ADVANCED COMPOSITES

ADVANCED COMPOSITES

Edited by

Ivana K. Partridge

School of Industrial Science,
Cranfield Institute of Technology
Bedford, UK

ELSEVIER APPLIED SCIENCE
LONDON and NEW YORK

ELSEVIER SCIENCE PUBLISHERS LTD
Crown House, Linton Road, Barking, Essex IG11 8JU, England

Sole Distributor in the USA and Canada
ELSEVIER SCIENCE PUBLISHING CO., INC.
655 Avenue of the Americas, New York, NY 10010, USA

WITH 46 TABLES and 184 ILLUSTRATIONS

British Library Cataloguing in Publication Data

Advanced composites.
1. Composite materials.
I. Partridge, Ivana K.
620.1′18

ISBN 1-85166-387-8

Library of Congress Cataloging-in-Publication Data

Advanced composites/edited by Ivana K. Partridge.
p. cm.
Bibliography: p.
Includes index.
ISBN 1-85166-387-8
1. Composite materials. I. Partridge, Ivana K.
TA418.9.C6A2675 1989
620.1′18—dc20

Phototypesetting by Tech-Set, Gateshead, Tyne & Wear.
Printed in Great Britain by Page Bros (Norwich) Ltd.

Preface

The idea for this book came from the collective enthusiasm of the lecturers and participants of a series of Seminars on Advanced Composites, organised by the editor at the Cranfield Institute of Technology. Several of the lecturers subsequently became contributors to the book.

Our understanding of the term 'advanced composites' at that time was primarily as covering organic matrix-based, continuous fibre reinforced materials. However, metal matrix and ceramic matrix composites are increasingly being considered in the same category and are therefore included here.

Two important topics relating to composites have been left out of this review: non-destructive testing and composite stressing. For the former, NDT techniques for composite materials and structures are already well covered, with several useful overview books in existence. In contrast to this the few books which deal with composite stressing tend to be limited to a laminate level design or specific to a particular composite structure; a generalised methodology similar to metallic design is, as yet, unavailable.

For the present book, each author was asked to present an overview of the state-of-the-art in his particular topic, to describe current problems and suggest, however tentatively, what the solutions may be. There was some reluctance to comply with the latter request. This was due to the recognition of the very rapid developments occurring in all aspects of composite materials, and an understandable wish on the author's part not to have been proved wrong by the time the book was published!

Of the many people deserving thanks for their part in the production of this book the first and foremost are the Seminars participants. Their encouragement and comments convinced me of the usefulness of bringing together information on materials, testing methods, production and performance of advanced composites, all in one publication. I am grateful to the School of Industrial Science at Cranfield for providing the environment necessary for this project to be completed. Finally, my personal thanks are due to W. Wright (RAE Farnborough), B. Bloch (ONERA, France), J. Bishop and I. Gurnell (Ciba-Geigy, UK) and A. Roulin (Tetra Pak, Switzerland) for their support.

IVANA K. PARTRIDGE

Contents

Preface v

Introduction 1
Willard D. Bascom
Department of Materials Science and Engineering, University of Utah, Salt Lake City, Utah 84112, USA

1 High Temperature Thermosetting Matrix Resins 7
I. K. Partridge
School of Industrial Science, Cranfield Institute of Technology, Bedford MK43 0AL, UK

2 Continuous Fibre Reinforced Thermoplastic Matrix Composites 43
David C. Leach
ICI Composites Inc., 2055 East Technology Circle, Tempe, Arizona 85284, USA

3 High Performance Fibres 111
W. Bishop
Aero Consultants (UK) Ltd, Stonehill, Stukeley Meadows Ind. Est., Huntingdon, Cambridgeshire PE18 6ED, UK

4 Approaches to Toughness Enhancement 145
C. B. Bucknall
School of Industrial Science, Cranfield Institute of Technology, Bedford MK43 0AL, UK

5 Reinforced Plastics in Anti-Corrosion Applications . . 163
Geoffrey Pritchard
School of Applied Chemistry, Kingston Polytechnic, Penrhyn Road, Kingston upon Thames, Surrey KT1 2EE, UK

6 Inorganic Matrix Composites 197
J. M. Quenisset
Laboratoire de Chimie du Solide du CNRS, Laboratoire de Genie Mecanique de L'IUT-A, Université de Bordeaux I, 351 Cours de la Libération, 33405 Talence Cédex, France

7 Advanced Composite Fabrication 251
Dave Woolstencroft
Composite Manufacturing Development, British Aerospace plc, Salmesbury Aerodrome, Balderstone, Blackburn, Lancashire B82 7LF, UK

8 Design of Bonded Structures 269
R. J. Lee[a] & J. C. McCarthy[b]
[a]Materials Development Division, [b]Engineering Division, Harwell Laboratory, UK Atomic Energy Authority, Didcot, Oxfordshire OX11 0RA, UK

9 Delamination 303
P. Davies
Division Polymères et Composites, Université de Technologie de Compiègne, 60206 Compiègne, France. Present address: *Polymer Laboratory, Swiss Federal Institute of Technology — Lausanne, 32 chemin de Bellerive, CH-1007 Lausanne, Switzerland*

10 The Fatigue of Organic Matrix Composite Materials . . 331
P. T. Curtis
Materials and Structures Department, Royal Aerospace Establishment, Farnborough, Hampshire GU14 6TD, UK

11 Damage Tolerance and Damage Assessment in Advanced Composites 369
Graham Dorey
Materials and Structures Department, Royal Aerospace Establishment, Farnborough, Hampshire GU14 6TD, UK

12 Certification of Civil Aircraft Composite Structures . . 399
John W. Bristow
Civil Aviation Authority, Aviation House, South Area — Gatwick Airport, Gatwick, West Sussex, RH6 0YR, UK

Index 431

Introduction

Willard D. Bascom

University of Utah, Salt Lake City, Utah, USA

It is indeed an honor to introduce this book which brings together timely reviews that cover a wide spectrum of composite technology ranging from materials, engineering, processing and quality assurance.

The development of new materials such as fiber reinforced composites is an essential factor for economic growth. Composites have spawned new industries for the production of raw materials and products ranging from aerospace vehicles to household appliances. However, the development of a new or advanced form of a material does not usually have an immediate economic impact. A number of factors influence its acceptance and viability. In the case of composites a primary factor is cost reduction in terms of energy consumption. Kelly[1] has emphasized this fact by pointing out that the need to reduce energy consumption has been a strong inducement for the introduction of advanced materials into engineering applications. The essential criteria have been reduction in specific weight and increased specific modulus. Other criteria such as the fracture resistance and durability are important but these secondary factors affect *which* material gains acceptance in the energy-lean market.

Composite materials have novel properties that offer new engineering opportunities. However, the development and commercialization of advanced composites is an evolutionary process. Materials scientists are continuously inventing new matrix materials and fibers but these are likely to remain in the laboratory unless they can hurdle a succession of 'acceptance' barriers. The first of these barriers is the development of an extensive data base, a data base that establishes the statistical variation in properties and the trade-off penalties that must be paid to gain a desired performance.

The next acceptance barrier is that the new material should meet engineering requirements and this has often been the slow, rate-determining step. Of necessity, the engineering community is very conservative. They ask tough questions about the reproducibility of properties, processability, cost, and merit relative to other material types. This engineering concern has been the major barrier to the acceptance of organic, matrix composites as replacements for metals. In this writer's experience, surmounting the engineering acceptance hurdle by organic composites was not accomplished in the airframe industry until the beginning of this decade. This acceptance came as a result of numerous trials of small composite components on aircraft. Unwilling to accept coupon or substructure tests, the engineering community requires that any new material be tested as non-critical structural elements; applications where its failure would not jeopardize the survival of the vehicle. Seemingly endless trials of composite wing flaps on aircraft or drive shafts on automobiles are in progress and until they are complete the economic fate of the composite material is undetermined.

If a material passes these in-service tests, then the next hurdle is cost. Does a new material offer a real cost advantage in terms of energy reduction or is this advantage lost in raw materials costs or processing costs? This hierarchy of materials development from invention to commercialization is shown in Fig. 1.

This text on advanced composite materials touches all the acceptance barriers indicated in the figure. State-of-the-art as well as new fiber and matrix materials are described. The balance of mechanical properties that composites offer is discussed along with areas of concern, notably delamination. Nearly every chapter discusses in detail or at least mentions cost considerations, especially processing. There is considerable attention given to military aircraft and space vehicles, where low weight and high modulus are critical and may override cost. The discussion of these applications is well balanced by reviews on applications where materials and processing costs are critical or where reliability is the governing and often overriding consideration.

The material of choice as reinforcement appears to be carbon fiber, which offers a balance of elongation, modulus and cost. Glass fiber is very cost-effective and aramid fibers offer unique damage tolerance properties. On the horizon, silicon carbide fibers show promise for

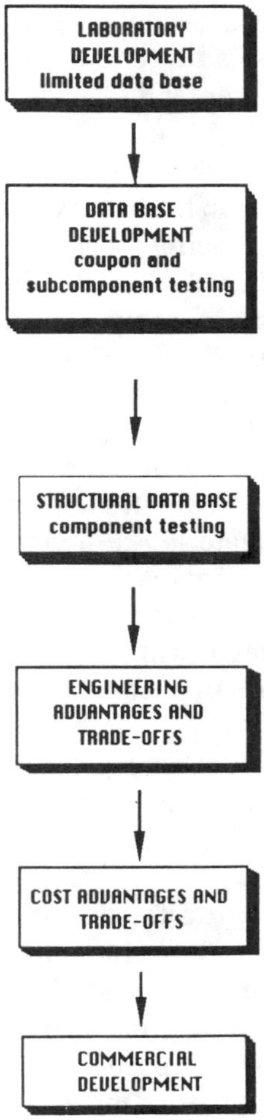

Fig. 1. Hierarchy of materials development from laboratory to commercialization.

thermal resistance applications and high modulus but are costly and in early stages of development (Chapter 3).

The principal matrix materials are organic polymers which for state-of-the-art structural applications have been the tetrafunctional, high modulus epoxies. These highly crosslinked polymers have met the needs of the aerospace industry for more than a decade but are deficient in two respects: upper temperature limit and toughness. For high temperature applications (200–400 °C) the thermosetting polyimides and bismaleimides are the leading candidates but, like the epoxies, have poor damage tolerance (Chapter 1).

Damage tolerance refers primarily to delamination resistance and two chapters in this book address this issue directly (Chapters 9 and 11). Delamination is dependent on the matrix fracture energy and so there is considerable interest in replacing thermosetting matrix polymers with the inherently tougher thermoplastic polymers. New thermoplastic polymers are on the market, notably PEEK, which offer high fracture energy and creep resistance, the latter property having deterred the use of thermoplastics as matrix materials in the past. However, the processing of these new thermoplastics requires high temperatures and large shearing forces and cannot be done on equipment already in place for processing thermosets. Nonetheless, new developments in the chemistry of these matrix materials and in processing methods indicate that they will capture an increasing share of the organic matrix composite market (Chapter 2). At the same time, advances in elastomer-toughened epoxies (Chapter 4) will keep the thermosets in the running, especially since they obviate the need for investment in new processing equipment.

Polymers are inherently more corrosion resistant than metals and so are used for low load applications such as pipe and small-scale storage in corrosive environments. Glass fiber has been the principle reinforcement in these applications. Advanced glass and carbon fiber reinforcement has opened up opportunities for large structural applications, especially in the chemical and marine industries. However, the design and production concerns of these materials are as severe as — and in some cases more so than — in aerospace applications (Chapter 5).

Ceramic and metal matrix composites offer some unique opportunities not attainable from organic matrix materials. Topping the list are high temperature applications especially in and around engines, but these materials also offer high dimensional stability and biocompatibility. Material costs/performance trade-offs are key considerations for these composites, as are the development of reliability data base and processing methods (Chapter 6).

From the engineering and design perspectives, damage tolerance poses some difficult questions. Delamination is unique to laminated composite structures so predictive engineering analyses need to be developed. In coupon and even substructure testing, the effect of impact loads is predominantly delamination (Chapter 9). The engineering challenge is scaling these test results to full-size structures.

Good fatigue resistance is one of the clear advantages of composites over metals (Chapter 10). Fatigue behavior is primarily determined by the matrix since the fibers are generally insensitive to fatigue loads (except the aramids). It does not appear that the thermoplastics offer any advantage over the thermosets in fatigue because high fracture energy is less critical in fatigue than for static loading.

Processing of composite raw materials into an end product is very often the cost-determining step. The traditional manual operations are slowly giving way to automated production (Chapter 7). Equally critical in the cost and design of composite structures is the joining of components — preferably by adhesive bonding rather than mechanical fastening which results in high stress concentration around drill holes (Chapter 8).

With all of the above considerations there is currently little reticence to the use of composites in small, private aircraft, including the home-built variety. This rapidly growing market presents some difficult problems to the civil airworthiness authorities worldwide. Qualification of aircraft components and all-composite planes must be done with limited design analysis capability. Also, the material supply chain, unlike that for metals, involves multiple manufacturers: chemical companies, fiber producers, processors and the end-item manufacturer. Quality control is therefore difficult and in many instances the QA parameters are undefined (Chapter 12).

REFERENCE

1. Kelly, A., *Composites Sci. Technol.*, **23** (1985) 171.

Chapter 1

High Temperature Thermosetting Matrix Resins

I. K. Partridge

School of Industrial Science, Cranfield Institute of Technology, Bedford, UK

1 INTRODUCTION

1.1 Background

Recent years have seen rapid developments in the area of heat resistant thermoset resins and blends, based on multifunctional epoxy resins or on polyimides. The ideal high temperature polymeric system for use as a laminating resin for structural composites should have the following characteristics: easy tape and prepreg preparation, good shelf-life, acceptable tack and drape, acceptable quality control procedures, no evolution of volatiles during processing, retention of mechanical properties over the required temperature–time ranges and environmental conditions, and acceptable repair.

Numerous commercially available systems exhibit many of these desirable characteristics; none has them all. Among the shortcomings of currently available matrix systems, those most frequently encountered are difficulty of fabrication, poor storage stability in the uncured condition, moisture pick-up and matrix cracking in the cured laminate.

For maximum use temperatures up to 180 °C, most problems can be overcome by a judicious use of blends of multifunctional epoxy resins, with suitable hardeners and modifiers. For applications requiring higher temperature stability than that obtainable with epoxy resins, polyimide-based matrices offer new possibilities. Current challenges in this area include improvement in the strain-to-failure of these resins and formulations for easier processing.

1.2 Review Scope

Before embarking on a description of the different thermosetting resin systems which are available for use in advanced composites it is useful to clarify what is meant by 'high temperature' in the present context. The figure of 93 °C is often quoted as a lower limit for continuous use temperature requirement for civil aircraft. This requirement rises to about 180 °C for skins of supersonic jets and military fighter aircraft. Thermally critical areas of aircraft such as engines have to withstand temperatures in excess of 300 °C and missile applications, albeit short-term, demand stability at 500 °C and above.

The above considerations effectively rule out the classical, well known polyesters and DGEBA-type epoxy resins and these materials will not, therefore, be included in this review.

An extensive account of synthetic resin matrices for service to 300 °C is provided in a book by Delmonte[1] and covers the developments in this field up to 1980. Detailed chemistry and evaluation of the thermo-oxidative properties of heat resistant polymers, including many of the experimental thermosets, are covered by Critchley *et al.*[2]

New information on the synthesis, characterisation and applications of polyimides became available in the mid-1980s.[3–5] Recently published books relating to the present topic include NASA conference proceedings[6] and an extensive, very up-to-date *Handbook of Thermoset Resins*.[7]

This chapter sets out to complement the above-cited reviews in those areas for which information still exists only in research or patent literature. In particular, attention is concentrated on formulated thermosetting blends and on some of the new types of thermoset resins. Chemical formulae are given only as part of the basic description of any particular resin. The emphasis is on highlighting the main advantages and disadvantages of each resin type and in comparing their processing characteristics and thermomechanical properties.

Many of the materials described in the above reviews have already become obsolete. However, new, increasingly specialised matrix resins are being added to the list of available systems at a rapid and growing rate. In such a situation, a designer faced with a material selection problem in this field would probably do well to become acquainted with the basic capabilities and limitations of just the main types of high temperature thermosets and thereafter to specify his requirements as accurately as possible to one or more pre-preg manufacturers.

It is hoped that the information presented here will facilitate the user–supplier dialogue.

2 HIGH TEMPERATURE RESIN TYPES

2.1 Phenolics

2.1.1 Conventional Phenolics

Historically, phenolics take their place as the first type of synthetic thermoset used in aircraft applications. To this day they excel over other matrix resins in terms of resistance to fire and the very low smoke and toxic evolution and for this reason have been widely employed in aircraft interiors[8] (see also Table 2 in Chapter 5).

Increasing awareness of safety requirements, coupled with recent developments of special paint systems, is stimulating rapidly increasing utilisation of phenolic-based composite panelling in underground installations, mass transportation and public buildings.[9]

Phenolic resols are cured by heating at between 120 and 180 °C; novolacs require the addition of a crosslinking agent such as hexamine. Water is evolved during cure and consequently very high pressures are required in processing, to bond pre-preg layers together and to minimise voiding.

Existing systems mainly use glass reinforcement. The composites exhibit excellent initial retention of strength (1000 h at 200 °C) but if long-term heat ageing is carried out then a catastrophic loss in properties is observed.[10] Nevertheless, the short term heat resistance of phenolics is good enough to enable their use in missile applications.

2.1.2 Phenol-aralkyl Resins

These phenol-based resins represent a much more recent class of thermosets. Developed in the late 1960s, they were subsequently made commercially available under the trade name of Xylok.[10] The pre-polymer is obtained by the condensation of a dimethoxy compound with phenols:[11]

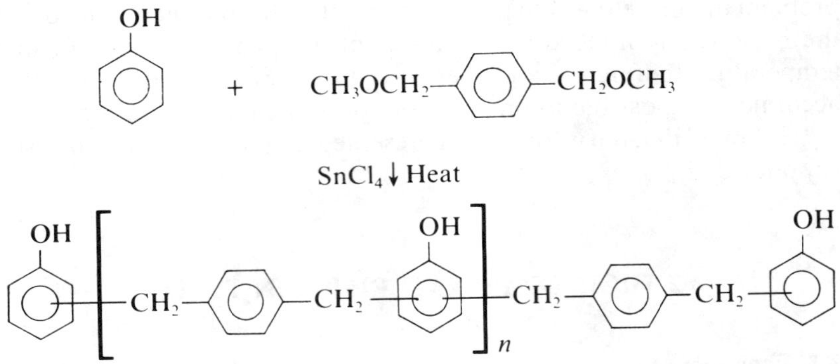

Phenol–aralkyl pre-polymer

Crosslinking is achieved by heating with hexamine or a di- or poly-epoxide. The disadvantage of the hexamine cure is the evolution of ammonia during the reaction, which leads to difficulties in the fabrication of thick sections. The evolution of volatiles may be minimised by a chemical modification of the pre-polymer, but only at the expense of heat resistance.

Glass fibre can be impregnated from a solution of the resins in methyl ethyl ketone and the pre-preg is consolidated by a short pre-cure at between 125 and 140 °C. Laminates are compression moulded at temperatures between 150 and 175 °C for 1–2 h. Pressure required can be between 0·7 and 7 MPa depending on application. A prolonged post-cure at 175 °C is essential for optimum property development.[10]

In common with conventional phenolic resins, phenol–aralkyls possess very good flame resistance and low smoke-generating characteristics. They can be blended with phenolic novolacs without any problems of compatibility and without necessitating any changes in processing. The main attraction of such blends is the possibility of improving the long-term property retention at elevated temperatures.[12]

2.2 Epoxies

2.2.1 Tetrafunctional

Multifunctional, i.e. tetra- and tri-functional, epoxy resins are used in most of the existing aerospace composites. For example, Narmco 5208 is *N,N'*-tetraglycidyl-4,4'-diaminodiphenylmethane (TGDDM) epoxy-cured with 4,4'-diaminodiphenylsulphone (DDS) hardener:

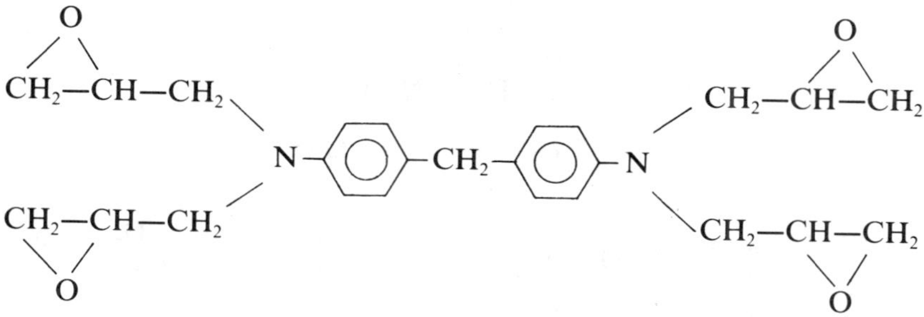

N,N'-Tetraglycidyl-4,4'-diaminodiphenylmethane TGDDM

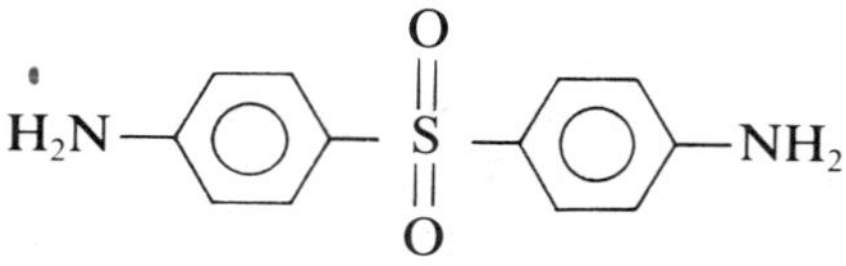

4,4'-Diaminodiphenylsulphone DDS

Another 'standard' epoxy resin of the last decade, Fiberite 934, is an accelerated version of the above. In fact, this particular glycidylamine epoxy monomer forms the basis of most of the currently available advanced CF/epoxy pre-pregs and is, therefore, the most utilised resin component of advanced composites. Several companies market the resin, in forms which at room temperature range from a highly viscous liquid to a semi-solid. Examples of trade names are MY720 (Ciba-Geigy), Rütapox TGMDA (Rütgers-Werke, Bakelite), Glyamine 120 (FIC Corporation) and Lopox B3302 (CdF Chimie-Résines).

Aromatic diamine hardeners, typically the DDS hardener shown above, are used in combination with the TGDDM epoxy. The cure regime for the composites includes a 177 °C (350 °F) step, giving rise to the well known '350 °F epoxy' name for this type of material. With additional post-cure at 190–200 °C the TGDDM–DDS-based materials attain glass transition temperatures (T_g) of around 240 °C.[13, 14]

Very advantageous shelf-lives can be obtained if a 'latent' crystalline hardener, dicyandiamide (Dicy) is used in place of the DDS. However, the use of Dicy limits the maximum T_g obtained to about 210 °C, a significant drop compared with the DDS-cure.[15]

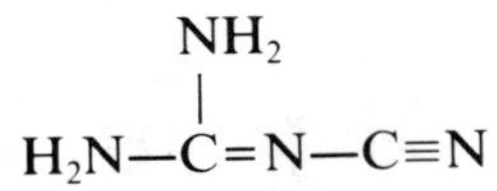

Dicyandiamide Dicy

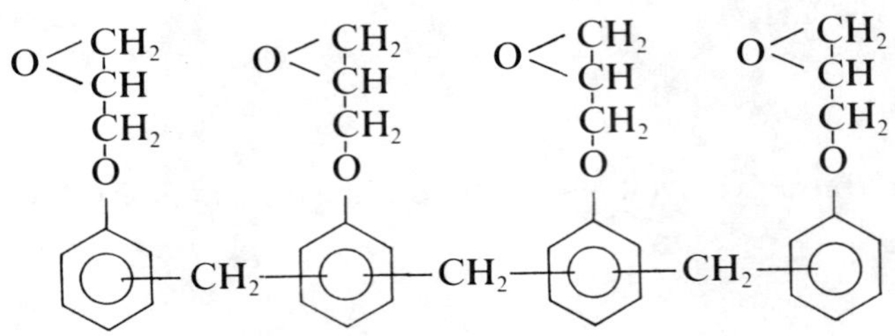

Tetrafunctional epoxy novolak
Trade names: Epikote 155 (Shell)
Rütapox 0300 (Bakelite)
DEN 438 (Dow)

Two different tetrafunctional epoxy resins are marketed by Shell: Epikote 155 is usually cured with an aliphatic diamine to a product with a T_g just above 180 °C; Epikote 1031 is capable of giving products with T_g in excess of 230 °C. Epikote 1031 is a hard solid at room temperature and is, therefore, often blended with the difunctional, liquid, Epikote 828 in order to assist processing characteristics.[16] A penalty in terms of thermal performance is necessarily incurred by such blending.

2.2.2 *Trifunctional*

Surprisingly little information on this class of epoxy resins is available in the open literature considering the vital part played by such resins in numerous adhesive formulations,[17] as well as in some well accepted blend systems.[15,18]

The two most frequently encountered trifunctional epoxies are triglycidyl-*p*-aminophenol (TGPAP) which is available from Ciba-Geigy (0510 resin) and a trisepoxy-novolak (TEN) developed by the Dow Chemical Company (XD-7342.02).

The Ciba-Geigy 0510 resin, TGPAP, is a low viscosity liquid at room temperature. It is highly reactive and autocatalytic, and therefore extreme care must be exercised when using resin systems which contain

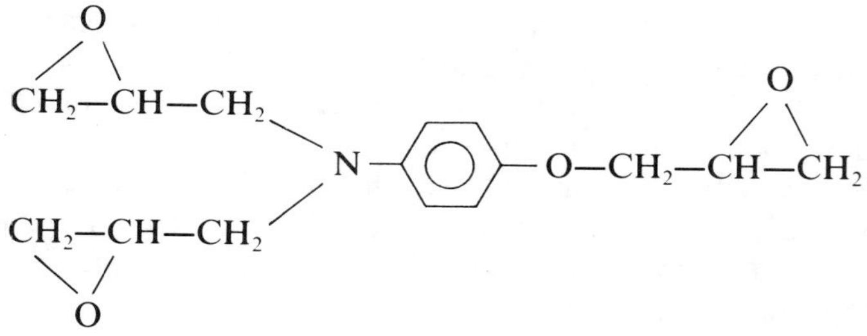

Triglycidyl-*p*-aminophenol TGPAP

a significant proportion of this epoxy. The monomer is less sterically hindered than its tetrafunctional version (TGDDM) and can give T_g values as high as 250 °C when cured with DDS.

The Dow TEN resin is a triglycidyl ether of tris(hydroxyphenyl)-methane, a semi-solid at room temperature.

This type of structure results in a very tightly crosslinked network upon curing and the resin has a very high heat distortion temperature and outstanding thermo-oxidative stability compared with other types of epoxy resins.[20] Glass transition temperatures of over 300 °C can be attained by curing with DDS.[21] The resin is, therefore, intended for adhesives and advanced composites in applications which demand property retention up to 177 °C 'wet' and 232 °C 'dry'.[22]

More recently a Japanese company has introduced a novel tri-functional amino-epoxy resin exhibiting a T_g of 261 °C when cured with DDS and having a good moisture resistance. The primary advantage of this resin appears to be its low viscosity which, in contrast to the case of

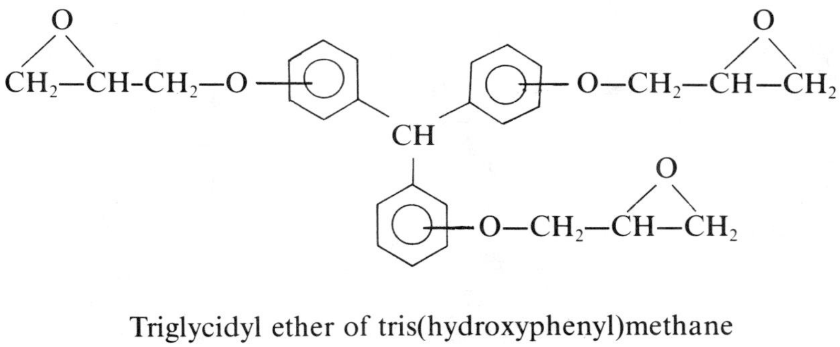

Triglycidyl ether of tris(hydroxyphenyl)methane

the TGPAP resin, can be achieved without having to carry out a laborious and expensive molecular distillation.[23]

Table 1 summarises the highest attainable 'dry' T_g values for the above-described multifunctional epoxy resins cured with amine hardeners. It must be remembered that the T_g of an epoxy resin saturated with moisture will be typically 50 °C lower than the 'dry' T_g value. Room-temperature dynamic viscosities, η, of the uncured resins are also given where appropriate.

2.3 Polyimides

2.3.1 Types of Polyimides

The vast amount of recent literature on the subject of polyimides bears witness to the dominance of these materials in current research and development programmes in the field of heat resistant polymers. They can be classified into three groups: condensation, addition and thermoplastic. A generalised polyimide structure is shown.

Only the addition type can be strictly classified as thermosetting resins. They polymerise through reactive end groups (maleimide, norbornene, acetylenic) on short chain pre-imidised oligomers. The resultant high crosslink density renders the materials extremely brittle.

The condensation types behave like thermosets, being almost completely infusible and insoluble. Their preparation involves a low temperature polycondensation of a dianhydride with a diamine in a polar solvent, to yield a soluble intermediate poly(amic-acid).

Several recent publications can be consulted for details of polyimide chemistry[1,3,7,24] and thermal stability[2,25] (Fig. 1). The attraction of

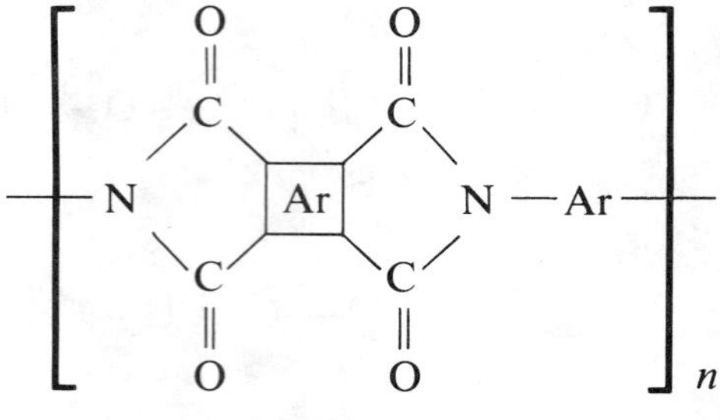

Polyimide

TABLE 1
Multifunctional Epoxy Resin Properties

Resin system [Manufacturer]	*Attainable 'dry' T_g [±5°C] (°C)*	*Ref.*	*$\eta_{25°C}$(Pa.s)*	*Ref.*
Tetrafunctional				
TGDDM type [Several]	240	19	>80	19
Epikote 1031 [Shell]	250	16	Semi-solid	
Novolak type [Several]	182		—	
Trifunctional				
TEN resin [Dow]	324	21	Semi-solid	22
ELM-100 [Sumitomo Chem. Co.]	261	23	10	23
TGPAP [Ciba-Geigy]	252	19	0·5–0·9	19
c.f. *Difunctional*				
DGEBA type	150		8–20	

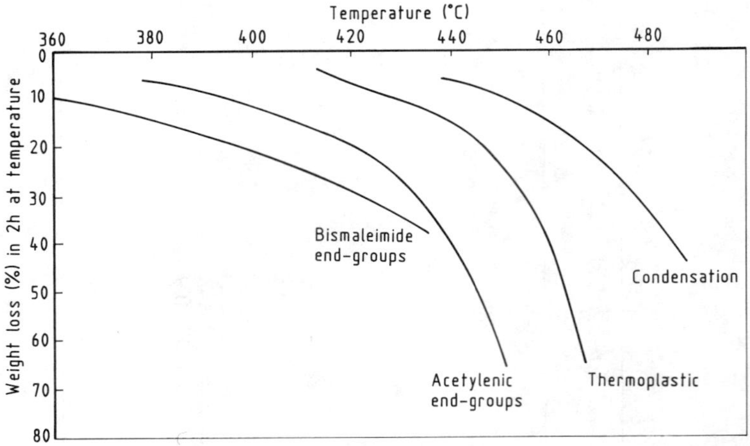

Fig. 1. Comparison of thermal stability in air of different types of polyimides. (Redrawn from Ref. 25.)

polyimides for advanced composite applications lies in their potential to operate successfully at temperatures above 250 °C and thus above the limit of performance of even the best multifunctional epoxy resins. Two major obstacles stand in the way of extensive exploitation of the 'classical' polyimides in composite structures. Firstly, the inherent brittleness of the tightly crosslinked polyimide structures leads to poor damage resistance in composites and to matrix cracking upon thermal cycling. Such cracking can lead to moisture pick-up and subsequent rapid deterioration of matrix integrity in a part subjected to freeze–heat cycles. The second problem is one of processability. For example, in the condensation types of polyimides the poly(amic-acid) precursor solution is used to impregnate fibres and the intermediate is then thermally dehydrated to give the fully cyclised polyimide. As with phenolics, this type of chemistry leads to considerable processing problems, particularly with evolution of volatiles. Moulding of laminates requires high temperature (280–300 °C) and pressure of about 1·5 MPa.

Both problems can be tackled and overcome to varying extents. Blending with different reactive modifiers can improve both processing characteristics and the toughness of polyimides (see Section 3.2, and also Chapter 4). Sections 2.3.2 and 2.3.3 describe two particularly successful methods of altering polyimides so as to make them more easily processable into continuous fibre reinforced laminates.

Recent international composites conference proceedings contain some excellent reviews of the status of polyimide resins in advanced composites.[26–29]

2.3.2 *The PMR Concept*

The PMR resins, developed at the NASA Lewis Research Center, are based on in-situ polymerisation of monomeric reactants, a process from which they derive their name. A mixture of the three monomers shown in Table 2 is dissolved in a low boiling point alcohol and this solution is used to impregnate the fibrous reinforcement. (N.B. A hot melt film process can also be used.) The heat-activated curing reaction is addition type, giving the opportunity for making high-quality void-free composites either by compression moulding or by a vacuum bag autoclave process. Laminates are typically cured for 1–2 h at 316 °C, followed by a free-standing post-cure in air, still at 316 °C, for between 4 and 16 h.[27]

The various members of the PMR family are designated by numbers, PMR-15 being the best known (see Table 2). The numbers relate to the so-called 'formulated molecular weight' of the polyimide and hence to the stoichiometry of the particular composition. The stoichiometry affects such parameters as resin flow and thermo-oxidative stability. The out-life of PMR pre-pregs is limited because of the high volatility of the alcohol solvents used in pre-pregging. Usually, tack and drape of pre-preg can only be maintained for a few days when exposed to atmosphere at ambient temperature.

A further development of the PMR concept was carried out at NASA Langley Research Center and is known as 'LARC-160' technology. The mixture of reactive monomers is obtained as a viscous liquid at room temperature and contains no residual solvent.

2.3.3 *Bismaleimides*

Bismaleimides (BMIs) have the generalised structure shown, where R may be —CH_2—, —O—, —SO_2 or other groups. The resins are cured by the application of heat, at temperatures between 150 and 250 °C. Numerous chemical modifications can be carried out and mixtures formulated to improve rheology and cured resin properties. Formulated bismaleimide resins are presently closest in processing characteristics to the conventional '350 °F' epoxy resins (Fig. 2). It is for this reason that an enormous amount of interest has arisen in these materials in the last few years, to such an extent that it is probably fair to say that as many research papers dealing with BMIs appear in any given period as those

TABLE 2
Monomeric Constituents of PMR-15 (After Ref. 27)

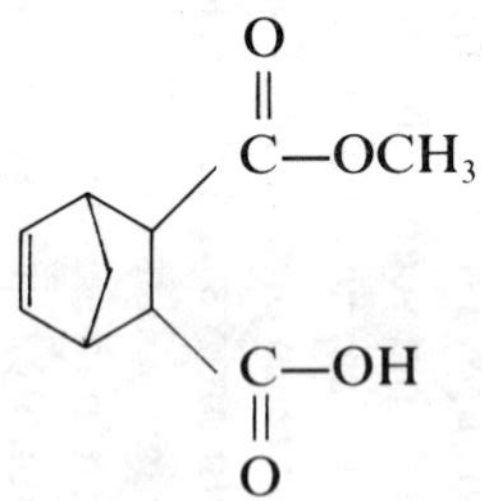	Monomethyl ester of 5-norbornene-2,3-dicarboxylic acid
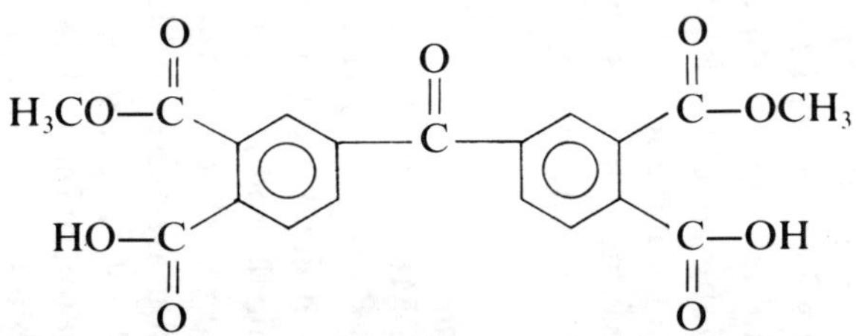	Dimethyl ester of 3,3′,4,4′-benzophenonetetracarboxylic acid
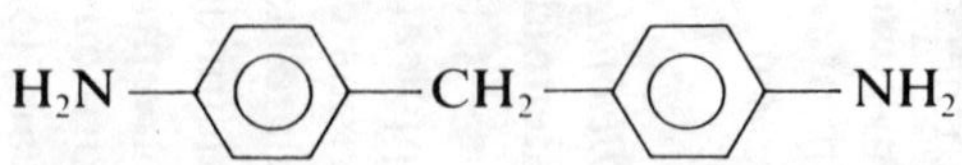	4,4′-Methylenedianiline

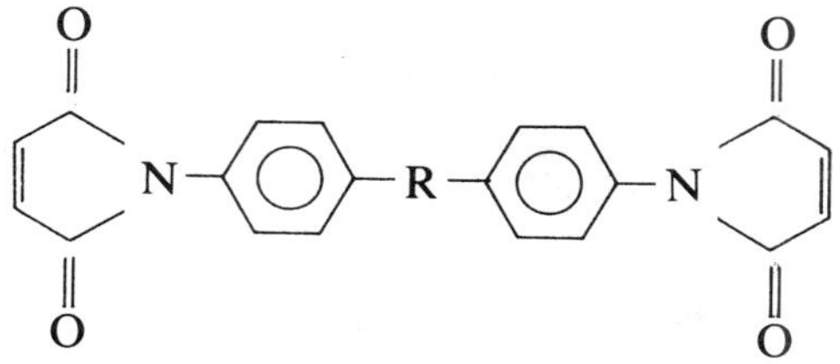

Bismaleimides

concerned with all the other types of high temperature thermosets together. It is, therefore, difficult to gain and maintain an up-to-date overview of the developments in this rapidly expanding area. Review chapters[24,30] and articles[31-33] are indispensable in such a task.

Most of the existing knowledge and experience of BMI systems originates in the USA but novel, highly specialised BMI resins (or BMI containing blends) are now commercially available from several leading Europe-based companies:

Technochemie GmbH–Shell	Compimide range
Ciba–Geigy	XU-295
Rhône–Poulenc	Kerimid range
British Petroleum	V378A
BASF	Narmco 5245C

Crosslinked BMI polymers are insoluble in all solvents, infusible, rigid and brittle. They have relatively high densities (1·35–1·4 g/cm^3) and exhibit T_g values generally between 250 and 300°C. Strain-to-failure is typically below 2%. Moisture absorption levels in BMI resins tend to be much the same as in epoxy resins (4–5% by weight) but saturation occurs faster than in epoxy resins. Table 3 lists some of the best properties attained so far by commercialised BMI resins.

Much research and development is still being directed to toughening of BMI resins.[34-38] New products, such as Narmco X5250,[39] aim for superior hot–wet performance, compression-after-impact in the composite, coupled with good processability and longer shelf-life over currently used epoxy and BMI composites.

One example of a recently developed formulated BMI is BP's V398 pre-preg, intended for use in wings, fuselage and stabilisers of the next generation of military aircraft. The material cures at 190°C and satisfactory performance under hot–wet conditions is expected between

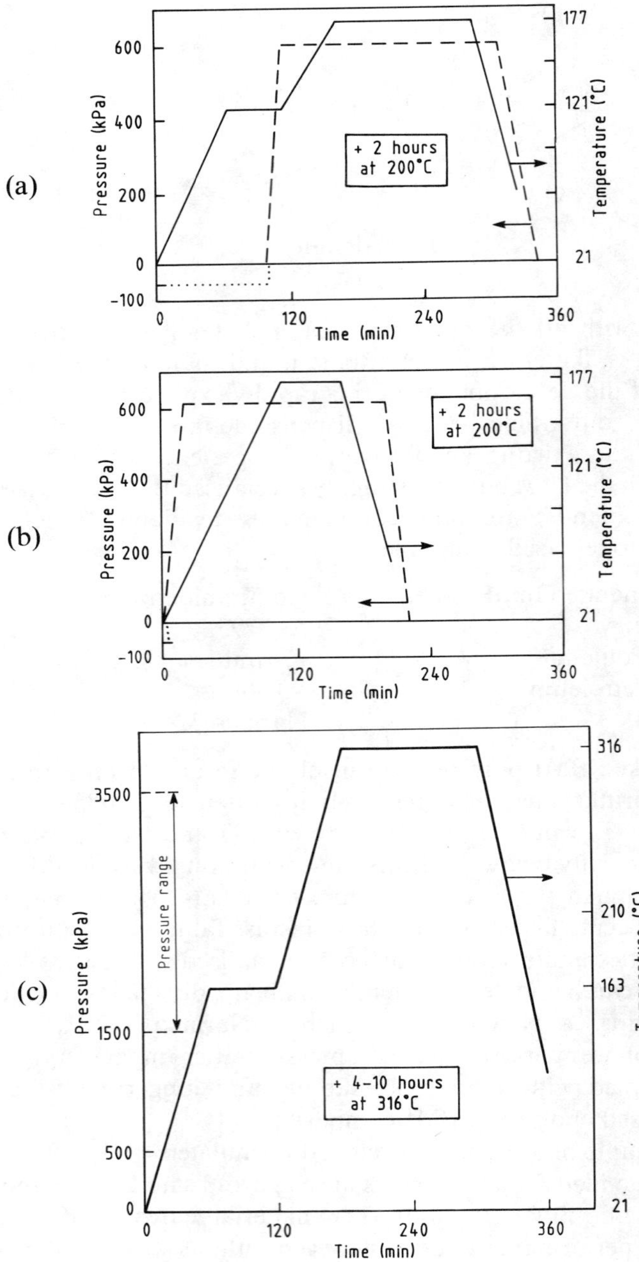
(a)
Pressure (kPa)
600
400
200
0
−100
Temperature (°C)
177
121
21
+ 2 hours
at 200°C
0
120
240
360
Time (min)
(b)
Pressure (kPa)
600
400
200
0
−100
Temperature (°C)
177
121
21
+ 2 hours
at 200°C
0
120
240
360
Time (min)
(c)
Pressure (kPa)
3500
1500
500
0
Pressure range
Temperature (°C)
316
210
163
21
+ 4–10 hours
at 316°C
0
120
240
360
Time (min)

(d)

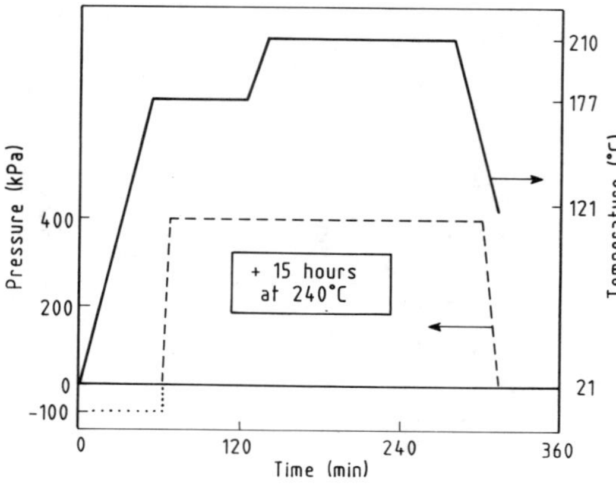

Fig. 2. Typical cure cycles for carbon fibre laminates based on high temperature epoxies and polyimides. (a) Epoxy-cure, including a 'dwell' step; (b) simplified epoxy-cure, suitable for 'controlled flow' systems; (c) PMR-15/LARC-160 cure cycle; (d) BMI-cure cycle.

175 and 200 °C. The technical literature claims good tack and drape, 21 days 'out-time' and, in addition, a promise is made of good micro-cracking resistance down to −196 °C.[40]

2.4 Polystyrylpyridine Resins

Polystyrylpyridine, PSP, resins could be classified as a type of new resins such as the ones described in Section 3.1. However, since carbon fibre/PSP pre-pregs are now commercially available, it seems more appropriate to include them here.

Initial developments were carried out jointly between the Office National d'Etudes et Recherches Aérospatiales (ONERA) and the Société Nationale des Poudres et Explosifs (SNPE) in France. SNPE now markets two series of the resins, PSP 6022 and PSP 7022. The 7022 type is a flexibilised version of the 6022 type.

The pre-polymer for this class of compounds is shown. It is obtained by a condensation reaction between aromatic aldehydes and methylated derivatives of pyridine.[41]

Subsequent crosslinking occurs by reactions of the unsaturated groups with the methyl groups. A typical press cure temperature regime may be 3 h at 200 °C, followed by 2 h at 250 °C. The shortest efficient

TABLE 3
Bismaleimide Resin Properties

	Highest published value	*BMI resin*	*Corresponding values for current 'standard' aerospace epoxy resins*
T_g (°C)			
Dry	400	Kerimid FE70003,	210–250
Wet	297	Ciba-Geigy XU-295	<200
Tensile strength (MPa)			
Dry, 25 °C	90	Technochemie H795	80–100
Wet, 25 °C	88	Ciba-Geigy XU-295	
Tensile strain-to-failure (%)			
Dry, 25 °C	2·9	Narmco 5245C	1–2
Dry, 177 °C	3·3	Hysol EA9102	
Fracture toughness (J/m²)	210	Ciba-Geigy XU-295	100–200

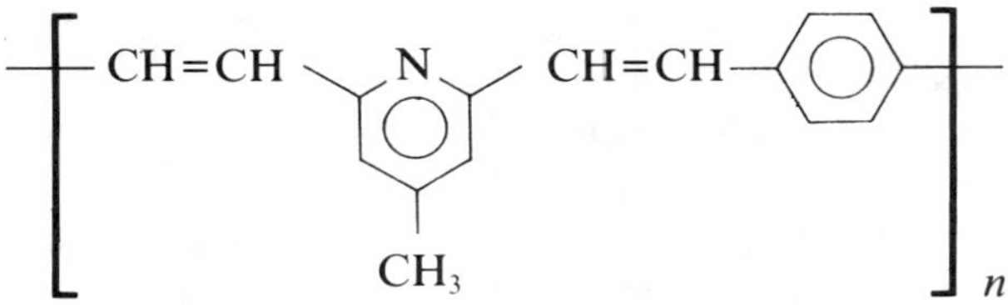

Polystyrylpyridine

autoclave cure schedule is 4 h at 200 °C. Autoclave pressure of around 500 kPa is required. In all cases, a post-cure at 250 °C for between 3 and 16 h is necessary to develop the potential of PSP resins for very high temperature resistance.[42] As in the case of polyimides, void formation can be a serious problem.[43] In comparison with the cure regimes of modern formulated epoxy resins, the cure conditions required by the PSP resins are complex and energy-demanding. However, the resulting property retention is impressive — the system appears capable of long-term stability at 250 °C. Further important advantages of PSP resins are good moisture resistance, very low flammability and low smoke toxicity.[42] Laminate mechanical properties are comparable with the performance of corresponding laminates based on the high temperature epoxies. Short-term retention of mechanical properties up to 400 °C enables their use in missile applications.[44]

Pre-pregs can be prepared either by hot melt impregnation or, more usually, from a solution in methyl ethyl ketone (MEK). Shelf-life is typically at least 3 months at 20 °C and more than 1 year at −18 °C.[41]

It should be noted that although the early problems associated with evolution of toxic pyridine volatiles have been overcome, there remains the need for good ventilation of any production spaces.

3 NEW APPROACHES

3.1 New Resins

Possibilities for synthesis of new types or varieties of thermosets with potential for high temperature applications are virtually unlimited. Most are unsuitable for large-scale production on grounds of economics, processability or toxicity. The number of real possibilities is further reduced if long-term thermo-oxidative stability is required at high temperatures.[2]

Of the dozen or so types of resins which have been evaluated with a view to applications as matrices for advanced composites, two are mentioned more frequently than others — polyphenylquinoxalines and polybenzimidazoles. Delmonte (Chapter 4 in Ref. 1) gives a brief description of and detailed references to the development of these two resin types up to 1980.

3.1.1 Polyphenylquinoxaline Resins

Further developments of the polyphenylquinoxaline polymers into thermosetting matrix resins have been carried out at the NASA Research Centers and at the Air Force Wright Aeronautical Laboratories.[45] The acetylene-terminated variety of the resin, shown, is currently the most favoured.

Such acetylene-terminated resins cure through a thermally induced, radical-initiated addition reaction, without evolution of volatiles. The ease of processability is inversely proportional to the oligomer chain length.[41–48] A disadvantage of the acetylene termination is its effect of reducing the inherent thermo-oxidative stability of the backbone oligomer, caused by the formation of polyene structure upon curing.

Nevertheless, the acetylene-terminated polyphenylquinoxalines exhibit good thermal stability to about 190 °C and short-term use is possible at 300 °C. Mechanical properties of the neat resins are very comparable with what would be expected from standard epoxy or BMI systems, but exhibit remarkable insensitivity to moisture (Table 4).

The combination of easy processing, excellent moisture resistance and the possibility of property tailoring by adjustment of the backbone chain length are the reasons for continued interest in this type of resin.

3.1.2 Polybenzimidazole Resins

The polymers are produced by reacting aromatic diamines with aromatic diacids. Several different synthetic routes can be followed,[49] resulting in relatively low molecular weight prepolymers. First applications of these resins as composite matrices were associated with glass cloth laminates and autoclave processing, which tended to give laminates with an unacceptably high void content (10–20%). More recently, interest has arisen in the application of polybenzimidazole/carbon fibre composites to missile systems. This is because the most thermally stable types of polybenzimidazoles outperform the PMR-15

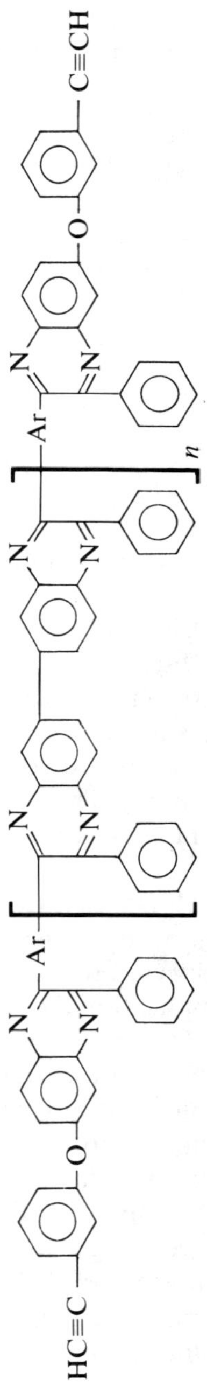

Polyphenylquinoxalines

TABLE 4
Polyphenylquinoxaline Resin Properties (After Ref. 45)

Tensile strength (MPa)	Dry, 25 °C	98
	Wet, 25 °C	98
	Dry, 230 °C	27
	Wet, 230 °C	32
Elongation at break (%)	Dry, 25 °C	3·5
	Dry, 93 °C	4·4
	Dry, 177 °C	6
K_{Ic} (MPa/m$^{1/2}$)		0·7–1·1 depending on backbone chain length

and LARC-160 bismaleimides in terms of their temperature resistance by some 50 °C.

Optimised autoclave cure cycles for benzimidazole-based composites are complex and expensive (Fig. 3). The isothermal dwell at about 166 °C is required, as in the case of the epoxy and polyimide cure cycles shown in Fig. 2, to allow the resin to completely wet-out the fibres. The cure reaction is accompanied by evolution of volatiles, giving rise to the above-mentioned void formation. The complicated procedure of gradual application of very high pressures is necessary in order to minimise this problem.

Recent work carried out at McDonnell Douglas Astronautics Co. on developing a low-cost, non-autoclave, benzimidazole laminate curing system has enabled high quality void-free laminates to be prepared.[50, 51]

In the high speed missile application the polybenzimidazole/carbon fibre system would be expected to perform well at short-term exposures to 650 °C. Despite such attractive properties, it is doubtful if polybenzimidazole resins can be put into conventional types of production because of the carcinogenicity potential of some of the benzidine compounds necessary for their manufacture. Special, safe, fabricating processes would need to be employed, limiting the utilisation of polybenzimidazoles to some very high added value applications, or, possibly, to their other use as raw materials for high temperature, flame resistant fibres, foams and fabrics.[48, 49] In spite of considerable activity in this field, the availability of information is limited by industrial secrecy.

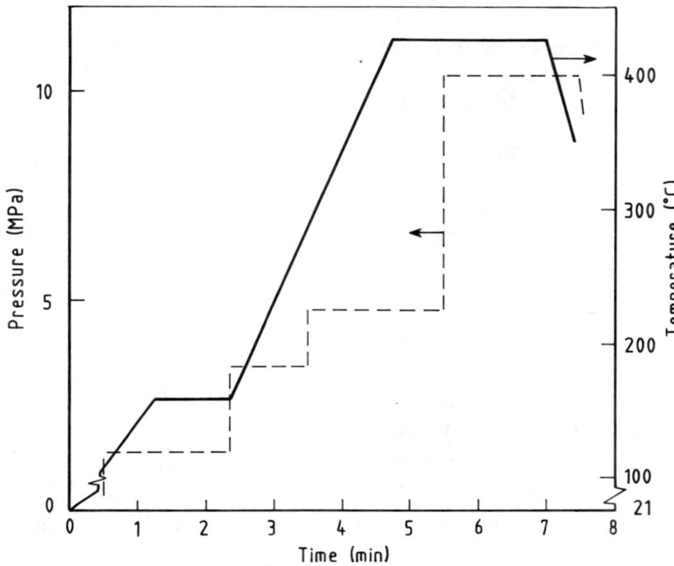

Fig. 3. Optimised autoclave cure cycle for polybenzimidazole/carbon fibre laminates (after Ref. 50).

3.1.3 Polyphthalocyanine Resins

This new type of potential matrix material for advanced composites has been developed at the Naval Research Laboratory in the USA. The processability of the resins is comparable with that of conventional epoxies but a major advantage is the inertness of the phthalocyanine monomers and pre-polymers at room temperature, thereby eliminating the need for cold storage.

The resin monomers are diamines with a generalised structure as shown.[52, 53] They have melting temperatures around 190 °C, depending on the length of the aliphatic chain. When heated for short periods at

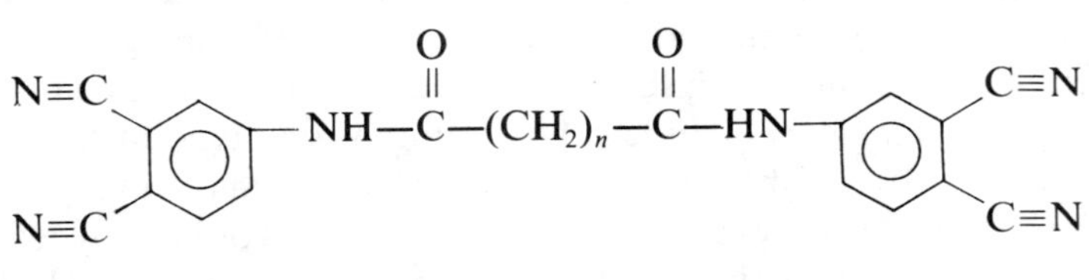

Phthalocyanine resins

TABLE 5
Dependence of T_g of Cured Polyphthalocyanine Resin on the Interaromatic Chain Length *n* of the Starting Monomer (After Ref. 52)

Chain length, n	T_g *(°C)*
4	350
8	280
20	130
34	90

temperatures above their melting point the resins become liquid and can be used for fibre impregnation. Continued heating activates polymerisation and a three-dimensional phthalocyanine structure is formed. The flexibility and hence the T_g of the final resin, following a 200 °C post-cure, are determined by the length of the interaromatic chain linkage (Table 5).

At present, insufficient data are available to enable an assessment of the future exploitation of these resins to be made. Nevertheless, the concept of material development indicated here is an elegant one, offering a chemically simple, pure system which can be tailor-made to requirements and which overcomes some of the practical manufacturing difficulties encountered with existing systems.

3.2 Blends

Preceding sections describe the essential characteristics of the more specialised 'chemical tools' put at the disposal of the composite matrix formulator by the research and development chemists. These products have been achieved by synthesis of new chemical structures and their subsequent development to achieve acceptable processing qualities.

An alternative and significantly cheaper approach to the development of a new high temperature thermoset is by blending of existing thermally stable materials in such a way as to obtain the desired balance of properties. This can include (a) blending of one type of thermoset with another, (b) suitable selection of hardeners, chain extenders or 'fortifiers'[54,55] and finally (c) the exciting possibilities offered by reactive blends of thermosets with elastomers or thermoplastics. The following sections describe these approaches in some detail.

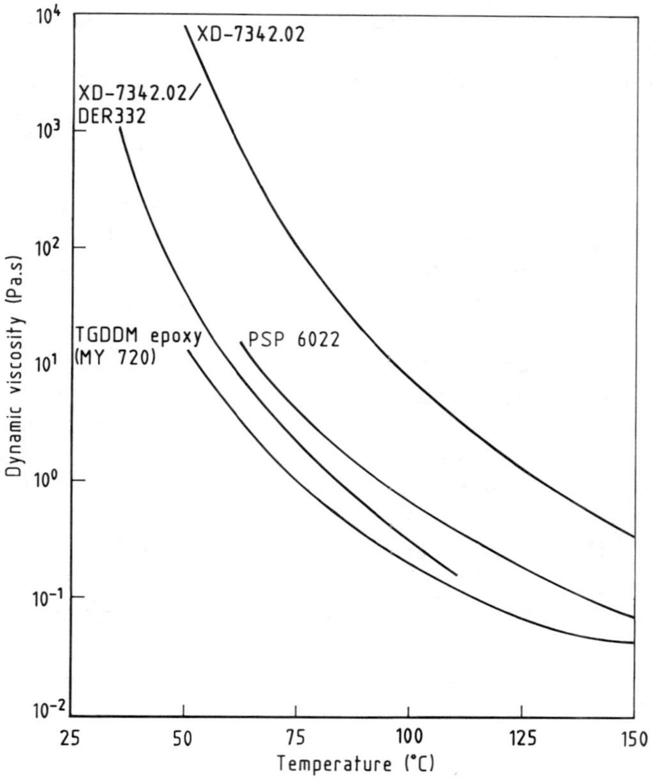

Fig. 4. Temperature dependence of dynamic viscosity of Dow trifunctional epoxy (XD-7342.02) and a blend of tri- and difunctional epoxies (XD-7342.02/DER 332). Corresponding curves for tetrafunctional epoxy (Ciba–Geigy MY 720) and for PSP 6022 resin are shown for comparison. (Data from manufacturers' literature.)

3.2.1 Mixtures of Different Monomers

The practice of combining different epoxy monomers in order to achieve good handleability at different stages of the fabrication process has already been mentioned in Section 2.2.1. Figure 4 shows the dynamic viscosity as a function of temperature for the Dow trifunctional epoxy resin XD-7342.02 which is solid at 25°C. Blending with the difunctional, liquid epoxy DER-332 reduces the viscosity to a level suitable for the pre-pregging operation. The accompanying drop in dry T_g, as a result of blending in about 25 wt% of the difunctional epoxy resin, is from 310 to 302 °C respectively, when the resins are cured with DDS hardener.[22]

Blending with epoxy monomer can also be used to impart good processing quality to more intractable thermosetting resins such as polyimides or to reduce the intrinsic brittleness of bismaleimides.[56, 57] The lower crosslink density of the epoxy component will result in the lowering of the thermal performance of the polyimide but is a necessary pre-requisite of a toughenable resin (see Chapter 4). Thus many of the commercial, formulated, toughened BMI resins can be expected to be based on a blend or a copolymer with a suitable epoxy resin.

3.2.2 Low Molecular Weight Modifiers

The best known way of altering the properties of an epoxy resin, both in terms of its cure kinetics and the properties in the cured state, is by changing the nature of the chemical agent used to polymerise the epoxy.

Figure 5 shows how the chemoviscosity of the TGDDM epoxy is affected by the choice of either DDS or Dicy hardener. In the cured state, TGDDM resin cured with 30 phr (parts per 100 parts of resin) of DDS exhibits a T_g of 240 °C whereas the stoichiometrically equivalent cure by 5 phr of Dicy results in a T_g of 200 °C.[19] In commercial practice it

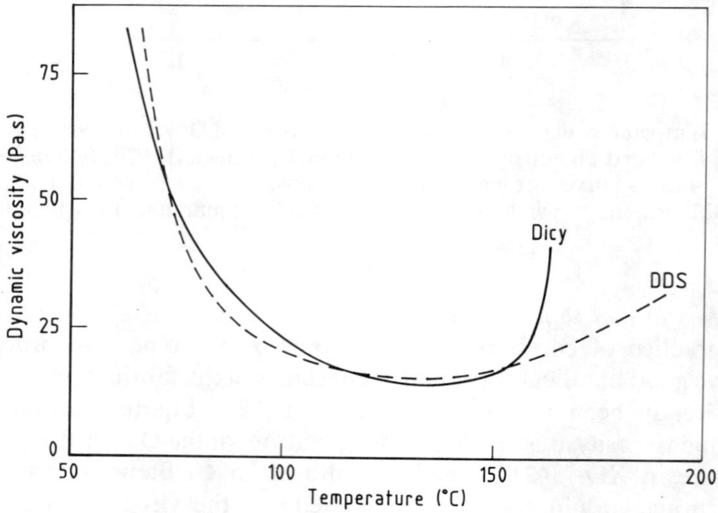

Fig. 5. Dynamic viscosity of TGDDM epoxy resin during cure with DDS or Dicy hardeners, at a heating rate of 2 °C/min. (Redrawn from Ref. 19.)

is not uncommon for a mixture of these two hardeners to be used, the Dicy then acting as an accelerator for the DDS cure.

The search is on for new hardeners which would enhance particular aspects of the classical or novel epoxies. For example, a whole new range of bisimide amine hardeners could become available, having the general formula:

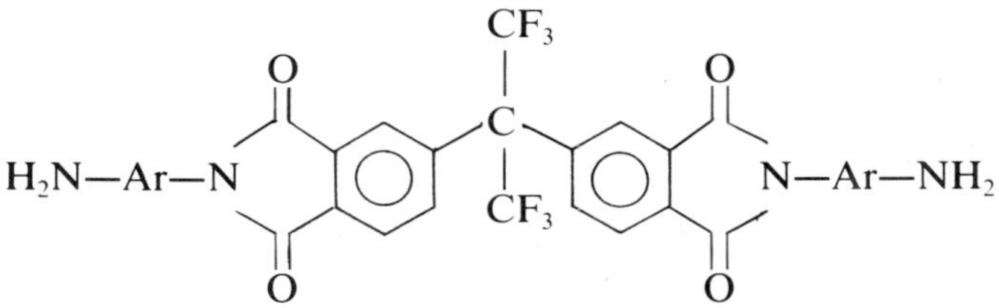

Bisimide amines

Depending on the choice of the Ar group, more or less flexibility can be introduced into the crosslinked network. Flexible chain linkages result in significant improvements in the tensile strength and the strain-to-failure of neat resins, whilst the presence of the hexafluoroisopropylidene group between aromatic rings ensures that such changes are achieved without sacrifice in the use temperature of the epoxy.[58]

An alternative approach to epoxy-cure, which avoids the introduction of flexible crosslinks, is to use an isocyanate compound. This type of cure gives rise to both chain extension and crosslinking via stiff heterocyclic structures, the oxazolidone and isocyanurate rings respectively. Nevertheless, epoxy resins cured in this way combine toughness and high T_g values. Further benefits may include low moisture absorption and low flammability.[59] Accompanying improvements in resin toughenability by conventional rubber toughening methods have also been found.[60]

Linear chain extension of BMIs by copolymerisation with various bis(benzocyclobutenes) is being explored in the same spirit of achieving improved fracture resistance with minimum loss in high temperature resistance.

Propenyl-terminated comonomers can be made to copolymerise completely with BMI formulations in a melt-blending process carried out at around 70 °C. The resulting flexibilised formulated BMI resin shows approximately 20% drop in stiffness and 10% drop in dry T_g compared with the base BMI. However, fracture resistance is increased

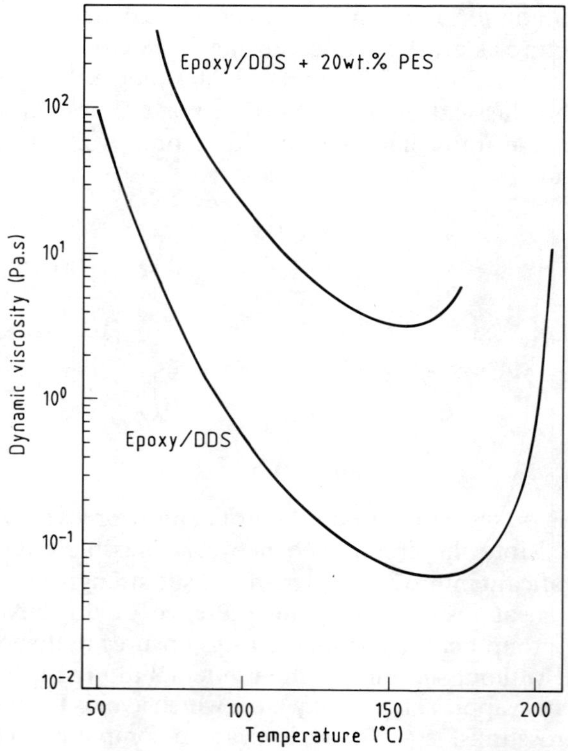

Fig. 6. Effect on the dynamic viscosity of addition of 20 wt% of polyethersulphone to a high temperature epoxy resin, during cure with DDS hardener. (Redrawn from Ref. 63.)

by nearly an order of magnitude, to values of G_{Ic} of 200–500 J/m^2. The accompanying reduction in equilibrium moisture absorption from 5 to 3% is another attractive feature of this approach.[61]

3.2.3 Elastomers and Thermoplastic Modifiers

The important role of rubbery and thermoplastic additives as toughening agents is covered in detail in Chapter 4. However, rubbers and low molecular weight thermoplastics may also be added to thermosets in order to control dimensional stability or, more importantly in the present context, viscosity and hence processing characteristics.[19,62,63] Essentially, low molecular weight thermoplastics which can form stable solutions in epoxy monomers lead to a significant increase in the viscosity of the blend (Fig. 6). This means that the usual 'dwell' step in the laminate cure can be eliminated and pressure can be applied at the start of the cycle (see Figs 2a and 2b). The resin flow is controlled, resulting in easy fibre management, low void content and generally good quality laminates. One of the first such commercial 'controlled flow' high temperature epoxy–carbon fibre pre-pregs was Ciba–Geigy's Fibredux 914. The success of the approach is shown by the fact that over the last decade Fibredux 914 has become 'the standard' high temperature epoxy system throughout Western Europe.

Furthermore, we are now witnessing very rapid developments in formulated high performance epoxies and bismaleimides based on the principle of dissolving a high temperature thermoplastic or rubber additive in the monomer, often with the aid of a common solvent. The thermoplastic is usually intended to phase-separate from the polymerising thermoset in the course of the reaction, resulting in morphologies such as that shown in Fig. 7(a). If the concentration of the thermoplastic is high enough, a complete phase inversion can take place (Fig. 7b). The thermoplastic is now the continuous phase and this will normally result in a tougher material than those where the thermoset is the continuous phase.

Good adhesion between the two phases is required in either case.[64,65] Solvent attack is clearly a potential problem, especially in the 'phase-inverted morphology' case. Reactive end groups on the additive can ensure sufficient grafting takes place between the additive and the monomer to avoid such problems.[66]

The exploitation of polymeric modifiers has not stopped at processing agents. The 'second generation' formulated epoxy resins utilise the technology to improve the toughness of the resins to a remarkable degree — from G_{Ic} values of about 200 J/m^2 typical of the best 'first generation' formulated epoxy resins to between 400 and 800 J/m^2 now being expected of commercial resin systems such as Fibredux 6376 (Ciba–Geigy),[67] Hercules 8551–7[68,69] or Torayca 3900–2.[70] Remarkable improvements in the toughness of bismaleimides have also been achieved, now falling into the 200–400 J/m^2 category[71] whilst maintaining advantages in high temperature performance over the epoxy systems.

Among the first polymeric modifiers investigated for the purposes of blending with high temperature thermosets were polysulphones. Better results were obtained with functionalised thermoplastics such as polyethersulphones.[15,18,19,72] Almost any engineering thermoplastic can

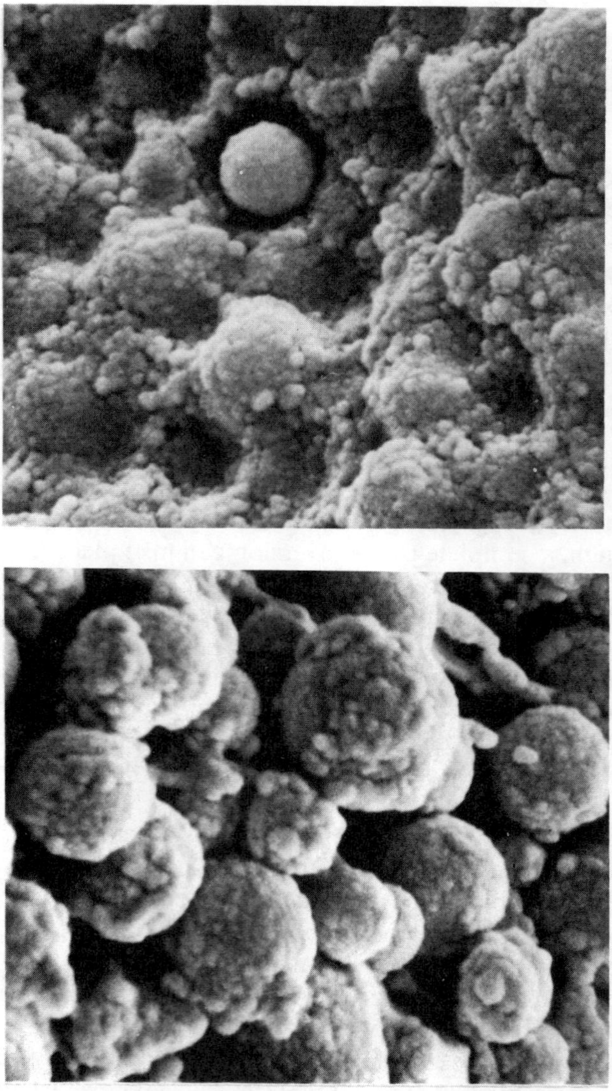

(a)

Fig. 7. (a) Scanning electron micrographs of fracture surfaces of cured thermoplastic-modified high performance epoxies. The thermoplastic is concentrated within the spherical inclusions (1 cm = 1 μm).

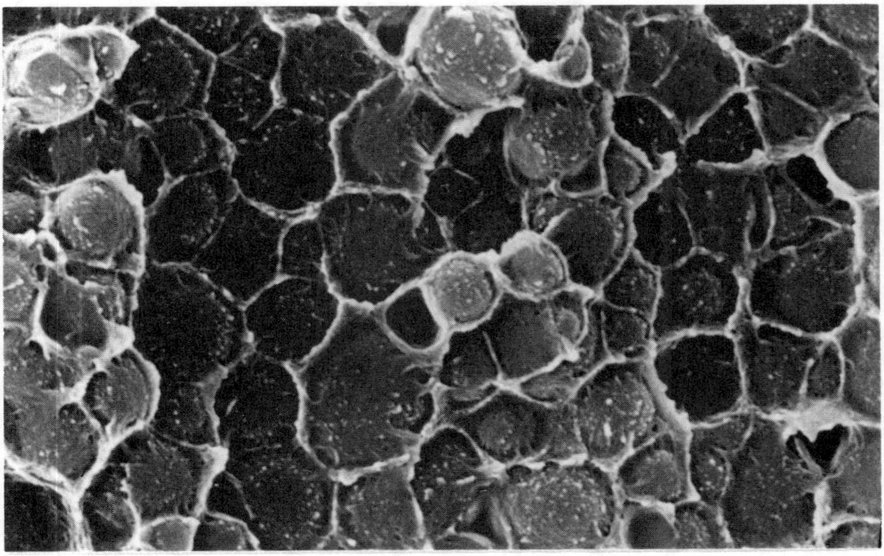

(b)

Fig. 7—*contd.* (b) Scanning electron micrograph of a fracture surface of a similar thermoplastic–epoxy blend, but showing the 'phase inverted' morphology. The thermoplastic here forms the continuous phase (1 cm = 1 μm). (Photo by courtesy of P. Davies.)

be considered (methacrylate, polycarbonate, poly(ether imide), etc.) provided it can be dispersed or, better still, dissolved into the resin monomers. Hence low molecular weight grades of the thermoplastics with reactive end groups are usually used.[64] Most recently chemical modifications of the thermoplastic backbone have allowed systematic changes in morphology of the cured blends to be produced. The toughness of 'semi-IPN' materials appears to be particularly promising.[73,74]

Polyimide-based blends employ the same tactics. Thermosetting polyimides can be blended with thermoplastic polyimides,[75] with the above-mentioned thermoplastics, silicone,[76] and silicone or fluorocarbon elastomers.[77,78]

Finally, the question of how the morphology and hence the properties of such blends in the cured state may or may not be affected by the presence of a high proportion of carbon, glass or aramid fibres remains a matter of further research and discussion.[79,80]

4 FUTURE DEVELOPMENTS

The available evidence suggests that polyimides and, more specifically, PMR resins and formulated bismaleimides are the most promising materials for the future, in the context of this chapter.[81, 82] The essential advantage of polyimides is their very high initial heat resistance which can thus be traded-off for improvements in toughness to a much larger extent than in epoxy systems.

In addition, the basic chemical blocks are easy to produce and relatively cheap. The processability of BMI resins has been improved to such an extent that it is now comparable with that of epoxy systems. They still require post-cures at somewhat higher temperatures than epoxies and tend to be more difficult to formulate into stable blends than the epoxy-based blends. Some problems of loss of tack can also be experienced. However, the 'hot–wet' performance of BMIs is sufficiently attractive to make any such disadvantages appear insignificant, especially in demanding applications such as structural adhesives for 'hot' parts in aircraft.

For any new formulations, addition type polymerisation is preferred and the use of solvents should be avoided. Cost-effectiveness of the entire process leading to the production of complete structures also needs to be taken into account. Any 'improved' matrix resin system should thus also offer good cure control and be suitable for automated, faster processing.[83]

To this end, faster cures may be offered by materials which can be made light-crosslinkable.[84] Along similar lines, the penetration limits of industrial high-energy electron beam accelerators are well in excess of usual laminate thickness, making radiation processing a realistic alternative to conventional heat cure.[85] Resin systems which can be conveniently cured by solar radiation in space, or by microwaves, are also being developed.

Of all the thermosets mentioned in this chapter, polybenzimidazoles offer the highest maximum operating temperature, in the region of 400 °C. This represents the upper limit for any organic-based polymer. If this barrier needs to be surpassed, then inorganic polymers have to be developed to a stage where they offer a viable alternative to other inorganic matrix composites (Refs 86–88; see also Chapter 6).

Finally, since the appearance of high temperature thermoplastic matrix composites (described in the following chapter), there has been lively speculation on the likely outcome of some sort of competition between high temperature thermosets and thermoplastics for advanced composite applications. The relative advantages and disadvantages of both types of polymers, in terms of their balance of properties and their processability, have come to a much clearer focus in the last few years. The best informed current opinion maintains that each class of material will find its own particular and cost-effective applications.[89]

REFERENCES

1. Delmonte, J., *Technology of Carbon and Graphite Fiber Composites*. Van Nostrand–Reinhold, New York, 1981.
2. Critchley, J. P., Knight, G. J. & Wright, W. W., *Heat Resistant Polymers — Technologically Useful Materials*. Plenum Press, New York and London, 1983.
3. Bessonov, M. I., Koton, M. M., Kundryavtsev, V. V. & Lains, L. A., *Polyimides — Thermally Stable Polymers*. Consultants Bureau, New York, 1987.
4. Fitzer, E., ed., *Carbon Fibres and Their Composites*. Springer-Verlag, Berlin, 1985.
5. *Proceedings of the 5th International Conference on Composite Materials (ICCM V), San Diego, July 1985.*
6. Serafini, T. T., ed., *High Temperature Polymer Matrix Composites*. Noyes Data Corporation, Park Ridge, NJ, 1987.
7. Goodman, S. H., *Handbook of Thermoset Plastics*. Noyes Publications, Park Ridge, NJ, 1986.
8. Forsdyke, K. L., Phenolic resin composites for fire and high temperature application. In *Fibre Reinforced Composites*. Plastics and Rubber Institute, London, 1984.
9. Forsdyke, K. L., Phenolic resin — the high temperature, fire resistant matrix. In *Polymers for Composites*. Plastics and Rubber Institute, London, 1987.
10. Ref. 2, Chapter 2.
11. Harris, G. I. & Coxon, F., British Patent 1 150 203 (1969).
12. Harris, G. I., Phenol-aralkyl and related polymers. In *Developments in Reinforced Plastics — 1*, ed. G. Pritchard. Applied Science Publishers, London, 1980, Chapter 4.
13. Morgan, R. J., O'Neale, J. E. and Miller, D. B., *J. Mater. Sci.*, **14** (1979) 109–24.
14. Munns, T. E. & Seferis, J. C., *J. Appl. Polym. Sci.*, **28** (1983) 2227–33.
15. Bucknall, C. B. & Partridge, I. K., *Polymer*, **24** (1983) 639–44.
16. *Shell Resins — EPIKOTE 1031*. Shell Chemicals, 1984.
17. Pearce, P. J., Davidson, R. G. & Morris, C. E. M., *J. Appl. Polym. Sci.*, **27** (November 1982) 4501–17.
18. Bucknall, C. B. and Partridge, I. K., *Polym. Eng. Sci*, **26**(1) (1986) 54–62.

19. Partridge, I. K., Effects of phase separation in crosslinked resins containing polymeric modifiers. PhD Thesis, Cranfield Institute of Technology, UK, 1984.
20. Hawthorne, K. L., Henson, F. C. & Pinzelli, R., *Org. Coat. Appl. Polym. Sci. Proc.,* **46** (1981) 493.
21. Chan, L. C., Nae, H. N. & Gillham, J. K., *J. Appl. Polym. Sci.,* **29** (1984) 3307–27.
22. *Experimental Tris-epoxy Novolak Resin*, Dow Epoxy Resins Data Sheet, XD-7342.02.
23. Saito, Y., Kamio, K., Morii, A. & Adachi, T., A novel tri-functional epoxy resin with low viscosity and high heat resistance. In *Proc. 32nd Int. SAMPE Symp., Anaheim, 6–9 April 1987*, pp. 1119–25.
24. Landman, D., Advances in the chemistry and applications of bismaleimides. In *Developments in Reinforced Plastics — 5*, ed. G. Pritchard. Elsevier Applied Science Publishers, London, 1986, Chapter 2.
25. Wright, W. W., *Brit. Polym. J.,* **16** (December 1984) 258–70.
26. Harvey, J. A., A survey of high temperature properties of polyimide composites. In Ref. 5, pp. 1623–32.
27. Serafini, T. T., Aerospace applications of PMR polyimide composites. In Ref. 5, pp. 1007–23.
28. Scola, D. A., The chemistry, processing characteristics and properties of commercial polyimide resins — a review. *Proc. 31st Int. SAMPE Symp., Las Vegas, 7–9 April 1986,* pp. 1844–56.
29. Driscoll, S. B. & Walton, T. C., Adhesion and characterisation of prepreg processable polyimides. *Proc. 32nd Int. SAMPE Symp., Anaheim, 6–9 April 1987*, pp. 1371–85.
30. Stenzenberger, H. D., Development of thermosetting polyimide matrix resins. In Ref. 4, Chapter 2, pp. 118–39.
31. Scola, D. A., The status of bismaleimides for composites. In Ref. 5, pp. 1601–19.
32. Parker, J. A., Kourtides, D. A. & Fohlen, G. M., Bismaleimides and related maleimido polymers as matrix resins for high temperature environments. In Ref. 6, pp. 54–75.
33. Boyd, J. D. & Biermann, T. F., Advanced bismaleimide composites, Pt II. Conversion of high performance structural fiber properties into bismaleimide composite properties. In *Proc. 32nd Int. SAMPE Symp., Anaheim, 6–9 April 1987*, pp. 1386–98.
34. Kinloch, A. J., Shaw, S. J. & Tod, D. A., Rubber-toughened polyimides. In *Rubber Modified Thermoset Resins*, ed. C. K. Riew & J. K. Gillham. Advances in Chemistry Series No. 208, ACS, Washington DC, 1984, Chapter 8.
35. Hsu, M. S., Chen, T. S., Parker, J. A. & Heimbuch, A. H., *SAMPE J.* (July/August 1985) 11–16.
36. Potter, K. D. & Robertson, F. C., Bismaleimide formulations for resin transfer moulding. *Proc. 32nd Int. SAMPE Symp., Anaheim, 6–9 April 1987*, pp. 1–12.
37. Chaudhari, M., King, J. & Lee, B., A new bismaleimide matrix resin for high performance advanced composites. *Proc. 32nd Int. SAMPE Symp., Anaheim, 6–9 April 1987*, pp. 24–31.

38. Steiner, P. A., Browne, J. M., Blair, M. T. & McKillen, J. M., *SAMPE J.*, **23**(2) (March/April 1987) 8–14.
39. Boyd, J. D. & Biremann, T. F., Advanced bismaleimide prepreg for up to 400°F service. *Proc. 31st Int. SAMPE Symp., Las Vegas, 7–10 April 1986*, pp. 977–95.
40. *BP Advanced Composites Ltd Prepreg Systems*. Advance Technical Information Sheet on US Polymeric BMI Prepregs, February 1988.
41. *Properties of the PSP 6022 and 7022 Resin Systems*, Parts I to VI. SNPE Technical Literature, 1986.
42. Malassine, B., PSP 6022 resin: a matrix for thermostable, fire resistant, high performance composite materials with epoxy-type manufacturing methods. *Proc. ICCM III, Paris, August 1980*, pp. 1483–97.
43. Hancox, N. L., Some properties of carbon fibre reinforced bismaleimide and polystyrylpyridine laminates. In *Polymers for Composites*. Plastics and Rubber Institute, London, 1987, Paper 10.
44. Bloch, B. & Maquart, J. G., PSP resins and their applications as heat and fire resistant matrices. *Proc. 14th Congress Int. Council of the Aeronautical Sciences (ICAS), Toulouse, 1984*, pp. 106–10.
45. Goldfarb, I. J., Lee, C. Y.-C., Arnold, F. E. & Helminiak, T. E., Acetylene terminated matrix resins. In Ref. 6, pp. 2–10.
46. Lee, C. Y.-C., Curing processes in acetylene terminated resins. In *Developments in Reinforced Plastics — 5*, ed. G. Pritchard. Applied Science Publishers, London, 1986, Chapter 4, pp. 121–50.
47. Lau, K. S. Y., Research polymers and future directions. In Ref. 7, Chapter 11, pp. 367–91.
48. Hergenrother, P. M., Heat resistant polymers. In *Encyclopedia of Polymer Science and Engineering*, ed. Mark, Bikales, Overberger, Menges & Keoschwitz, Vol. 7, John Wiley & Sons, New York 1987, pp. 639–55.
49. Ref. 2, Chapter 5.
50. Jones, J. F., Waldrop, J. C. & Fountain, R., High temperature resistant polybenzimidazole composites for missile applications. In *Proc. 29th Nat. SAMPE Symp., 3–5 April 1984*, pp. 777–83.
51. Jones, J. F. & Fountain, R., Polybenzimidazole composites. In Ref. 5, pp. 1591–9.
52. Ting, R. Y., Nash, H. C., Cottington, R. L. & Rall, C. J. N., Phthalocyanine resins — new matrix materials for advanced composites. In *Proc. ICCM III, Paris, August 1980*, Vol. 2, pp. 1501–13.
53. Ref. 2, Chapter 5.
54. McLean, P. D., Scott, R. F. & Garton, A., *Brit. Polym. J.*, **15** (March 1983) 66–70.
55. McLean, P. D., Garton, A., Stevenson, W. T. K. & MacPhee, K. E., *Polym. Composites*, **7**(5) (1986) 330–6.
56. Woo, E. M., Chen, L. B. & Seferis, J. C., *J. Mater. Sci.*, **22** (1987) 3665–71.
57. Gotro, J. T., Appelt, B. K. & Papathomas, K. I., *Polym. Composites*, **8**(1) (1987) 39–45.
58. Scola, D. A., *Polym. Composites*, **4**(3) (1983) 154–62.
59. Hodd, K. A. & Lee, Y. S.-K., Cross-linked polyoxazolidones: one approach

to higher performance epoxies. In *Polymers for Composites.* Plastics and Rubber Institute, London, 1987.
60. Clarke, J. A., Rubber toughening of oxazolidinone-modified epoxy novolacs. In *Rubber Modified Thermosetting Resins*, ed. C. K. Riew & J. K. Gillham. Advances in Chemistry Series No. 208, ACS, Washington, DC, 1984, Chapter 5.
61. Stenzenberger, H. D., Konig, P., Herzog, M. & Romer, W., Advances in addition polyimide matrix materials. In *Polymers for Composites*. Plastics and Rubber Institute, London, 1987.
62. Hayes, B., *Composites,* **14**(2) (1983) 93–8.
63. Partridge, I. & Gurnell, I., Modification of epoxy resins with PES: effects on fabrication and properties of composites. In *Proc. 5th French Nat. Conf. on Composite Materials (JNC 5), Pluralis, Paris, 1986*, pp. 429–41 (in French).
64. Cecere, J. A., Hedrick, J. L. & McGrath, J. E., Morphology and properties of amine terminated poly(arylene ether ketone) and poly(arylene ether sulfone) modified epoxy resin systems. In *Proc. 31st Int. SAMPE Symp., Las Vegas, 7–10 April 1986*, pp. 580–8.
65. Muraki, T., Takeo, A. & Hirata, M., Tough resin/carbon fibre composites. In *High Tech — The Way into the Nineties*, ed. K. Brunsch, H.-D. Goelden & C.-M. Herkert. Material Sci. Monographs 35, Amsterdam, Elsevier SC BV, 1986, pp. 163–76.
66. Hercules Inc., US Patents 4 656 207 and 4 656 208 (7 April 1987).
67. Jonas, T. F. & Siegmund, R. F., New high strain to failure prepreg. In *Proc. SAMPE, 5th Technology Conference of the European Chapter, Montreux, 12–14 June 1984*, Paper 5.
68. Gawin, I., Tough thermosetting resins with superior damage tolerance. Hercules 8551 series. In *Proc. 31st Int. SAMPE Symp., Las Vegas, 7–10 April 1986*, pp. 1204–13.
69. Boll, D. J., Bascom, W. D., Weidner, J. C. & Murri, W. J., *J. Mater. Sci.,* **21** (1986) 2667–77.
70. Tobukuro, K., Odagiri, N., Ito, Y. & Nishimura, T., High impact resistant CFRP from thermosetting system of TORAYCA T800H/3900 prepreg. Paper presented at SAMPE, 9th Technology Conference of the European Chapter, Milan, 14–16 June 1988.
71. Stenzenberger, H. D., Konig, P., Herzog, M., Romer, W., Pierce, S. & Canning, M. S., Bismaleimide resins. Improved novel toughening modifiers for BMI resins. In *Proc. 32nd Int. SAMPE Symp., Anaheim, 6–9 April 1987*, pp. 44–57.
72. Raghava, R. S., Secondary transitions and fracture toughness of cured epoxy resins and their blends with polyethersulfones. In *Proc. 28th Nat. SAMPE Symp., 12–14 April 1983*, pp. 367–73.
73. Sefton, M. S., McGrail, P. T., Peacock, J. A., Wilkinson, S. P., Crick, R. A., Davies, M. & Almen, G., Semi-interpenetrating polymer networks as a route to toughening of epoxy resin matrix composites. In *Proc. 19th Int. SAMPE Tech. Conf., 13–15 October 1987*, pp. 700–11.
74. Hsiue, E. S., Miller, R. L. & Segal, L., A new matrix system for advanced composites: dicyanate semi-IPN. In Ref. 5, pp. 1655–65.

75. Yamamoto, Y., Satoh, S. & Etoh, S., Preparation and characterisation of thermo-plastic/thermo-setting polyimide blends. In *Proc. 30th Nat. SAMPE Symp., 19–21 March 1985*, pp. 903–11.
76. Viot, J.-F. & Seferis, J. C., *J. Appl. Polym. Sci.,* **34** (1987) 1459–75.
77. St Clair, A. K. & St Clair, T. L., *Int. J. Adhesion Adhesives* (July 1981) 249–55.
78. Mijovic, J., Pearce, E. M. & Foun, C.-C., Fluoroelastomer-modified thermoset resins. In *Rubber Modified Thermoset Resins*, ed. C. K. Riew & J. K. Gillham. ACS Advances in Chemistry Series No. 208, ACS, Washington, DC, 1984, Chapter 18.
79. Cecere, J. A., Senger, J. S., McGrath, J. E., Steiner, P. A., Wong, R. S. & Sacheva, Y., Fabrication and characterisation of amine terminated poly(arylene ether sulfone) modified epoxy–carbon fiber composites. In *Proc. 32nd Int. SAMPE Symp., Anaheim, 6–9 April 1987*, pp. 1276–87.
80 Bucknall, C. B. & Partridge, I. K., *Composites,* **15**(2) (1984) 129–33.
81. de Abajo, J., Modern trends in the synthesis and application of heat-resistant polymers. Plenary lecture. In *Proc. 1st European Symposium on Polymeric Materials, Lyon, 14–18 Sept. 1987.*
82. Balme, Barthelemy, Camberlin & Lopez, Impact improvement of bis-maleimide resins by siloxane modification. *Proc. 1st European Symposium on Polymeric Materials, CPA 01, Lyon, 14–18 Sept. 1987*, pp. 1–5.
83. Hewitt, R. W. & Bonner, D. C., New epoxy resins for economical fabrication of advanced composites. In Ref. 5, pp. 401–7.
84. Berger, J., *Polym. Commun.,* **26** (January 1985) 11–14.
85. Saunders, C. B., Dickson, L. W., Singh, A., Carmichael, A. A. & Lopata, V. J., Radiation curable carbon fiber prepreg composites. In *Composites '87*, NRCC, Boucherville, Quebec, 5–6 Nov. 1987, Paper 17 (to appear in *Polymer Composites*).
86. Ref. 2, Chapter 9.
87. Prewo, K. M. & Minford, E. J., Graphite fiber reinforced thermoplastic glass matrix composites for use at 1000 °F. *SAMPE J.,* **21**(2) (1985) 26–33.
88. Phillips, D. C., High temperature fibre composites. In *Proc. ICCM VI–ECCM 2*, ed. F. L. Matthews, N. R. C. Buskell, J. M. Hodgkinson & J. Morton. Elsevier Applied Science Publishers, London, 1987, pp. 2.1–2.32.
89. Johnston, N. J., Thermoset and thermoplastic composites — room for both? Plenary lecture presented at ICCM VI–ECCM 2, Imperial College, London, 20–24 July 1987.

Chapter 2

Continuous Fibre Reinforced Thermoplastic Matrix Composites

David C. Leach

ICI Composites Inc., 2055 E. Technology Circle, Tempe, Arizona, USA

1 INTRODUCTION

Thermoplastic composites have emerged in recent years as serious candidates for use as advanced materials in applications such as aerospace and other high performance industries. These applications had been dominated by thermosetting matrix composites until recently. Thermoplastic polymers and their short fibre composites have dominated the field of moulding materials where the ability to fabricate complex shapes rapidly by techniques such as injection moulding have made them very attractive materials.

The emergence of thermoplastic composites as advanced materials has occurred as a result of materials development and technological demand. The US Office of Technology Assessment has identified toughness improvement and reduction of fabrication costs as two of the major requirements for composite materials used in aerospace and military applications.[1] This is not a new target and materials scientists have been working towards these goals in thermosetting matrix composites. However, there have been parallel developments in thermoplastic polymers and impregnation technology which have lead to the point where thermoplastic matrix composites can now fulfill the requirements of aerospace applications.[2]

The field of advanced thermoplastic composites is expanding rapidly and can already be confusing. Advanced thermoplastic composites are taken to mean composites containing at least 50% by volume of continuous fibres. Injection moulding and sheet stamping materials will not be discussed.

A large number of polymers are available as composite matrices. They may be grouped by morphology into amorphous, semi-crystalline and liquid-crystalline polymers. Composites are available in the form of fully impregnated tapes or fabrics (pre-pregs) and in other forms including hybrids of polymer and reinforcing fibres, and powder-impregnated forms. In some cases the same polymers are available impregnated in a number of different forms with differences in handling, fabrication and properties. Some pre-pregs contain a fully polymerised, 'true' thermoplastic, whilst others contain a pre-polymer in which case the polymerisation chemistry must be completed during the processing/fabrication stage. These materials are normally referred to as 'pseudo-thermoplastics'. The aim of the chapter is to summarise the polymers, composite forms, properties and fabrication of this rapidly growing class of materials.

2 BACKGROUND

2.1 Historical Development

The principal reason for using thermoplastic composites is economic and this will be the subject of a separate section. Thermoplastic matrix composites offer potential for considerable reduction in the initial and in-service costs of components. The US Air Force and Navy sponsored programmes during the 1970s to examine the relative costs of finished parts made from thermoset and thermoplastic composites. Work by Hoggatt *et al.*[3] demonstrated that thermoplastic composites could offer significant cost reductions compared with thermosets for a range of aircraft parts. This is due to the rapid fabrication of true thermoplastic composites. The work used polysulphone/carbon fibre composites and this represented a practical limitation to the implementation of the work as these materials did not have adequate resistance to aerospace environments. From these initial studies some applications were developed, principally for small volume, specialist components such as radomes.[4]

At the time polysulphone represented the state-of-the-art in polymer development but during the 1970s several new polymers became available which combined the required temperature performance and good environmental resistance. These included poly(phenylene sulphide) (PPS), poly(ether ether ketone) (PEEK), and poly(amide imide) (PAI). These polymers require higher processing temperatures and in some cases annealing periods to give optimum properties. It was difficult or impossible to impregnate these polymers by the techniques that had been developed for thermosetting matrices.

The final stage was the development of impregnation technologies. Thermoset polymers have relatively low viscosities during the impregnation stage and can be readily impregnated into the reinforcing fibres using a variety of techniques. In addition the polymers undergo a low viscosity stage during fabrication which enables further impregnation to occur. Thermoplastic polymers are normally fully polymerised and have relatively high melt viscosities making it difficult to impregnate them into the small diameter fibres (5–15 μm) at the required fibre contents (>50% by volume). During the early 1980s several groups developed techniques to give good penetration of the polymer to wet the fibres uniformly. These techniques are generally proprietary but the available information is reviewed later in the chapter.

By the early 1980s thermosetting matrix composites were being used in significant portions of the primary structure of military aircraft and secondary structure of commercial aircraft. Composites are used because of their weight savings compared with metals, and hence reduced fuel consumption, longer range, higher performance or greater payload. It was recognised that thermoset composites were sensitive to damage, and in particular low energy and low velocity impact which occurs in routine service. This is most commonly characterised by a post-impact compressive test.[5] The sensitivity of thermosetting composites is due to their low toughness, which in turn is due to the highly crosslinked nature of the polymer. Although it was possible to improve the toughness of thermosetting matrix composites this was accompanied by a reduction in other critical properties such as compressive strength or processability (see Chapter 4). The low toughness represented a major constraint and limited design strains to 0·40% whereas it was generally believed that design strains of 0·60% were needed to achieve the desired further weight savings.

The initial proprietary thermoplastic composites based on PEEK,

PPS and PAI demonstrated significant improvements in toughness compared with the available thermoset composites while retaining good environmental resistance. This balance of properties stimulated considerable interest in these materials, which led to the rapid characterisation and development of fabrication processes. This has further stimulated material suppliers, and in recent years many more composites have become available based on a wider range of polymers and product forms.

2.2 Reasons For Using Thermoplastic Matrix Composites

Some of the advantages and problems of thermoplastic composites are evident from the previous section, and are summarised in Table 1. Thermosetting matrix composites have a 'shelf-life' due to the chemical reactions which occur in the resins. The pre-pregs must be stored below

TABLE 1
Advantages and Problems of Thermoplastic Matrix Composites

	Advantages	*Problems*
Storage and handling	Indefinite shelf life. No special storage conditions required.	Pre-preg stiff and boardy with no tack. (Pseudo-thermoplastics may have some tack and drape.)
Properties	High toughness. Good environmental resistance (semi- and liquid-crystalline matrix polymers only).	Limited evaluations of long-term properties (creep, fatigue) and environmental resistance.
Processing/ fabrication	No chemistry during fabrication (true thermoplastics only). Potential for rapid, low cost fabrication. Potential to re-process (true thermoplastics only). Scrap recovery. Potential for easy quality control.	Fabrication processes still being developed. High temperature processing. Tooling, bagging materials, sealants expensive. May need capital investment for new equipment.

room temperature and the handling time for lay-up of parts is limited. Careful inventory control, scheduling and special storage conditions are therefore required; this adds to the cost. This is not the case with true thermoplastic matrix composites which normally have an indefinite shelf-life and do not require any special storage conditions. Thermoset and some pseudo-thermoplastic pre-pregs are drapable and tacky which can be an advantage in the handling and lay-up. In contrast fully impregnated, true thermoplastic pre-pregs are stiff and boardy which can make handling and lay-up of complex parts more difficult.

Thermoplastic composites generally offer a combination of high toughness and good environmental resistance. The toughness comes from a combination of the inherent toughness of high molecular weight, linear chain polymers, combined with good impregnation and interfacial characteristics. This high toughness can also translate into reduced operating costs as there is a reduced requirement for costly, non-destructive inspection and a lower frequency of repairs. Epoxy resins are plasticised by hot water but have excellent resistance to more aggressive environments encountered in aerospace applications. Most of the thermoplastic polymers have little sensitivity to water though some may be affected by environments such as paint strippers and hydraulic fluids.

Thermosetting matrix materials require long cure cycles, typically taking many hours, because of the chemical reactions in the resin. True thermoplastic matrix composites can be processed very quickly as it is only necessary to heat, shape and cool. Fabrication temperatures generally are higher for thermoplastics than for thermosets, but the processing time is usually minutes rather than hours. Polymerisation takes place during fabrication for thermosetting matrix composites and therefore it is impossible to verify the quality of the part until the final stage. In this respect the pseudo-thermoplastics are similar to thermosets as chemical reactions take place during final fabrication, and long processing times are usually required. True thermoplastics may be processed using a wide variety of fabrication techniques, but further work is needed to prove the practicality and economics of the techniques in production. For true thermoplastic matrix composites the chemistry is complete and therefore it is possible to verify the quality of the polymer before making the component. There is also the potential for re-processing imperfect parts and recovering unused material for other applications, neither of which is possible with thermosets.

3 THERMOPLASTIC POLYMERS

3.1 Polymer Chemistry

An understanding of the chemical structure of the polymers and how this affects properties is useful, though it is not intended here to examine the polymer chemistry in detail. The structures of some of the thermoplastic polymers used as composite matrices are shown in Fig. 1. They are arranged in order of ascending glass–rubber transition temperature (T_g) as this usually determines the upper service temperature of advanced composites. The upper service temperature requirements are ~100 °C for transport aircraft and helicopters, up to 175 °C for fighter aircraft, and much higher temperatures can be encountered in specific areas, e.g. close to engines.

All of the polymers are heavily aromatic in nature, usually with flexibilising groups in the backbone. The first three polymers in Fig. 1 are semi-crystalline: poly(phenylene sulphide) (PPS), poly(ether ether ketone) (PEEK), and poly(ether ketone) (PEK). The polymer chains are able to align themselves forming physical linkages between adjacent chains to form a crystalline morphology. This is possible because of the repeat sequences and the absence of bulky side groups. The semi-crystalline polymers retain some stiffness and strength above T_g, and the retention depends upon the level of crystallinity. PPS can develop up to 50% crystallinity, PEEK and PEK up to 40% under normal processing conditions. Another feature of semi-crystalline polymers is that it is necessary to process above the crystalline melting point (T_m).

The remaining polymers in Fig. 1, with the exception of some of the polyimides, are all amorphous, that is they do not develop a defined morphology. Several of the polymers contain sulphone linkages. These provide a stiff unit in the polymer backbone and hence increase T_g, but it is also a bulky group, which prevents crystallisation. The T_g values of the sulphone containing polymers range from 190 °C for polysulphone (PS) to 230 °C for poly(ether sulphone) (PES).

Poly(ether imide) is the first of the imide containing polymers with a T_g of 216 °C. The imide link is very stiff, giving higher T_g values, and is very thermally stable. It is therefore commonly found in the higher temperature thermoplastic and thermoset polymers. In the case of PEI the imide is present in relatively low proportions and therefore the polymer has a relatively low T_g but is easy to process. Where the imide is

Polymer	Glass Transition Temperature °C	Structure
Poly(Phenylene Sulphide)	85	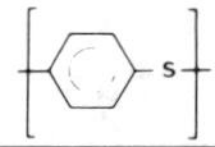
Poly(Ether Ether Ketone)	143	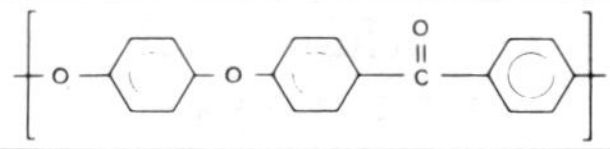
Poly(Ether Ketone)	165	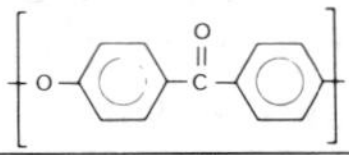
Poly(Sulphone)	190	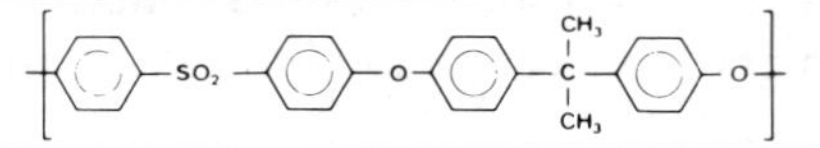
Poly(Ether Imide)	216	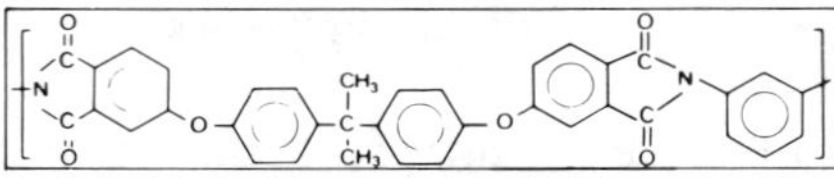
Poly(Phenyl Sulphone)	220	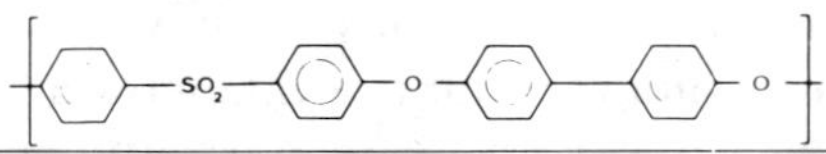
Poly(Ether Sulphone)	230	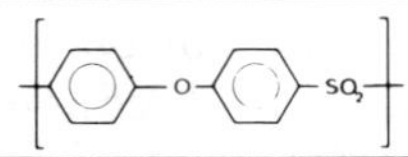
Poly(Amide Imide)	249-288	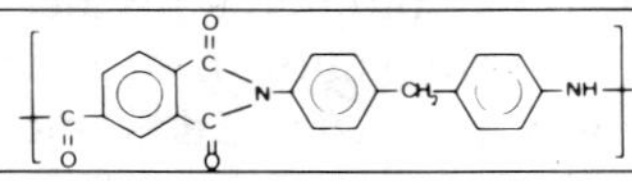
Poly(Imide)	256	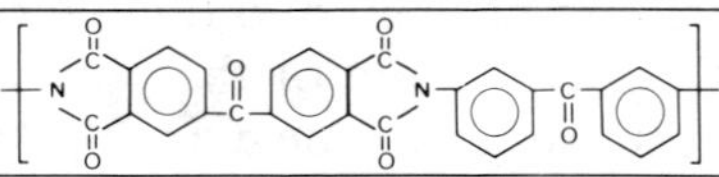

Fig. 1. Chemical structures of thermoplastic polymers used as composite matrices.

present in higher proportions the polymers may have very high T_g values but can also be difficult to process, due to the higher viscosities. For example poly(amide imide) (PAI) has a very high viscosity and must be annealed to achieve the higher T_g and optimum properties. Thermoplastic polyimides are frequently impregnated as monomers or pre-polymers due to the high viscosity in their fully polymerised state. The polymerisation is completed during the fabrication process or during a post-fabrication stage. Several different polyimides are available and the one illustrated in Fig. 1 is that of LARC-TPI (Langley Research Center — Thermoplastic Polyimide). Most of the polyimides are amorphous but it is possible to crystallise some of them. For example crystallinities of up to 30% have been reported in LARC-TPI.

A relatively new class of polymers are the liquid-crystalline polymers. These have not been included in Fig. 1 as the chemical structures of the commercially available polymers have not been disclosed. The polymers have rigid backbones, frequently an aromatic polyester, which become highly aligned during processing to produce very ordered structures in the solid state. The polymers are therefore intrinsically anisotropic. A detailed description of liquid crystalline polymers has been made by Calundann & Jaffee.[6]

3.2 Commercially Available Polymers

The manufacturers and trade names of the more common high performance thermoplastic polymers are listed in Table 2. The information in this table is taken from the manufacturers' literature.[7–16] In some cases there are several manufacturers of the same polymer but the polymers may not be identical in thermal, mechanical or other properties. Table 2 only contains polymers which have been announced as being made on a commercial scale.

The glass–rubber transition, melt and typical processing temperatures for the polymers are listed in Table 2. Melt temperature is only applicable for the semi- and liquid-crystalline polymers.

The chemical nature of the polymer in terms of true or pseudo-thermoplastic is also noted in Table 2. A true (or 'conventional') thermoplastic is taken to mean one in which the chemistry is complete, and processing involves a reversible physical change of heating to melt and cooling to solidify. A pseudo-thermoplastic is one in which a chemical reaction takes place during processing to increase molecular weight and/or expel volatiles to achieve the desired properties. The polymers may retain some thermoplasticity after processing but this is usually very limited. LARC-TPI was licensed to Mitsui–Toatsu and Rogers Corporation and both companies have developed polymers from the original chemistry.

A number of polymers are now being developed to meet high temperature applications. They have T_g values in the range 200–300 °C and include amorphous, semi- and liquid-crystalline polymers. In some cases the polymers may be available only in the reinforced form.

3.3 Polymer Properties

The properties of some of the thermoplastic polymers are summarised in Table 3 and have been reviewed by Johnston.[21] The target values are taken from Ref. 22 and have been determined to be the values needed to give the required composite properties for aerospace applications. The data on the polymers are taken from the manufacturers' published values. Most polymers are available in various grades and the properties may vary considerably. The particular grades used in the composites may not be available commercially but the grades selected for comparison in Table 3 are believed to be typical of grades used as composite matrices.

The target values are derived from comparisons of neat resin and composite properties, supported by theoretical assessments, and are minimum requirements. The resin modulus has a strong effect on composite compressive strength;[23] the higher the resin modulus the greater the composite compressive strength. A high yield strength is required to give adequate composite strength in matrix dominated lay-ups such as [±45°], and in transverse plies. The toughness requirement is to ensure good post-impact compressive strength. There is still considerable debate as to the relationship between T_g and the maximum use temperature. The heat distortion temperature has been included as this represents the temperature at which mechanical properties start to deteriorate significantly, and therefore may be a more useful assessment of upper service temperature than T_g. Achieving the target values does not ensure that the composite will be satisfactory, as there are many factors which determine the composite properties.

The semi-crystalline polymers have high moduli and yield strengths and in most cases exceed the target values for these properties. Limited toughness data are available for the semi-crystalline polymers but the values vary considerably, partially due to the different levels of crystallinity in the polymers. The heat distortion temperatures of the semi-crystalline polymers are similar to or greater than T_g, confirming that mechanical performance is retained at temperatures up to or above the T_g. Despite this there are concerns in using polymers above their T_g due to creep effects. The T_g values of all the semi-crystalline polymers,

TABLE 2
Manufacturers, Chemical Names and Morphology of Thermoplastic Polymers

Manufacturer	*Trade name*[a]	*Polymer*	*Polymer abbreviation*	*Glass transition temperature* T_g (°C)	*Melt temperature* T_m (°C)	*Processing temperature* T_p (°C)	*Morphology*	*Chemical nature*[b]
Amoco	UDEL	Polysulphone	PS	190	—	300	Amorphous	Conventional
	RADEL	Poly(phenyl sulphone)	PPYS	220	—	330	Amorphous	Conventional
	TORLON	Poly(amide imide)	PAI	249–288	—	400	Amorphous	Pseudo-
	XYDAR	Aromatic polyester	APE	350	421	400	Liquid-crystalline	Conventional
BASF	ULTRAPEK	Poly(ether ketone ether ketone ketone)	PEKEKK	175	375	400	Semi-crystalline	Conventional
DuPont	J-2	Polyamide	PA	160	—	300	Amorphous	Conventional
	AVIMID K-III	Polyimide	PI	250	—	350	Amorphous	Pseudo-
General Electric	ULTEM	Poly(ether imide)	PEI	216	—	360	Amorphous	Conventional

Hoechst–Celanese	FORTRON	Poly(phenylene sulphide)	PPS	85	285	340	Semi-crystalline	Conventional
	VECTRA	Aromatic polyester	APE	70	280	285	Liquid-crystalline	Conventional
	HOSTATEC	Poly(ether ketone)	PEK	167	360	400	Semi-crystalline	Conventional
ICI	VICTREX PEEK	Poly(ether ether ketone)	PEEK	143	334	380	Semi-crystalline	Conventional
	VICTREX PEK	Poly(ether ketone)	PEK	162	365	400	Semi-crystalline	Conventional
	VICTREX PES	Poly(ether sulphone)	PES	230	—	350	Amorphous	Conventional
Mitsui–Toatsu	LARC-TPI	Polyimide	PI	261	270–340	340	Semi-crystalline	Pseudo-
Phillips	RYTON	Poly(phenylene sulphide)	PPS	85	285	320	Semi-crystalline	Conventional
	PAS-2	Poly(arylene sulphide)	PAS	215	—	329	Amorphous Crystalline	Conventional
Rogers	DURIMID	Polyimide	PI	256	305	340	Semi-crystalline	Pseudo-

[a] All are registered trade names.
[b] See text for definitions.
Note: All data from manufacturers.

TABLE 3
Properties of Thermoplastic Polymers Used as Composite Matrices

Property	*Target*	*Semi-crystalline polymers*				*Amorphous polymers*						*Liquid-crystalline polymers*		*Typical current epoxy*
		RYTON[a] PPS	*VICTREX PEEK 450G[a]*	*VICTREX PEK[a]*	*LARC-TPI*	*VICTREX PES 4100G*	*UDEL P1700*	*ULTEM 6000*	*RADEL A400*	*AVIMID K-III*	*TORLON C[b]*	*VECTRA A-900[c]*	*XYDAR SRT-300[c]*	*HERCULES 3501-6*
Specific gravity	—	1·36	1·27–1·32	1·30	1·40	1·37	1·27	1·29	1·29	1·31	1·28–1·40	1·40	—	
Modulus (GPa)	2·8	3·4	3·7	4·0	3·5	2·6	2·6	3·0	2·1	3·7	3·8–4·8	45	17·2	3·8
Tensile yield strength (MPa)	90	79	92	105	166	84	70	103	72	101	152–193	614	138	46[e]
Elongation (%)	—	2–20	50	5	9	40–80	50–100	30	60	14	10–15	1·8	5	1[e]
Fracture toughness (kJ/m^2)	1·9–3·0	—	4·8	—	—	3·0	2·5	—	3·5	1·5	3·2–3·9	—	1·2	0·08
Izod notched impact strength (J/m)	—	21	83	—	21	84	69	40	—	—	80	530	125	—
Glass transition temperature (°C)	25°C above use temperature	85	143	162	261	230	190	230	220	250	249–288	70	350	190
Heat distortion temperature at 1·82 MPa (°C)	—	137	140	186	—	203	174	215	204	—	274	180	355	—
Limiting oxygen index (%)	—	44	35	40	—	38	30	44	—	50[d]	43	35	42	20–25
Source of data (Ref.)	17, 22	7, 18	8, 19	8	15	8	9	10	9	14	9	12	13	20

[a]Properties dependent on crystallinity.
[b]Properties dependent on post-cure.
[c]Properties highly anisotropic.
[d]AVIMID K-II.
[e]Brittle failure prior to yield.

except for PPS, appear adequate for most aerospace applications.

The amorphous polymers have lower moduli and yield strengths but higher toughnesses (as assessed by the notched Izod impact strength) than the semi-crystalline polymers. The moduli and yield stress of the sulphone based, amorphous polymers ('VICTREX' PES, 'UDEL' and 'RADEL') are lower than the requirements. This is of concern because compressive strength, shear strength and transverse ply strength are frequently limiting properties for composites. The imide containing polymers (ULTEM, AVIMID K-III and TORLON) have higher moduli and yield strengths than the other amorphous polymers due to the stiffness of the imide unit. These polymers meet the target values for all properties but have the processing disadvantages discussed earlier. The heat distortion temperatures of the amorphous polymers are lower than T_g by 14–30 °C. This indicates that the mechanical performance starts to fall significantly below T_g. Therefore for the amorphous polymers it is probably necessary to have a larger margin between maximum service temperature and T_g.

The liquid crystalline polymers have exceptionally high stiffness, strength and toughness. The polymers are highly anisotropic and it may not be possible to achieve these values in a typical composite moulding. It has not been demonstrated whether the liquid crystalline morphology can be achieved in a high fibre volume composite. If the polymers could be oriented to provide the maximum properties in the desired directions then the liquid-crystalline polymers would have a major property advantage over the other groups of polymers.

It is useful to compare the properties of the candidate thermoplastics with those of a widely used epoxy resin, Hercules '3501-6'. This has high modulus but is extremely brittle, fracturing prior to yield in tensile test and with a fracture toughness 15–60 times lower than those of the thermoplastics.

The limiting oxygen index of the thermoplastic polymers is relatively high for all the polymers for which data is available. The results indicate that it is necessary to have an atmosphere containing at least 30% oxygen to support combustion of the polymer.

Chemical resistance is an important aspect of performance. Chemical and environmental resistance data is available but it is difficult to compare it directly owing to the differences in exposure times, temperatures and tests performed. Some general comments may be made. The semi-crystalline polymers have high resistance to most environments due to the crystalline morphology. They even resist chlorinated solvents, strong acids and alkalis. The weight uptake and effect on properties due to moisture is negligible. Some of the amorphous polymers have a sensitivity to water with a weight uptake of 2·2% being reported for a thermoplastic polyimide immersed in hot water.[24] The effect of the absorbed moisture is to plasticise the resin, reduce the T_g and the elevated temperature mechanical properties. Polar, organic chemicals have a significant effect on amorphous polymers and in some cases are used as solvents for impregnation. Chlorinated solvents can readily dissolve many of the amorphous polymers. There have been attempts to improve the solvent resistance of certain amorphous thermoplastics through the introduction of crosslinks and crystallinity.[25] The environmental resistance of the liquid-crystalline polymers appears to be very good, owing to the crystalline morphology.

Current epoxy resins show exceptionally good resistance to most environments due to their crosslinked structure, but they are sensitive to water. Weight uptakes of 5% may occur in realistic exposure conditions, leading to a drop in T_g of up to 50 °C. Readers are also referred to Chapter 5 for a further discussion of environmental effects.

4 PRODUCT FORMS AND MANUFACTURERS

4.1 Product Forms

It is necessary to obtain good impregnation of the matrix into the reinforcing fibres in order to achieve the full properties of the composite. Good impregnation is also a prerequisite to creating the fibre–matrix interface, which is increasingly recognised as important in determining the composite behaviour. Several impregnation techniques may be used and the resulting differences in product forms affect the handling, processing and properties of the composite.

Impregnation of true thermoplastics is generally more difficult than of thermosets due to the relatively high melt viscosities of thermoplastics, typically 10^3–10^4 N s/m^2 at a shear stress of 10^5 N/m^2. Pseudo-thermoplastics may have lower viscosities as they are typically not fully polymerised at the time of impregnation.

The techniques may be divided into those which give a pre-impregnated material and those which give a post-impregnated form. Pre-impregnated forms are ones in which the reinforcing fibres are

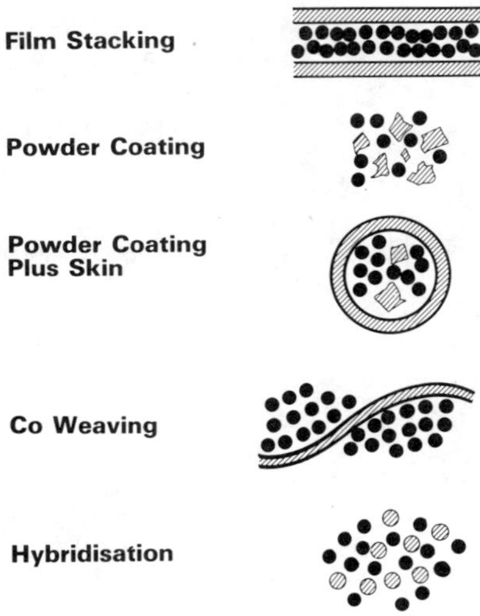

Fig. 2. Impregnation techniques for thermoplastic composites.

completely wetted and impregnated with the matrix. Post-impregnated forms achieve only a physical mixing of the matrix and fibre and do not give full wetting of the reinforcement. The techniques are discussed below and some of them are illustrated in Fig. 2.

4.1.1 Pre-impregnation Techniques

The main feature of the pre-impregnated materials is full wetting and impregnation of the reinforcing fibres, with no voids. In addition, it is desirable that the matrix be uniformly distributed without resin rich areas, and that the pre-preg should be of uniform thickness. Micrographs through a high quality thermoplastic/carbon fibre pre-preg tape and a laminate moulded from it are shown in Figs 3 and 4 respectively.

4.1.1.1 Melt impregnation. In principle the simplest method of impregnation is to pull the reinforcing fibres through molten polymer. The high viscosity of true thermoplastics makes it extremely difficult to achieve good impregnation by this technique. It may be possible to achieve good impregnation with pseudo-thermoplastics or very low molecular weight true thermoplastics. This method has the advantage of not requiring any solvents, which must later be removed, or costly intermediates, such as polymer films or fibres.

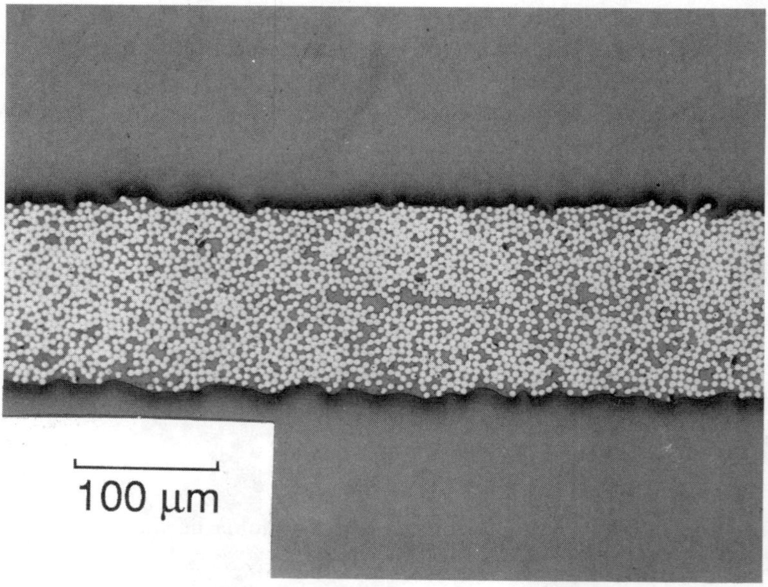

Fig. 3. Photomicrograph of fully impregnated carbon fibre/thermoplastic pre-preg tape. Note the uniform fibre/matrix distribution, absence of voids and uniform thickness.

4.1.1.2 Solvent-aided impregnation. The viscosity of the polymer may be reduced by partial or total solvation, and impregnating the polymer by techniques similar to those used for thermoset impregnation. Most thermoplastics cannot be dissolved in common, low boiling solvents and it is necessary to use higher boiling point solvents such as *N*-methyl pyrrolidone. It can be difficult to remove the solvent totally and if any is left in the composite it may plasticise the matrix. One limitation of the technique is that polymers which can be readily solvated also have poor resistance to those solvents in service. In particular the semi- and liquid-crystalline polymers cannot readily be dissolved for solvent impregnation. Some of the solvent may be deliberately left in the pre-preg to give some tack and drape. Solvent-aided impregnation was used to manufacture the early thermoplastic pre-pregs such as polysulphone matrix composites, and is the most common method of impregnation for amorphous thermoplastics and pseudo-thermoplastics.

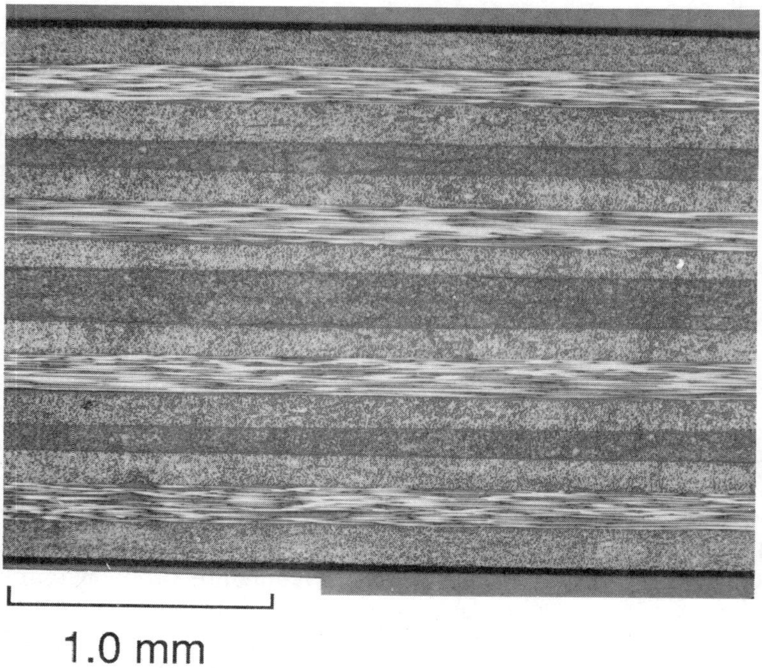

Fig. 4. Photomicrograph of 16-ply, quasi-isotropic carbon fibre/thermoplastic laminate.

4.1.2 Post-impregnation Techniques

4.1.2.1 Film stacking. Film stacking was one of the first techniques to be used and may be applied to any thermoplastic that can be converted into a film. Layers of polymer film are interlayered with the reinforcing fibres and impregnation is achieved by applying heat and pressure. The principle is illustrated in Fig. 2 and is discussed in Refs 4 and 26. Film stacking may be used to make flat sheets which can then be fabricated in a subsequent stage. Alternatively impregnation may take place during fabrication in a single stage process. It is difficult to achieve good impregnation due to the viscosity of the polymer and it is usually necessary to use long times or high pressures together with high temperatures. The use of high pressure can be self-defeating as it has the effect of compacting the fibres and hence making impregnation more difficult. It is usual for the reinforcement to be in a fabric form as this is less prone to distortion and 'wash' during the impregnation stage. It is also possible to carry out film impregnation on a continuous basis in which film and fibres are carried through a lamination process. Film stacking has an advantage in that it is possible to make components from combinations of fibres and polymers which are not available in commercial pre-preg form. Film stacking is also used to make specialist components such as radomes.[4]

4.1.2.2 Powder coating. Particles of the polymer are distributed into the reinforcing fibre and impregnation is usually achieved in the processing/fabrication stage. In order to achieve good impregnation it is necessary to have fine polymer particles, preferably the same size as the fibre diameter, and to have the particles uniformly distributed within the fibres. This ideal state is illustrated in Fig. 2. Segregation of the powder from the fibres may occur during normal handling, leading to non-uniform polymer distribution and changes in actual resin content. Three approaches have been used to overcome this problem. Gantt *et al.*[27] have reported a technique in which a direct current is applied to the carbon fibre reinforcement during impregnation of the powder. The current causes resistive heating and local melting of the polymer onto the fibres. A second approach is the use of a tackifying agent to give adhesion between the polymer and fibre.[28] The tackifying agent may also provide sufficient tack for pre-preg layers to adhere to one another during lay-up at room temperature. The third approach is the addition of a 'skin' around the powder-impregnated composite to prevent loss and movement of the polymer particles. This is illustrated in Fig. 2. It is usually difficult to achieve high fibre contents due to the polymer in the powder and the skin. The powder technique has the same disadvantages as film stacking, in that impregnation must be completed during the processing/fabrication stage. The times, temperatures and pressures required will be highly dependent on the size and distribution of the particles. An advantage of the technique is that it can provide a drapeable pre-preg and may also have tack if a suitable agent is used.

4.1.2.3 Co-weaving. Co-weaving consists of weaving tows of the reinforcing fibre and polymer fibre together. Figure 2 illustrates the usual technique in which one fibre is woven as the warp and the other as the weft. As an alternative the polymer could be a slit film which is co-woven with the reinforcing fibres. The technique has the simplicity of being a textile operation, and is therefore easy to perform, and the fabric has drapability. Disadvantages are that it is necessary to obtain the polymer in a fibre form and that impregnation takes place during the processing/fabrication stage. The ease of impregnation depends on the weave style and the size of the polymer and reinforcing fibre tows. A variation is to alternate reinforcing and polymer tows in each direction. Co-woven composites are made by a number of textile suppliers, and have the advantage of providing a route to materials not available in pre-preg form. Co-woven composites have been discussed further by Silverman & Jones.[29]

4.1.2.4 Hybridisation. Hybridisation is also known as co-mingling and involves an intimate mixing of polymer and reinforcing fibres into a single tow. The resultant hybrid tow is normally woven into a fabric but could be used in a unidirectional form.[29,30] Small diameter fibres are desirable and it is necessary to achieve good intermingling of the polymer and reinforcing fibres. Segregation of the fibres may occur during handling. As with powder coating and co-weaving, final impregnation is achieved during the processing/fabrication stage. The quality of the composite will depend upon the size and distribution of the polymer fibres. In principle it should be easier to achieve full impregnation in a hybrid system than in a co-woven one as there should be a greater degree of mixing in the hybrid.

4.2 Manufacturers of Thermoplastic Composites

The principal manufacturers of thermoplastic composites are summarised in Table 4. Most of the manufacturers are also suppliers of high performance thermoplastics, and in developing composite systems the suppliers have concentrated on their own polymers. In some cases the continuous fibre composite is sold under a separate trade name, and in other cases the composite is designated by the polymer name. The product form is also indicated, though in most cases the impregnation process is proprietary. ATO and Spie–Batignolles do not supply composites based on advanced polymers as defined earlier but have been included as impregnation technology can be applied to other polymers. For example Ciba–Geigy have licensed the ATO process in the USA.

There are also a number of other suppliers of solvent-impregnated

TABLE 4
Principal Manufacturers and Trade Names of Thermoplastic Composites

Manufacturer	*Matrix*	*Trade name of composite*	*Product form*
American Cyanamid	ULTEM—poly(ether imide)	CYPAC	Fully impregnated
Amoco	RADEL—poly(phenyl sulphone) TORLON—poly(amide imide)	—	Fully impregnated
ATO	Nylons—polyamide	FIT (fibre impregnation with thermoplastic)	Powder + skin
BASF	Various	—	Hybrid, powder
Ciba-Geigy	Various	—	Powder + skin (licensed from ATO)
DuPont	AVIMID K-III—polyimide J-2—polyamide	— —	Fully impregnated
ICI Fiberite	VICTREX PEEK—poly(ether ether ketone) Other aromatic polymers	APC (aromatic polymer composite)	Fully impregnated
Phillips	RYTON—poly(phenylene sulphide) Other aromatic polymers	—	Fully impregnated
Spie-Batignolles	Nylons—polyamide	SPIFLEX	Fully impregnated

composites and co-woven/hybrid composites. These techniques are readily available and can be applied to a range of different polymers and reinforcements.

5 PROCESSING AND PROCESSING EFFECTS

5.1 Pre-impregnated, True Thermoplastics

The processing conditions are defined by the temperatures, pressures and times that are required to consolidate the pre-impregnated composite into a laminate. These conditions are usually dictated by the polymer matrix and the product form of the composite. The manufacture of complex shapes by different fabrication processes is discussed in Section 7. The recommended conditions for processing various pre-impregnated composites are summarised in Table 5. In order to process a fully impregnated, true thermoplastic it is only necessary to melt the polymer and apply moderate pressure to consolidate the pre-preg layers. This is a rapid process requiring typically only a few minutes at the processing temperature and pressures of 0·7–2·0 MPa (100–300 psi) as shown for the APC-2, RYTON and CYPAC composites. It is necessary to process above the melt temperature for semi-crystalline polymers (Table 2). There is no melting process in amorphous polymers and typically it is possible to process over a wider temperature range, though longer times will be needed at lower temperatures. Semi-crystalline polymers may be typically processed at 200 °C above T_g and amorphous polymers at 150 °C above T_g. For true thermoplastics the heat-up rate is not critical.

A wide range of processing conditions were examined for APC-2 by Iaconis[33] including temperatures of 343–393 °C, pressures of 350–4150 Pa (50–600 psi) and consolidation times of 1–10 min. The conditions which produced good quality laminates were at temperatures of greater than 357 °C, pressures of 350 Pa (50 psi) or higher, and times of 5 min or more.

No change in the properties of RYTON PPS composites were found with moulding temperatures in the range 291–360 °C and times of 7–20 min.[34] The thermal stability of PPS and PEEK matrix composites at processing temperatures has been assessed by Ma *et al.*[35] PPS was found to have lower thermal stability than PEEK and additional

TABLE 5
Processing Conditions for Pre-impregnated Thermoplastic Matrix Composite

	True thermoplastics			*Pseudo-thermoplastics*		
	APC-2	*RYTON*	*CYPAC*	*TORLON*	*AVIMID K-III*	*LARC-TPI*[a]
Polymer	PEEK	PPS	PEI	PAI	PI	PI
Volatiles during processing	No	No	No	Yes	Yes	No
Moulding technique	Press	Press	Press	Press	Autoclave	Press
Heat-up						
Rate (°C/min)	Any	Any	Any	1–28	2	8
Pressure	Not critical	Not critical	Not critical	—	Vacuum	—
Consolidation						
Temperature (°C)	380–400	300–343	304–343	349	350	350
Time (min)	5	5	5	5–20	60	60
Pressure (psi)	100–200	150	100	200–500	200	300
(MPa)	0·7–1·4	1·0	0·7	1·4–3·5	1·4	2·0
Cool-Down						
Rate (°C/min)	(10–700)	—	Not critical	3–36	0·5°C/min to 250°C; 3°C/min below 250°C	Not critical
Pressure (psi)	100–300	150	100	200–500	100–200	300
(MPa)	0·7–2·0	1·0	0·7	1·4–3·5	0·7–1·4	2·0
Post-Cure						
Time (days)	—	—	—	3–12	—	—
Temperature (°C)				204–266		
Source of data (Ref.)	31	7	32	9	14	44

[a] Pre-preg 'B'-staged prior to moulding.

crystallisation processes were observed during thermal cycling of PPS. The viscosity of both polymers was observed to rise slightly with time in the melt, though the effect was less pronounced for PEEK. This rise is consistent with the 10% rise in molecular weight of PEEK held at 400 °C in air for 2 h reported by Cogswell.[36] It is interpreted that the changes in molecular weight, thermal properties and viscosity did not represent a limitation in processing PEEK matrix composites.[36]

It is not usually necessary to control the cooling rate for amorphous matrix composites but it may be necessary for semi-crystalline thermoplastics, to ensure that the appropriate crystallinity and microstructure is achieved. For example it has been shown that for PEEK matrix composites the crystallinity and properties are relatively invariant in the cooling rate range 5–700 °C/min.[37–40] Toughness is expected to be the property most affected by variations in crystallinity and crystalline morphology. No change in interlaminar fracture toughness of APC-2 was observed in the cooling rate range of 3–19 °C/min, and the toughness at 1 °C/min was observed to be about 20% lower than at the higher cooling rates.[40] Berglund[38] has examined higher cooling rates ranging from 15 °C/min to 1000 °C/min, and the effects of annealing. Crystallinity varied from 0% (amorphous) at the higher cooling rates to 33% at the slowest cooling rate. Transverse fracture toughness varied from 2·0 kJ/m^2 at the fastest cooling rate to 1·0 kJ/m^2 at the slowest cooling rate. The toughness of the quenched and annealed sample followed the same behaviour of toughness versus crystallinity as the other samples indicating that the level of crystallinity, rather than processing history, is the controlling factor in determining toughness. The interlaminar fracture toughness of PPS composites was examined as a function of crystallinity by Davies *et al.*[41] The level of crystallinity ranged from 5%, giving an interlaminar toughness of 0·92 kJ/m^2, to 31% which gave a toughness of 0·76 kJ/m^2. For PPS matrix composites it has been shown that a rapid quenching followed by annealing at 200 °C for 2 h produced optimum properties.[34] This was attributed to smaller crystallites being formed during this sequence, resulting in a tougher polymer. These results indicate that the effects of cooling rate and crystallinity on toughness of PEEK and PPS matrix composites are relatively small over a wide range of processing conditions.

The variation of cooling rate through the thickness of laminates under different cooling conditions has been examined and was found not to result in significant differences except for thick laminates at extremely high cooling rates.[42]

5.2 Pre-impregnated, Pseudo-thermoplastics

The processing cycles of the pseudo-thermoplastics have less flexibility than those for the true thermoplastics. This is due to the need to control the chemical reactions which take place during the processing of the composite. TORLON, AVIMID K-III and LARC-TPI undergo a condensation polymerisation and therefore it is necessary to ensure that the polymerisation by-products and any residual solvents can be removed during processing. Post-curing of TORLON causes an increase in molecular weight and glass transition temperature. Solid-state polymerisation and removal of reaction water occur during the post-curing, although no significant crosslinking takes place.[9,43]

Two different cure cycles have been presented for AVIMID K-III laminates, the one in Table 5 being the simplest.[14,24] It is necessary to ensure that vacuum is applied during the moulding cycle to remove the condensation polymerisation products (water and ethanol) and to remove the excess solvent (*N*-methyl pyrrolidone). The cycle in Table 5 produces laminates with less than 0·5% voids. An alternative cycle involving several holds has been recommended as having advantages in ensuring that the laminate is equilibrated within the moulding temperature range and to allow sufficient time for consolidation.[24] The cycle incorporating the holds gives a total moulding cycle time of 12 h compared with 8 h for the cycle in Table 5.[24]

The cycle for LARC-TPI is for a material which has been 'B-staged', i.e. the imidisation reaction is complete. Hartness[44] has described two alternative cycles which involve holds at 260 °C during heat-up, and application of consolidation pressure at different times to ensure that wetting, voidage and mechanical properties are optimised. Although these cycles involve holds they do include more rapid heat-up and so the total cycle time is similar to the cycle in Table 5.

It is apparent that processing of the pseudo-thermoplastics is closer to thermosets than to true thermoplastics. Longer cycles are needed for pseudo-thermoplastics and it may be necessary to carry out a post-curing stage to optimise the properties. The cycles described here could be used to produce final components, or to produce sheets which could be subsequently fabricated into components as the pseudo-thermoplastics retain some thermoplasticity after initial processing.[14] The long cycles and high temperatures represent a limitation and it is unlikely that the pseudo-thermoplastics will be amenable to the use of rapid fabrication techniques which are one of the advantages of thermoplastic composites.

5.3 Post-impregnated Thermoplastics

Processing cycles for various post-impregnated PEEK matrix composites are shown in Table 6, compared with a pre-impregnated PEEK composite. The cycles for the post-impregnated composites typically require longer times than for a pre-impregnated material, and it may be necessary to use higher pressures. Silverman & Jones[29] have reported that it is possible to obtain laminates of similar quality in terms of voidage and resin distribution in co-woven and hybrid composites, to those made from a pre-impregnated form. Despite this the reported values of void content in the post-impregnated laminates were up to 4·4% but only 1·9% for the pre-impregnated laminates. The particular conditions used may be dependent on the number of carbon fibres in the tow. A moulding temperature of 416°C was needed for 6 k tow and 399°C was sufficient for 3 k tow.[29] Considerable differences in the fibre–matrix interface were observed, which was attributed in part to the different sizing agents applied to fibres.[29] Hartness[45] made PEEK composites by the film stacking technique and found that a 2-h consolidation time was needed to achieve adequate wet-out of the fibres, compared to the 5 min used by Iaconnis[33] to produce good quality laminates from pre-impregnated tape.

In another study Chang[46] compared the properties of unidirectional composites made by various routes in a matrix of 'J-Polymer' polyamide copolymer. The results from the study are summarised in Table 7. The co-wound composites were made by winding the reinforcing fibres with the thermoplastic fibres to produce a hybrid yarn. The highest mechanical properties were obtained with the melt coated composites. The number of filaments in the carbon fibre tow had an effect on mechanical properties and the interlaminar shear strength was three times higher for the smaller tow size. These results further show that the product form can have a large effect on the properties of the composite.

It is necessary to use longer moulding times and possibly higher temperatures for a post-impregnated composite compared to a pre-impregnated material. There may be considerable differences in quality of laminates and in mechanical properties of laminates made from

TABLE 6
Processing and Properties of Pre- and Post-impregnated PEEK Matrix Composites (All Data from Ref. 29)

	Pre-impregnated APC-2/AS4 (Fibre: AS4)	*Post-impregnated*		
		Plain Weave[a] *(Fibre: G30-500)*	*No crimp structure*[a] *(Fibre: AS4)*	*Co-woven plain-weave*[a] *(Fibre: AS4)*
Processing				
Consolidation temperature (°C)	400	416	416	399
Consolidation pressure (psi)	200	200	200	200
Consolidation time (min)	30	60	60	60
Properties				
Fibre content (volume %)	60	56	58	62
Void content (%)	1·9	2·1	3·2	4·4
0° Flexural modulus (GPa)	108	98	106	65
0° Flexural strength (MPa)	1 687	1 514	1 222	1 150
90° Tensile strength (MPa)	91	64	26	Not tested

[a]Unidirectional reinforced fabrics.
Fibres: AS4 — Hercules Magnamite; G30-500 — BASF Celion.

TABLE 7
Effect of Product Form on Mechanical Properties of Polyamide Matrix Composites (Unidirectional Reinforced J-Polymer)[a]

Process	*Number of filaments in carbon tow*	*Interlaminar shear strength (MPa)*	*Compressive strength (MPa)*
Melt coating	3 000	86	1 110
Film impregnation	12 000	78	932
Yarn co-winding	3 000	21	—
Yarn co-winding	1 000	76	966

[a]Reinforcement Hercules Magnamite AS4 fibres; 60% fibre volume.
Data from Ref. 46.

post-impregnated forms due to fibre–resin distribution, voids and the fibre–matrix interface.

6 PROPERTIES OF THERMOPLASTIC MATRIX COMPOSITES

6.1 Room Temperature Mechanical Properties

The mechanical properties of thermoplastic matrix composites have been extensively characterised and data on unidirectional carbon, woven carbon, glass and aramid fibres are presented in Tables 8, 9, 10 and 11 respectively. Wherever possible the comparisons have been made with the same grades of reinforcing fibre and at the same fibre content.

The properties of unidirectional carbon fibre composites are summarised in Table 8. The stiffness properties are largely determined by the type of fibre and the fibre content. There are significant differences in strength between the various composites and this cannot be accounted for entirely by the fibre differences. This is more evident for flexural, compressive and short-beam shear strengths. The 0° compressive strength of the thermoplastic composites is generally lower than that of typical epoxy matrix composites. This is usually attributed to the lower modulus of the thermoplastic polymers[21] but this does not totally account for the differences. For example PEEK and 3501-6 epoxy have almost identical resin moduli (Table 3) but the 0° compressive strength of pre- and post-impregnated PEEK matrix composites is only

TABLE 8

Mechanical Properties of Unidirectional, Carbon Fibre/Thermoplastic Composites (All Properties Measured at Room Temperature)

	Pre-impregnated true thermoplastics				Pre-impregnated pseudo-thermoplastics			Post-impregnated true thermoplastics		Typical current epoxy resin
	VICTREX PEEK	*RYTON PPS*	*J-polymer*	*Liquid crystal polymer*	*TORLON*	*AVIMID K-III*	*LARC-TPI*	*PEEK*[a]	*PPS*[a]	
Manufacturer	ICI Fiberite	Phillips	DuPont	Celanese	Amoco	DuPont	Mitsui–Toatsu	BASF	BASF	Hercules
Composite	APC-2/AS4	—	—	—	—	—	—	—	—	AS4/3501-6
Fibre[b]	AS4	AS4	AS4	6 000	AS4	IM6	HTA-7	AS4	G30-500	AS4
Physical properties										
Fibre content (volume %)	61	60·5	—	45–57	—	—	68	63	62	62
Void content (%)	<0·5	—	—	—	—	<0·5	—	—	—	—
Unidirectional mechanical properties										
0° Tensile modulus (GPa)	134	134	—	143	—	158	—	140	138	136
0° Tensile strength (MPa)	2 130	1 828	—	1 492	—	—	—	2 040	1 860	2 210
0° Compressive strength (MPa)	1 100	938	1 110	809	1 379	990	—	1 100	—	1 371
0° Flexural modulus (GPa)	121	117	—	107	128	124	124	135	132	—
0° Flexural strength (MPa)	1 880	1 897	—	1 515	2 070	1 590	1 590	2 280	1 790	—
90° Tensile strength (MPa)	80	—	—	—	—	47	—	75	19	52
0° Short-beam shear strength (MPa)	105	69	86	52	110	99	80	—	62	110
Interlaminar fracture toughness (kJ/m^2)	2·7/2·1	0·8	2·0	—	1·1–1·2	1·4	—	—	—	0·22
Source of data (Ref.)	31/47	7, 48	46	49	9, 43	14/24	15	50	50	51, 52

[a]Hybrid.
[b]Fibre: AS4, IM6 — Hercules Magnamite; HTA-6 — Toho Besfight; G30-500, 6000–BASF Celion.

TABLE 9
Mechanical Properties of Pre-impregnated Woven Carbon Fibre/Thermoplastic Composites
(All Properties Measured at Room Temperature)

	VICTREX PEEK	*RYTON PPS*	*TORLON PAI*	*ULTEM PEI*	
Manufacturer	ICI Fiberite	Phillips	Amoco	American Cyanamid	American Cyanamid
Composite	—	—	—	CYPAC X7005	CYPAC X7005
Fibre/weave	—/3k	—/Satin	Celion 3000/4-HS	—/3k 70P	—/3k 135-8H
Physical properties					
Fibre content (volume %)	60	60	—	54	54
Mechanical properties					
Tensile modulus (GPa)	70	77	62	56	52
Tensile strength (MPa)	769	690	676	724	676
Compressive modulus (GPa)	—	—	63	55	57
Compressive strength (MPa)	694	400	566	503	572
Flexural modulus (GPa)	61	50	59	55	52
Flexural strength (MPa)	1 052	745	1 007	828	621
Short-beam shear strength (MPa)	80	—	90	66	46
Interlaminar fracture toughness (kJ/m^2)	—	—	1·75	4·0	—
Source of data (Ref.)	54	48	9	32	32

Fibres: Celion 3000 — BASF Celion.

TABLE 10

Mechanical Properties of Pre-impregnated Glass Fibre/Composites (All Properties Measured at Room Temperature)

	Unidirectional glass				*Woven glass*		
	VICTREX PEEK		*RYTON PPS*		*RYTON PPS*	*ULTEM PEI*	
Manufacturer	ICI Fiberite		Phillips		Phillips	American Cyanamid	American Cyanamid
Composite	—	—	—	—	—	CYPAC X7005	CYPAC X7005
Fibre/weave	E-glass/—	S-2 glass/—	E-glass/—	S-2 glass/—	E-glass/Satin	E-glass/120E	E-glass/7781
Physical properties							
Fibre content (volume %)	62	61	57	55	60	53	52
Mechanical properties							
Tensile modulus (GPa)	55	51	52	49	24	21	26
Tensile strength (MPa)	1 126	1 158	835	1 117	248	338	469
Compressive modulus (GPa)	—	—	—	—	—	24	29
Compressive strength (MPa)	1 044	871	800	—	297	572	524
Flexural modulus (GPa)	53	54	42	21	21	21	26
Flexural strength (MPa)	1 213	1 246	1 145	952	366	414	579
Short-beam shear strength (MPa)	95	70	43	32	—	70	76
Interlaminar fracture toughness (kJ/m^2)	3·3/3·1	2·5	1·3	1·3	—	—	—
Source of data (Ref.)	55	31	48	56	48	32	32

TABLE 11
Mechanical Properties of Pre-impregnated Aramid Fibre/Thermoplastic Composites (All Properties Measured at Room Temperature)

	VICTREX PEEK	*RYTON PPS*	*J-polymer PA*
Manufacturer	ICI Fiberite	Phillips	DuPont
Fibre	Aramid[a]	Kevlar 49[b]	Kevlar 49[b]
Physical properties			
Fibre content (volume %)	60	53	60
Mechanical properties			
Tensile modulus (GPa)	63	64	73–79
Tensile strength (MPa)	976	1 103	1 200–1 400
Compressive strength (MPa)	230	193	248–276
Flexural modulus (GPa)	47	48	70–73
Flexural strength (MPa)	520	593	675–700
Short-beam shear strength (MPa)	44	—	55–62
Interlaminar fracture toughness (kJ/m^2)	—	0·7	2·0
Source of data (Ref.)	55	48	57

[a]Fibre not defined.
[b]Fibres: Kevlar 49 — DuPont.

80% that of the epoxy matrix composite. It has been suggested that the low values may be due to fibre damage, fibre misalignment or non-optimised fibre–matrix interface.[21] The latter theory is supported by Fife *et al.*[53] who showed that PEEK matrix composites with a non-optimised interface had lower 0° compressive, 0° tensile and 90° flexural strengths than composites with an optimised interface, and the observations of Chung & McMahon[49] that low compressive strength in liquid-crystalline matrix composites was due to poor interface rather than fibre misalignment. The low 90° tensile strengths of most of the composites is further indication of a poor fibre–matrix interface. The short-beam shear test is used to assess interlaminar shear strength, but as has been pointed out by several authors the failure mode is rarely by interlaminar failure, and therefore these values are of limited use in comparing materials.

The properties of woven carbon fibre thermoplastic composites are summarised in Table 9. The properties show similar effects to those noted in the unidirectional composites. Good fibre utilisation is shown by the 0° tensile properties, but in most cases the 0° compressive properties are low. The interlaminar fracture toughness is notably high for the ULTEM PEI matrix composite. It is expected that thermoplastic matrix woven composites should have superior mechanical properties to thermosetting matrix composites, as the tougher matrix is less likely to fracture at the stress concentrations produced by the fibre crimps in the weave. At present there are insufficient data to confirm that this is indeed the case.

Table 10 summarises the properties of pre-impregnated, glass fibre composites. Results have been presented on PEEK and PPS composites with both E- and S-glass fibres. Glass fibres must have a sizing agent applied to the surface of the fibre as soon as they are manufactured in order to protect them during subsequent handling operations, and a coupling agent is usually applied to give improved bonding between the fibre and matrix. The sizing and coupling agents which withstand the thermoplastic processing temperatures are only now becoming available, and therefore the fibre–matrix interface is probably not optimised. This almost certainly affects the strength properties of the composites.

Some information on aramid fibre reinforced composites has been published and this is summarised in Table 11. Aramid fibres are organic polymers and it is possible that some degradation occurs during impregnation and processing in thermoplastic matrices.

In summary the thermoplastic composites show good fibre utilisation in stiffness. The strength properties vary considerably and are dependent on the composite type.

6.2 Toughness

The toughness of thermoplastic composites is higher than that of thermosets and for this reason the toughness has been extensively examined.[41, 52, 58–67] The property most commonly characterised is the interlaminar fracture toughness (G_{Ic}), obtained using a double cantilever beam (DCB) test. A more extensive description of the test methods and fracture behaviour is found in Chapter 9. The toughness of various thermoplastic matrix composites is given in Tables 8–11 and is summarised in Table 12.

A typical current thermosetting matrix composite such as Hercules AS4/3501-6 has an interlaminar fracture toughness of 0·2 kJ/m^2,

TABLE 12
Toughness of Carbon Fibre/Thermoplastic Composites

Composite	*Polymer/ fibre*	*Interlaminar fracture toughness G_{Ic} (kJ/m²)*	*Post-impact compressive strength (MPa)[a]*	*Source of data (Ref.)*
Unidirectional				
APC-2/AS4	PEEK/AS4	2·7/2·1	338	31
RYTON	PPS/AS4	0·8	—	7
J-polymer	PA/AS4	2·0	283–400	46
TORLON	PAI/—	1·1–1·2	327	9/43
AVIMID K-III	PI/—	1·4	274	14/24
Woven				
TORLON	PAI/plain weave	1·75	296	9/43
CYPAC X7005	PEI/plain weave	4·0	320	32
Typical current epoxy				
3501-6	Epoxy/AS4	0·22	165	51/52

[a]Test described in Ref. 5. Impact energy per unit thickness 6·7 J/mm (1500 in lb/in).
Fibres: AS4 — Hercules Magnamite.

whereas the thermoplastic matrix composites have toughnesses in the range 0·8–4·4 kJ/m². The high toughness imparts a much greater resistance to delamination, which is one of the most common failure modes of laminated composites. The low interlaminar fracture toughness of thermosetting matrix composites has been one of the main limitations to their use in service, due to the ease with which delaminations can initiate and propagate.[5, 68] The most commonly cited problem is that of a low energy impact, such as might happen during routine service, causing a delamination which is propagated during subsequent compressive loading,[5] see Chapter 11.

The high toughness of thermoplastic matrix composites is normally attributed to the high toughness of the thermoplastic polymers as shown in Table 3. The interface also plays a significant role and a strong interface is needed in order to achieve the potential toughness of the polymer matrix.[53]

The fracture behaviour of carbon fibre/PEEK composites has received considerable attention.[52, 53, 58–65] Two types of failure have been observed in DCB testing of APC-2, PEEK matrix composites. Stable crack growth is observed in some regions with values of fracture toughness in the range 1·5–2·5 kJ/m^2, and a very high matrix ductility as shown in Fig. 5(a). Unstable fracture is observed in other regions with toughness in the range 1·0–2·0 kJ/m^2 and reduced matrix ductility as shown in Fig. 5(b). The crack has been observed to slow during stable propagation prior to a transition to unstable fracture, and the G_{Ic} rises locally due to blunting at the crack tip.[63] The two regions of fracture have been examined in detail by Crick *et al.*[64] and Purslow.[65] The differences in fracture morphology are attributed to the change in crack speed. Both authors note the tendency for crazing to occur on multiple planes during fracture, and that the pattern of ductility during unstable fracture is related to the spherulitic structure of the PEEK. Several authors have noted the strong fibre–matrix interface in APC-2, PEEK matrix composites.[53, 64, 65] Polymer is seen adhering to the fibres on the fracture surface shown in Fig. 5 in both fracture modes, implying a strong fibre–matrix interface.

The effects of temperature and test rate have been examined.[38, 58, 60, 62, 66] The propagation interlaminar toughness of APC-2/AS4 increased from 1·9 kJ/m^2 at −30 °C to 3·0 kJ/m^2 at 80 °C and the arrest toughness remained relatively constant at 1·0 kJ/m^2 over the temperature range. The proportion of stable fracture increased with temperature and totally stable fracture was observed at 120 °C.[62] The effect of rate on APC-2/AS4 has been examined by Berglund[38] and Smiley & Pipes,[58] who also tested a thermoset for comparison. The toughness of both materials decreased at high rates, however the thermoplastic was less sensitive to rate and had a considerably higher toughness than the epoxy matrix composite.

Failure in PPS matrix composites has been observed to occur predominantly at the interface rather than in the matrix.[41] Even so, the reported values of fracture toughness were in the range 0·76–0·92 kJ/m^2 covering a range of specimen configurations and crystallinities.

Several authors have discussed Mode II interlaminar fracture toughness tests.[58, 60, 62] In general the observations are similar to Mode I; however the increase in toughness compared to current thermosets is not so great.

Fatigue toughness has also been examined.[59, 60, 67] This has involved crack growth measurements (da/dn)[59, 60] and the number of cycles required to initiate delamination.[67] The initial results have been

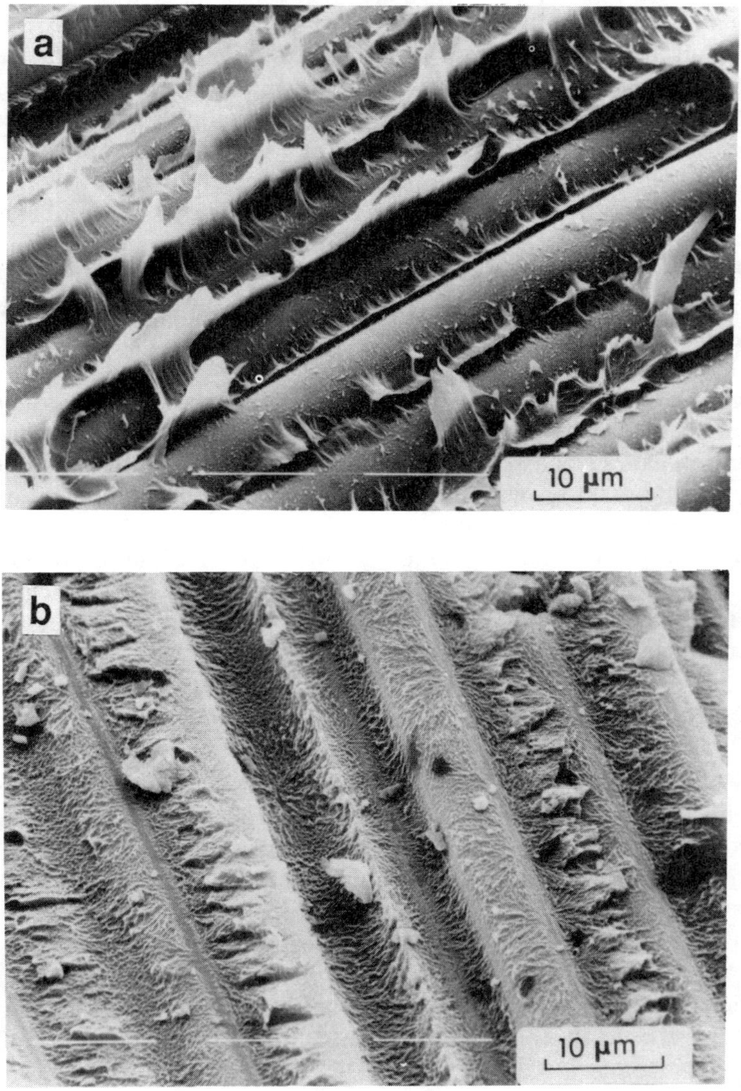

Fig. 5. Interlaminar fracture surfaces of PEEK/carbon fibre composite. (a) Stable fracture, (b) unstable fracture.

contradictory, with different thresholds and crack growth rates being reported for PEEK matrix composites. Prel *et al.*[59] concluded that there was a lower bound of 1 kJ/m^2 below which crack propagation will not occur. O'Brien[67] showed that the threshold toughness for delamination initiation decreases markedly with increasing numbers of fatigue cycles for all the tougher composites including thermoplastics. However the initiation toughness was still higher than that of a current thermoset system over the range of composites examined.

Another test frequently used to assess the toughness of composites is a post-impact compression (PIC) test in which a laminate is subjected to a low energy impact followed by a compression test to fracture.[5] This simulates the in-service requirement to withstand low energy damage. The impact energy is usually expressed as an energy per unit thickness and a typical level is 6·7 J/mm (1500 in lb/in). The post-impact compressive strengths of various thermoplastic matrix composites are summarised in Table 12. The higher interlaminar fracture toughness of thermoplastics leads to a reduced area of delamination during impact,[5,61] and improved resistance to the propagation of the delamination during compressive loading.[61]

Thermoplastic matrix composites offer considerable improvements in toughness and damage resistance over current thermosets. This represents an advantage in service as this is a limiting property in the current use of advanced composites.

6.3 Effect of Temperature on Properties

The effect of temperature on properties is important in determining the usable temperature range of the composite. The effect of temperature on flexural modulus is shown in Fig. 6. The data are plotted as percentage retained compared to 23 °C for each material. Most of the composites retain a high proportion of the room temperature modulus to at least 200 °C. The effect of temperature on retention of flexural strength is shown in Fig. 7. Another useful reference point is the temperature at which 70% of the room temperature flexural strength is retained and this is summarised in Table 13. The temperatures range from 124 °C for RYTON PPS to 213 °C for TORLON PAI. The value for PPS is above the T_g of the polymer due to the high crystallinity that can be achieved. For the PEEK composite the 70% retention temperature is at the T_g of the polymer, but for the amorphous matrix composites the 70% retention is below T_g. This indicates that the margin between polymer T_g

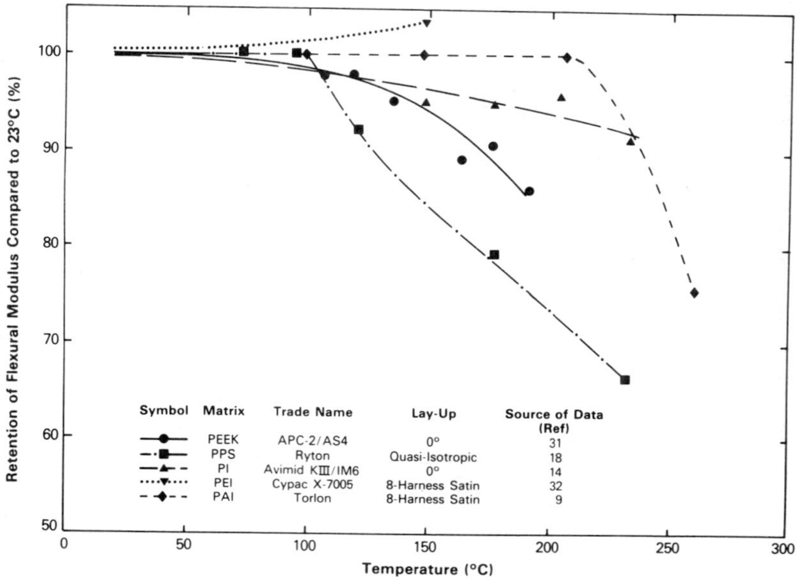

Fig. 6. Retention of flexural modulus at elevated temperatures relative to 23°C.

and use temperature varies with the polymer morphology and it indicates that it is possible to use semi-crystalline matrix composites at temperatures much closer to their polymer T_g values than the amorphous matrix composites.

The effects of temperature on toughness were discussed in Section 6.2.

There has been little examination of other properties at low temperatures, though the properties of PEEK matrix composites have been examined at 4·2 K for cryogenic applications.[69] The stiffness and strength of PEEK matrix composites increased by 20% at 4·2 K compared to room temperature. The fracture surfaces showed evidence of matrix ductility despite the low temperature.[69]

The characterisation of temperature effects is limited and further information is needed to determine the temperatures over which thermoplastic composites may be used.

6.4 Environmental Resistance

The effect of fluid environments on thermoplastic composites is reported by several authors[14, 24, 31, 32, 70–75] and in Chapter 5.

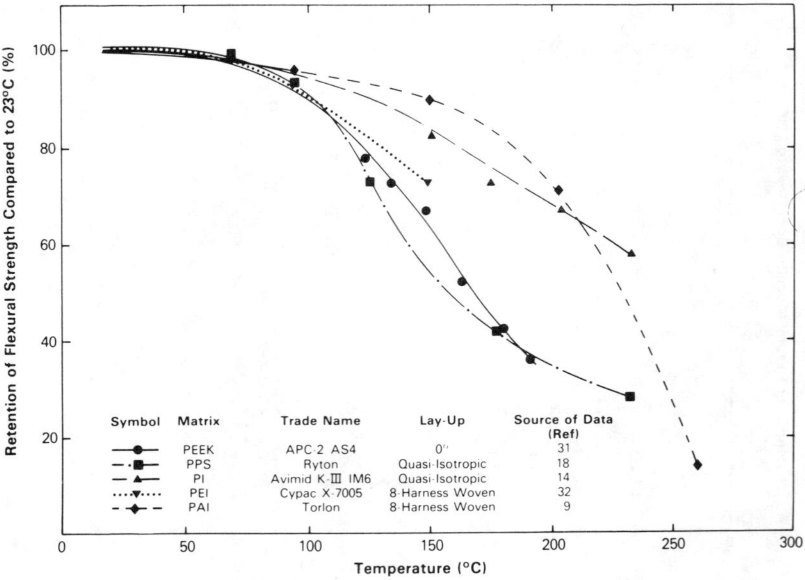

Fig. 7. Retention of flexural strength at elevated temperatures relative to 23 °C.

The most common environment is moisture. In order to meet the requirements of high performance applications, composites must also have good resistance to a range of environments including organic solvents, chlorinated solvents, acids, alkalis, fuels and hydraulic fluids. The most common effect of an environment is to plasticise the polymer, leading to a reduction in T_g of the matrix (Chapter 5). The weight uptake and retention of short-beam shear strength of some thermoplastic composites in various fluid environments are shown in Table 14.

There are considerable differences in the weight uptakes in moisture environments. The equilibrium weight uptake of AVIMID K-III/IM6 in water is 0·67% and this leads to a reduction in T_g of about 50 °C.[14] Moisture has little effect on room temperature properties but at elevated temperatures there may be significant effects. The retention of 0° flexural strength at various temperatures after moisture conditioning is shown in Fig. 8. These data are normalised to the room temperature/ unconditioned strength. The strength in the moisture conditioned state is lower even at room temperature for several materials and the retention at elevated temperatures is considerably reduced for several of

TABLE 13
Retention of 0° Flexural Strength at Elevated Temperatures

Composite	*Matrix*	*Polymer dry T_g (°C)*	*Temperature at which 70% of room temperature flexural strength is retained (°C)*		*Water conditioning*	*Source of data (Ref.)*
			Dry	*Wet*		
RYTON PPS	PPS	85	124	100	71°C/14 days	48
APC-2/AS4	PEEK	143	140	138	100°C/4 days	31
CYPAC X7005	PEI	216	153	—	—	32
AVIMID K-III/IM6	PI	250	192	160	71°C/14 days	14
TORLON	PAI	249–288	213	—	—	9

Fibers AS4, IM6 — Hercules Magnamite.

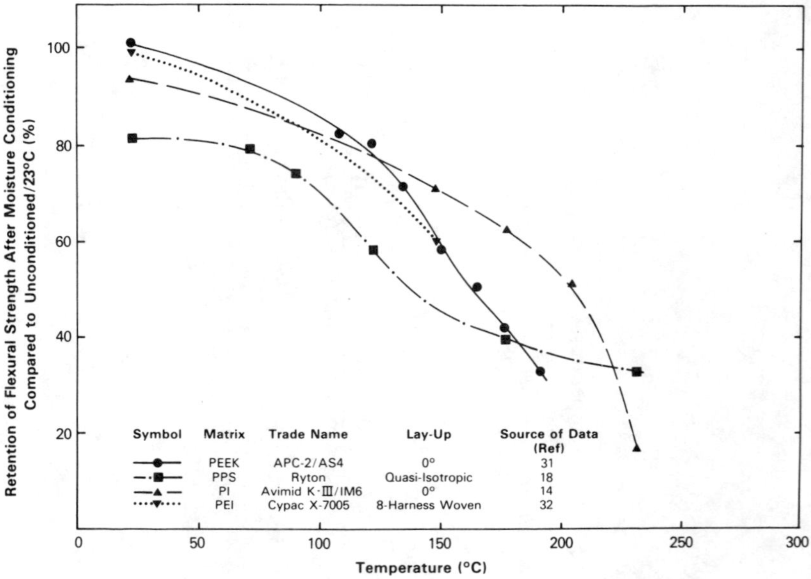

Fig. 8. Retention of flexural strength at elevated temperatures after moisture conditioning. Retention is relative to 23°C/dry for each material.

the materials (Fig. 8; cf. Fig. 7). The temperature at which 70% of room temperature/unconditioned flexural strength is retained is shown in Table 13.

The weight uptakes in the other environments shown in Table 14 are relatively small except for dichloromethane with APC-2, though it must be noted that this fluid dissolves many of the amorphous polymers. The interaction of PEEK and dichloromethane has been examined by Stober *et al.*[74] and Grayson *et al.*[75] Dichloromethane is of relevance as it is a constituent of many paint strippers, however it is being phased out because of health and safety aspects. The retention of short-beam shear strength in the environments is very high, the lowest retention of 80% being for CYPAC PEI in dichloromethane.

An extensive study examined the effects of water, salt water, hydraulic fluid, ethylene glycol, jet fuel and paint stripper on PEEK/carbon fibre composites tested in several ways.[72] Specimens were immersed without stress and with up to 20% of ultimate stress for 60 days. The largest effect on properties was after immersion in paint stripper which reduced the

TABLE 14
Environmental Resistance of Thermoplastic Composites

Environment	*Immersion conditions*		*Weight uptake (%)*		*Retention of short-beam shear strength (%)*[a]		
	Temperature (°C)	*Time (days)*	*APC-2/ AS4*	*AVIMID K-III/IM6*	*APC-2/ AS4*	*AVIMID K-III/IM6*	*CYPAC X7005*
Water	82	14	0·28	0·67	101	—	—
Methyl ethyl ketone (MEK)	23	14	0·03	0·10	103	114	—
Dichloromethane	23	1	2·41	0·43	99	112	80
Monsanto low density hydraulic fluid	82	14	−0·08	0·09[b]	102	121[b]	100
Jet fuel JP4	82	14	0·10	0·041	114	110	100[b]
Source of data (Ref.)			31	14, 24	31	14, 24	71

[a]Compared to unconditioned control.
[b]Immersion temperature 23°C.

open-hole compressive strength by 14% and the tensile strength by 11%.

Some limited data have been published on the effects of environments in the presence of stress. The flexural creep of CYPAC PEI composites over 100 days shows little change in hydraulic fluid, jet fuel and methyl ethyl ketone.[32] Pritchard & Randles[73] have shown that stressed APC-2 transverse tensile specimens immersed in water, ethylene glycol and jet fuel showed no change in properties. Dichloromethane had a significant effect, and specimens loaded at 65% of ultimate strength failed on contact with the fluid.

The information available to date indicates that thermoplastic composites have sufficient resistance to typical environments encountered in aerospace applications; however, further information is needed.

6.5 Long-term Properties

Composite materials must perform for extended periods, possibly up to 20 years, and therefore creep and fatigue are of concern.

Tensile creep testing has been conducted on PEEK matrix composites,[47,76,77] and flexural creep on PEI matrix composites.[32] No creep was observed in tensile loading of fibre dominated lay-ups.[47] Creep tests on matrix dominated lay-ups of PEEK matrix composites were conducted in transverse tension and in shear at various temperatures.[47,76,77] This enabled creep compliance master curves to be obtained[76,77] and such a curve for the shear creep is shown in Fig. 9. Xinran[77] noted that the glass transition temperature appeared to be depressed by approximately 15 °C in the composite compared to neat resin with consequently greater creep at temperatures approaching T_g, but was essentially temperature independent in the range 25–90 °C.

The fatigue properties of thermoplastic composites have been discussed by a number of authors[57,67,78–82] and have recently been reviewed by Moore.[83] Kim & Hartness[78] and Dickinson *et al.*[79] examined the behaviour of PEEK/XAS composites in comparison with epoxy matrix composites, the work of Kim being in flexural fatigue and that of Dickinson *et al.* being in tensile fatigue. Both studies concluded that the PEEK composites had superior fatigue properties to the epoxies examined. The more recent evaluations have included tensile fatigue on [±45 °] and quasi-isotropic laminates,[80,81] and a summary of the results is shown in Fig. 10. There is a considerable difference in the fatigue behaviour of the [±45 °] lay-up from Refs 80 and 81 and this has been

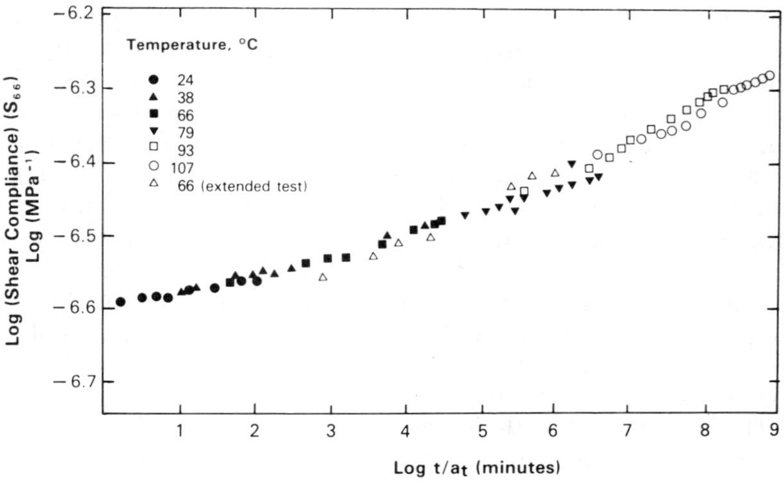

Fig. 9. Master shear creep compliance curve for PEEK/AS4 composite. Data derived from [±45°] creep tests. Reference temperature = 24°C. (Reproduced from Ref. 76.)

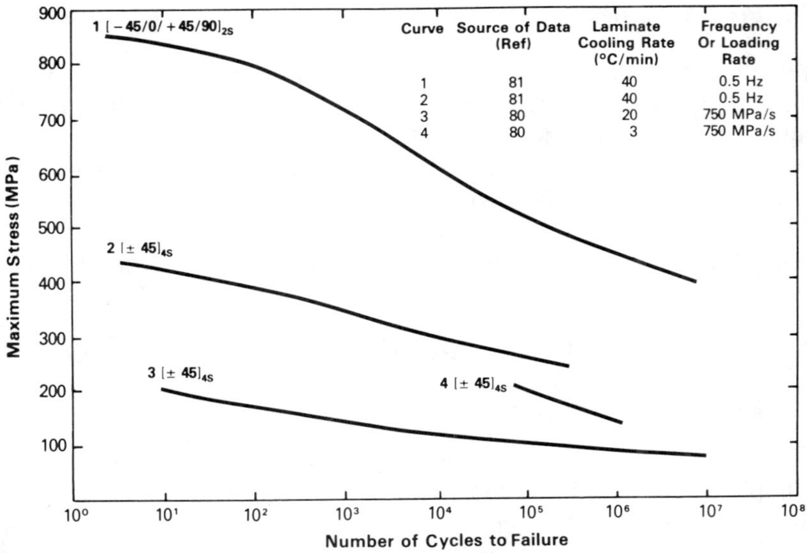

Curve	Source of Data (Ref)	Laminate Cooling Rate (°C/min)	Frequency Or Loading Rate
1	81	40	0.5 Hz
2	81	40	0.5 Hz
3	80	20	750 MPa/s
4	80	3	750 MPa/s

Fig. 10. Tensile fatigue behaviour of PEEK/carbon fibre composites.

discussed by Moore.[83] Significant heating effects were observed in fatigue testing of [±45°] laminates with temperature rises of up to 150°C being observed in specimens tested at 5 Hz.[81] The differences in the data may be due to heating effects, to the processing of the laminates or other reasons. Krueger *et al.*[57] and Croman[82] have studied the fatigue behaviour of J-2 polyamide matrix composites. The fatigue limit of a current epoxy composite (AS4/3501–6) was higher than that of a J-2 composite reinforced with the same fibre. Delamination failure was observed in the epoxy and not in the J-2, and the superiority of the epoxy was attributed to the extra energy absorbed in delamination. Tensile fatigue properties of J-2/aramid fibre were found to be very similar to those of epoxy/aramid.[57]

Toughness fatigue is discussed in Section 6.2

The issue of whether thermoplastics have superior fatigue resistance to thermosets is unresolved. Some authors have reported that thermoplastic matrix composites are superior in fatigue[78,79,81] whilst others have reported that thermoset composites are superior.[80,82] The evaluations have covered various different material systems and fatigue modes, and to date the information available is limited. It is anticipated that further data will become available in the near future.

6.6 Fire Properties

Fire properties have assumed increasing importance in recent years for aircraft interiors and for structural applications where the material must provide a fire barrier. The fire properties of PEEK[84] and PPS[85] matrix composites have been published and these are summarised in Table 15. The fire resistance of both materials is exceptionally good and easily meets typical industry requirements. The toxic gases given off by both composites are also very low and within aerospace requirements.

7 FABRICATION OF THERMOPLASTIC MATRIX COMPOSITES

7.1 Fabrication Techniques

Fabrication of advanced composites is discussed by Woolstencroft in Chapter 7. Thermoplastic composites offer the possibility of using novel and rapid fabrication techniques which cannot be applied to conventional, thermosetting matrix composites. The ability to fabricate complex shapes more economically is one of the principal reasons for

TABLE 15
Fire Properties of Thermoplastic Composites

Polymer/fibre	PEEK/CF	PPS/CF
Composite	APC-2/AS4	RYTON AC 31-60
Manufacturer	ICI Fiberite	Phillips
Limiting oxygen index (%)	65	48
Flash ignition temperature (°C)	575	493
Self ignition temperature (°C)	595	493
Smoke density (Ds) in NBS smoke density chamber		
Smouldering mode	0	0·7
Flaming mode	2	0·9
Source of data (Ref.)	84	85

using thermoplastic composites. The cost reduction is usually achieved through reducing the fabrication time. More than twenty fabrication techniques have been identified for thermoplastic composites.[86, 87] The principal techniques are summarised in Table 16 and have been reviewed by Cattanach & Cogswell.[88]

Thermoplastics may be laid-up into the final shape using conventional techniques, but the stiffness of the pre-preg and the cost of hand lay-up of complex shapes does not make this attractive. It is preferable to start with a flat lay-up or pre-consolidated sheet and perform the shaping when the polymer is in the molten state. This requires an understanding of the rheology involved in the shaping process, and this is discussed in Section 7.3.

It is possible to make some general comments as to which techniques are appropriate for certain types of parts but further experience is needed to identify the optimum processes. This must include both technical and economic input. Griffiths *et al.*[89] discussed a variety of techniques used to manufacture a flight-worthy helicopter tailplane. Their work included manufacture of components by shaping flat laminates, press-claving of pre-preg and heat forming of pre-preg into a preform for molding into complex shapes.[89] This diversity of techniques opens up many new possibilities for manufacturing and assembling structures.

The techniques described may be applied to fully impregnated, true thermoplastic pre-pregs. They may not all be applicable to post-impregnated or pseudo-thermoplastic pre-pregs due to the need to

TABLE 16
Fabrication Techniques for Thermoplastic Composites

Forming principle	*Fabrication technique*
Matched moulding	Platen pressing Compression moulding Roll-forming Pultrusion
Stretch–draw	Hydroforming Rubber pressing Double-action matched-die moulding
Compliance moulding	Autoclaving Press-clave moulding Vacuum forming Cushion platen moulding Piston matched-die moulding Diaphragm forming
In-situ consolidation/lamination	Tape laying Filament winding Tow placement Continuous belt lamination

impregnate and complete chemical reactions respectively during the fabrication of these pre-pregs.

7.1.1 Matched-mould Techniques

In this technique the material is melted and then shaped between two matched dies. The dies must be carefully machined and aligned, and the pre-preg must have a uniform thickness to achieve success with this approach. The simplest technique is to mould a flat panel using a single platen press. More complex shapes may be made using matched male–female tools. The technique requires a high temperature press and there is a practical limitation to the size of parts that can readily be made owing to the temperature uniformity and clamping forces required. Heating and cooling a single press may be time consuming.

An alternative technique to the one press technique is to melt the composite in one press and transfer it to a cooler mould for forming. Pre-heating may be in either another press or an oven. It is preferable to start with a sheet which is at least partially consolidated.

Two other matched-mould techniques are roll-forming and pultrusion.

Fig. 11. Roll-formed sections.

It is possible to make constant cross-section shapes at high rates using both techniques, and therefore they are ideal for manufacturing channels, beams and similar articles. In roll-forming several plies of aligned pre-preg are heated on a continuous basis, usually using an oven, and then drawn through a series of rolling stations to gradually form the shape required. Some roll-formed parts are illustrated in Fig. 11. Roll-forming has been demonstrated with sections such as C-channels and 'top-hat' sections. It is also possible to make sections which are profiled along the machine direction such as 'sine wave' spars. Further information on the roll-forming technique may be found in Ref. 88.

Pultrusion offers more flexibility in the shape of sections that may be produced but at present they must be of constant cross-section. In pultrusion the material is pre-heated and then drawn through a die or series of dies to make the required shape. A variety of shapes including rods, channels and beams have been demonstrated using a range of polymers.[90,91] To date pultruded sections have been relatively small, up to a few centimetres in size, however there is no reason why larger sections could not be pultruded.

Roll-forming and pultrusion have an advantage in that it is possible to manufacture intermediate-stage products in standard shapes. The ability to have standard shapes such as beams 'off-the-shelf' has the potential of providing considerable cost savings particularly for larger transport aircraft and industrial applications.

7.1.2 Stretch–Draw Processes

The term stretch–draw is applied to processes which involve a blank of material being pre-heated and then transferred for forming. Cooling usually takes place *in situ* but the part may be removed from the forming station for cooling. The techniques normally use a blank which is at least partially consolidated. Although this is an extra step in the fabrication process, the speed of the stretch–draw processes more than offsets the cost of the extra step.

Hydroforming is a metal working technique which uses hydrostatic pressure via an oil reservoir to form the part over the tool. The blank is pre-heated, usually in an infra-red oven, and then rapidly transferred to a male or female tool for forming. Another similar process involves pressing with a rubber block rather than the oil reservoir. A further variant is the double-action matched die, in which the pre-heated blank is transferred to a press with a double-action die to form the part.

The tooling for the stretch–draw techniques is normally cooler than the processing temperature for the material, and may be at room temperature. The type of cooling and its temperature will affect the cooling rate of the part, which may in turn affect the properties. For example, metal tooling and the oil and metal diaphragm used in hydroforming all have high thermal conductivity and therefore the part will be cooled rapidly, whereas rubber forming will give a lower cooling rate. Control of the cooling rate may be achieved from understanding the thermal properties of the composite and tooling material, and by control of the tooling temperature.[42]

Parts may be made in minutes using the stretch–draw techniques, but their size and complexity is limited by the size of presses available and the speed of moulding. The stretch–draw processes are very attractive due to the high production rates that can be achieved.

7.1.3 Compliance Moulding

The compliance moulding techniques all involve using a tool surface and a compliant surface. The simplest technique is vacuum forming/

consolidation in which only atmospheric pressure is used for consolidation. Initial work has demonstrated successful vacuum consolidation and forming, and the technique provides an inexpensive feedstock for other processes.

Thermoplastic parts may be formed using conventional autoclave technology, but this is not an attractive route owing to the difficulty of laying up complex shapes and the long cycle times involved in heating and cooling the autoclave. These difficulties may be overcome by using an autoclave with a moving platen for forming, and by using heated tooling, eliminating the need to heat the whole autoclave to the processing temperature of the thermoplastic.

A recent development in compliance forming is diaphragm forming.[87] This technique is derived from the superplastic forming of metals, and uses diaphragms which have superplastic properties at the processing temperature of the thermoplastic. It is illustrated in Fig. 12. The composite lay-up is placed between two diaphragms in a suitable temperature/pressure chamber and the part is formed by applying a differential pressure on the two parts of the chamber to draw the diaphragms and lay-up across the tool.[87] The deformation in the composite is controlled by the diaphragms and the plies are always held in tension, eliminating the 'kinking' which occurs if the fibres go into compression during forming. Initially superplastic aluminium diaphragms were used but more recently polymer films have been employed.[92] It is possible to use male or female tooling and to adapt a conventional autoclave for the process. Complex parts may be made using diaphragm forming and some of these are shown in Fig. 13. Total cycle times of less than 1 h have been used in making complex parts[93]

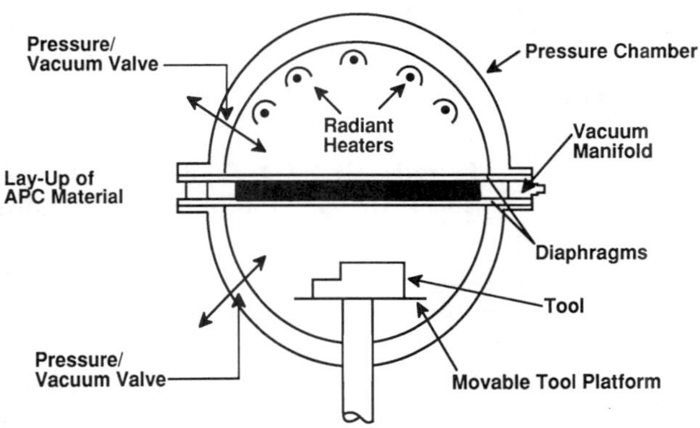

Fig. 12. Diaphragm forming technique.

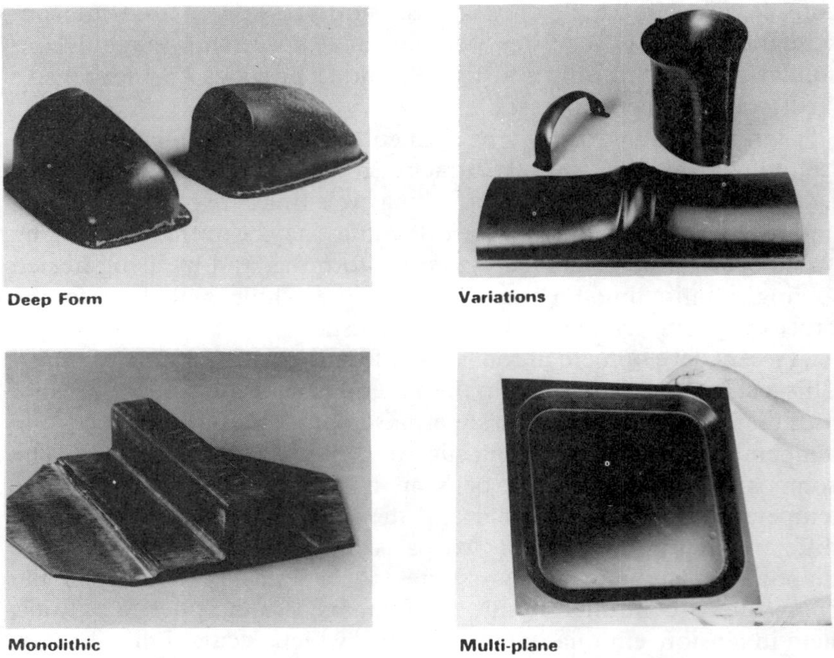

Fig. 13. Diaphragm formed components.

and the ability to use conventional autoclaves opens up the possibility of forming large parts using this technique.

7.1.4 Ply Assembly

These techniques involve assembling plies individually to make an intermediate or final product. The techniques are tape laying, filament winding, tow placement and double-belt lamination. The goal with these techniques is to melt the pre-preg tape or tow and consolidate it locally in a single operation. This is illustrated in Fig. 14. Several heat sources have been used including laser, infra-red, flame, hot gas and heated shoe.[94] In addition the pre-preg or the substrate may be heated.

Tape laying is ideally suited to making large area, gently contoured shapes, and may also be used to make flat sheets for use in other fabrication techniques.[95] PEEK and PPS matrix composites have been laid at speeds of 2·5–3·8 m/min, giving 80% consolidation. It is anticipated that the speeds can be doubled as the technique is

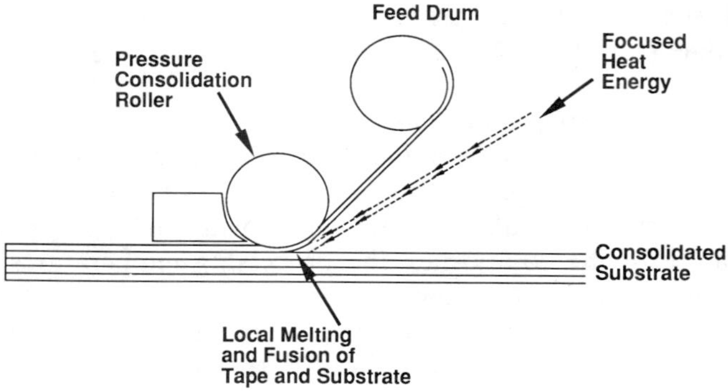

Fig. 14. Tape laying of thermoplastic composites using in-situ consolidation.

developed. Even higher speeds have already been achieved but with a reduction in consolidation. The partial consolidation is satisfactory if the panel is to be used as the feedstock for another process. Panel sizes up to 1·5 m × 6 m with up to 96 plies have been demonstrated.[95]

Filament winding has been extensively used with thermosetting matrix composites and is now being applied to thermoplastics. Some of the aspects of filament winding with thermoplastics have recently been discussed by Egerton & Gruber.[94] The key steps are identified as tow pre-heating, tow guiding, contact point heating, mandrel heating and post-consolidation, with a range of options available for each stage. One interesting feature with thermoplastics is the ability to wind shapes containing concave areas, due to the local consolidation during winding. Ideally no post-consolidation is used but in some cases it may be necessary to achieve the desired part quality. Void contents as low as 2·5% can be achieved with in-situ consolidation, and this can be reduced to 1% if a post-consolidation step is used.[94,96] The properties of in-situ consolidated components are 70–100% of autoclave formed parts.[94,96]

Tow laying is a process which lies between tape laying and filament winding. It is similar to tape laying but the use of the narrower tow enables more complex parts to be laid.

Another ply assembly technique is double-belt lamination. This has been used for some time to make sheet moulding compounds and reinforced thermoplastic sheets, and could be readily applied to continuous fibre reinforced thermoplastics. Pre-cutting and arrangement of the off-axis plies enables sheets of any lay-up to be made. The pre-preg may be pre-heated, or heated by conduction as it passes through the laminators. This is another route to a full- or partially-consolidated sheet feedstock.

7.1.5 Bagging Materials and Sealants

Some of the processing techniques such as vacuum, autoclave and press-clave, require a pressure tight bag. This is normally made with polymer film and sealants which will withstand the processing temperature of the composite. The processing temperatures usually necessitate the use of high temperature films and specialist sealants. The current sealants have poor room temperature tack, making it difficult to get a good seal, and the films usually have limited room temperature drapeability. Some sealants and bagging films with improved properties are now available, but they are expensive compared with those for thermosetting matrix composites. Further improvements in materials and reduction in cost are expected in the near future.

7.2 Assembly

Some of the assembly techniques for thermoplastic composites are summarised in Table 17 and are discussed in Ref 87. Conventional mechanical fastening may be used with thermoplastics, though it has the usual limitations of requiring many holes, giving a local strength reduction, and being expensive, due to the extra operations and fasteners.

Adhesive bonding using thermosetting adhesives has been successfully applied.[97,98] The most critical stage has been identified as the surface

TABLE 17
Assembly Techniques for Thermoplastic Composites

Approach	*Technique*	*Source of further information (Ref.)*
Fastening	Mechanical fastening	—
Bonding	Adhesive	97, 98
	Resistance	99
	Vibration	99
Welding	Hot knife	—
	Induction	98, 99, 100
	Ultrasonic	99

preparation. High strength joints have been demonstrated with PEEK matrix composites using either chromic acid etching or dry media abrasion,[97] and with PPS matrix composites using flame treatment, corona discharge and plasma treatment.[98]

Other techniques utilise the thermoplastic nature of the matrix to melt the polymer locally and create a joint. This may be performed to join the composite together directly (welding) or through the use of additional material (bonding). The various techniques for fusion joining have been reviewed by Benatar & Guttowski.[99]

Resistance bonding may be performed by applying an electrical current to a sacrificial ply (if the fibre is conductive) or to a metallic grid. Resistive heating causes the polymer to melt and fuse locally in order to form the joint. It has been possible to achieve short-beam shear strengths of 70% of the original laminate strength for polyamide (PA) composites by using an extra ply of polymer film at the joint.[99] Vibration bonding is another technique used to heat the polymer locally and is normally used with additional polymer film at the joint.

Welding may also be accomplished by direct heating of the two surfaces and then bringing them together under pressure. This may be achieved by a hot knife or infra-red heating.[99]

Induction welding may be performed directly on the composite, if the fibres are conductive, or may require a susceptor, usually a metallic grid. Stein *et al.*[100] have described the induction bonding of PEEK/CF to itself using a steel screen as a susceptor. Cohesive failure was observed and overlap shear strengths similar to the shear strength of the PEEK matrix were reported.[100] Similar experiments using a PA composite with a susceptor and additional film at the interface gave joint strengths of 55% of the laminate strength.[99]

Ultrasonic welding of PEEK and PA composites has been carried out and bond strengths equivalent to 80% of the laminate strength have been achieved.[99]

Benatar and Guttowski[99] conclude that the most promising techniques are ultrasonic, induction and resistance heating. It is apparent that bond strengths almost as great as the composite strength can be achieved by a number of welding and bonding techniques.

7.3 Fabrication Science

There are two major differences in fabricating thermosetting and thermoplastic matrix composites:

(1) Thermosets are normally laid-up into the final shape, whereas thermoplastics are usually formed to shape in the melt.
(2) Fabrication of thermosets normally takes place over long periods of time, typically hours, whereas fabrication of thermoplastics often takes place in minutes.

An increased understanding of the rheological processes is required in order to fabricate parts from thermoplastic composites successfully. The methods of processing frequently lead to a re-distribution and re-orientation of the fibres and this can affect the mechanical and physical properties of the component.

The principal mechanisms of flow have been identified as resin percolation, transverse flow, interply slip and intraply slip.[36] The mechanisms are illustrated in Fig. 15. The reinforcing fibres are essentially inextensible but the plies may spread transversely and in shear. Rheological properties have been obtained for PEEK matrix composites in each of the four modes of deformation.[36, 101] A hierarchy of flow processes has been identified in which the more complicated the shape, the greater the number of flow processes that are required. For a double curvature component all four mechanisms are needed to form the shape successfully.

An extensive study of the diaphragm forming process was conducted by Smiley[102] including a comparison of the predicted and observed flow. It was possible to predict successfully the slippage and re-orientation of the plies using finite element analysis.

The mechanism of deformation during the manufacture of two-dimensional parts has been examined by Guttowski *et al.*[103, 104] This included a theoretical and experimental approach to assess the effect of forming conditions on fibre misorientation, surface quality and mechanical properties.

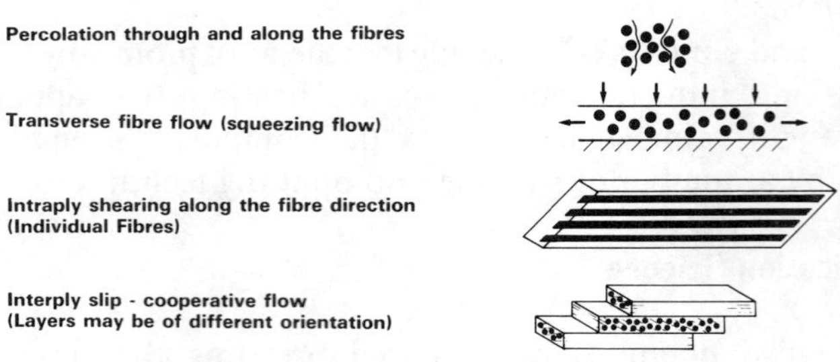

Fig. 15. Flow mechanisms in thermoplastic composites.

The understanding of the rheological processes is in its early stages. The initial work has shown that it is possible to determine experimentally the rheological properties and to predict successfully the deformation processes under various conditions. Considerable effort is directed at this area and it is expected that further progress will be made in predicting the performance of the finished parts. This will reduce significantly the time involved in defining the optimum conditions for manufacturing a particular part.

8 ECONOMICS, APPLICATIONS AND FUTURE TRENDS

8.1 Economics

Economic factors must be taken into account in the use of any new material, and economic benefits are one of the major reasons for using thermoplastic composites.

Thermoplastic resins are typically more expensive than current thermosets, although the cost will fall as production increases. Thermoplastic impregnation is generally more complex than thermoset impregnation and there may be considerable differences in manufacturing costs between the various techniques for thermoplastics.[105] For example, manufacturing small diameter fibres can double the cost of the polymer and performing a hybridisation can multiply the cost 1·5-fold. Current prices for thermoplastic pre-pregs are probably not a good guide to the long-term price owing to the small production quantities. Most predictions are that in the longer term the prices of thermoplastic pre-pregs will be comparable with those of the higher performance thermosets.

The elimination of special storage conditions and the infinite shelf life of thermoplastics leads to a cost benefit to the user. The most significant advantage is the reduction in part manufacturing costs. Cost estimates were made for manufacturing various parts for the YC-14 aircraft in thermoplastic and thermoset composites.[3] Setting-up costs were higher for the thermoplastic because of the more expensive tooling but fabrication costs were considerably lower for the thermoplastic. The total part cost in thermoplastic varied from 57% to 76% of the thermoset part cost over 100 production units. The greatest savings were for fuselage body sections and the lowest savings were in more complex structure such as the horizontal stabiliser. More recently an engineering

TABLE 18
Relative Costs and Weights of Westland 30-300 Helicopter Tailplane (Information from Ref. 106)

	Metallic aluminium alloy	*Thermoset CF/epoxy*	*Thermoplastic APC and CF/PEI*	*Optimised thermoplastic APC and CF/PEI*
Weight	100%	70%	71%	68%
Manufacturing cost	100%	76%	52%	44%

substantiation has been published on manufacturing a helicopter tailplane from metal, thermoset and thermoplastic composites.[106] A summary of the relative costs and weights is given in Table 18. The cost of the thermoplastic tailplane is 52% of that of the aluminium tailplane, and this can be reduced to 44% when the design is optimised for thermoplastics. This represents 58% of the cost of the thermoset tailplane. The cost reduction is achieved through rapid fabrication and parts consolidation.

The realisation of reduced manufacturing costs depends on the fabrication process and the type of composite. For example some techniques cannot be applied to pseudo-, or post-impregnated thermoplastic composites. Single-stage manufacturing techniques such as tape laying and filament winding are potentially the lowest cost techniques. The relative costs of various manufacturing processes have been estimated by Harvey.[105]

Additional savings may come from an increased success rate in manufacturing components, due to the absence of chemical reactions during processing, and from the ability to re-work and reclaim defective parts.

The final area of economic benefit is in service and this is the least characterised aspect. Thermoplastic composites are very resistant to in-service damage due to their high toughness. This reduces the need for replacement and repair of damaged parts, and it may also be possible to reduce the inspection frequency. The cost saving over the life of an aircraft can be considerable and several programmes are currently examining the supportability and in-service cost benefits of thermoplastic composites.

8.2 Applications

There is a relatively long lead time in implementing a new material into aerospace structural applications. This involves property characterisation, development of manufacturing processes, sub-scale and full scale evaluation, leading to in-service testing of a demonstration part prior to production. Some parts have been in service for several years, notably the poly(ether sulphone)/glass radome used on the RAF Hercules transport aircraft. This is shown in Fig. 16 and the background is described further in Ref. 4.

There are also many demonstration components currently under test. Boeing has evaluated access panels from PEEK/CF and PPS/CF.[7, 107] The latter won an award of excellence from the Society of the Plastics Industry in 1986 and is shown in Fig. 17. Boeing has also manufactured a complete demonstration wing for the proposed US Advanced Tactical Fighter from thermoplastic composites. Westland started flight tests on an access panel on the Lynx III helicopter and on the rudder of an AP1-88 hovercraft in 1985, and are evaluating a thermoplastic tailplane for the 30-300 helicopter.[90, 106] The tailplane utilises both PEI/CF and PEEK/CF and is shown in Fig. 18. Other thermoplastic parts which have undergone flight test include Lockheed C-130 Transport belly-

Fig. 16. Radome for Hercules C-130 transport aircraft made from film stacked VICTREX PES/glass fibre fabric. (Photo courtesy Specmat, Whitney, UK.)

Fig. 17. Prototype aircraft access door made by Boeing Aerospace from RYTON PPS/woven carbon fibre pre-preg. Left, external view of door; right, internal substructure. (Photo courtesy Phillips Petroleum, Bartlesville, OH, USA.)

Fig. 18. Tailplane for Westland 30–300 helicopter using VICTREX PEEK/unidirectional carbon fibre tape and ULTEM PEI/woven carbon fibre. (Photo courtesy Westland Helicopters, Yeovil, UK.)

skin, Northrop F-5 landing gear doors and McDonnell Douglas F-18 outer wing skin.

It is believed that thermoplastic composites are already being used in production applications and they will feature highly in the next generation of aircraft such as the US Advanced Tactical Fighter. The initial applications are likely to be parts such as access panels, undercarriage doors and under-body fairings which suffer considerable in-service damage. The flight tests of structural components such as the Lockheed, Northrop and McDonnell Douglas demonstrators are expected to lead to a rapid transition of thermoplastics into primary structural components.

8.3 Future Trends

Thermoplastic composites are still relatively new but the technology is maturing quickly. The areas for future progress are:

(a) *Property evaluation*. Further data on creep, fatigue, environmental resistance and temperature effects. Increased understanding of property translation in components.
(b) *Fabrication/manufacturing*. Increased experience and knowledge of processing science leading to better understanding of which techniques are appropriate for each type of part. Characterisation of part costs and cost savings relative to thermosets.
(c) *Materials development*. Emergence of 'Industry Standard' materials through experience and qualification. Development of higher temperature performance composites and greater diversity in reinforcing fibres available in pre-pregs.
(d) *Applications*. Production applications in aircraft primary structure. Applications in other areas such as space structures, naval applications, advanced industrial and biomedical applications.

9 CONCLUSIONS

Thermoplastic composites have developed rapidly in recent years to the point where they are now being used in aircraft structural applications. Thermoplastic polymers are available which combine good mechanical properties, environmental resistance, temperature performance and processability. Impregnation techniques have been developed to give full impregnation of the fibres and to ensure that an adequate fibre–matrix interface is created. The properties of most thermoplastic composites are equivalent to or exceed those of thermoset composites. One notable difference is toughness, with thermoplastic composites possessing much higher toughness and damage tolerance than current thermoset composites.

Thermoplastic composites provide the opportunity to use low cost fabrication techniques due to the speed with which fully impregnated, true thermoplastics may be formed. This results in a considerable reduction in the final cost of the component.

The extensive characterisations of properties and fabrication techniques have laid the ground for the transfer of thermoplastic composites into production applications. Therefore a rapid expansion in the use of advanced, thermoplastic matrix composites is anticipated in the future.

ACKNOWLEDGEMENTS

The author is indebted to his many colleagues in ICI Advanced Materials at Wilton, Cleveland, UK, and Tempe, Arizona, USA, whose activities are reflected in this text.

REFERENCES

1. US Congress Office of Technology Assessment, New structural materials technologies: opportunities for the use of advanced ceramics and composites. Report PB87-118253/KG, US Technical Information Service, Springfield, VA, USA, 1987.
2. National Materials Advisory Board, The place for thermoplastic composites in structural components. US National Research Council Report NMAB-435, 1987.
3. Hoggatt, J. T., Oken, S. & House, E. E., Advanced fiber reinforced thermoplastic structures. US Air Force Report AFWAL-TR-80-3023, April 1980.
4. Hogan, P. A., The production and uses of film stacked composites for the aerospace industry. In *Proc. 14th SAMPE Technical Conference, New York, 1980.*
5. Byers, B. A., Behavior of damaged graphite/epoxy laminates under compression loading. NASA Contractor Report 159293, August 1980.
6. Calundann, G. W. & Jaffee, M., Anisotropic polymers — their synthesis and properties. In *Proc. Robert A. Welch Foundation Conferences on Chemical Research, Research Conference XXVI, 1982.*
7. Product Literature, Phillips Petroleum Company, Building 71-C, PRC, Bartlesville, OK 74004, USA.

8. Product Literature, Victrex Group, ICI Advanced Materials, PO Box 6, Bessemer Road, Welwyn Garden City, Herts, UK.
9. Product Literature, Amoco Performance Products Division, Ridgefield, CT 06877, USA.
10. Product Literature, General Electric Plastics, Ultem Products Operation, One Plastics Avenue, Pittsfield, MA 01201, USA.
11. Product Literature, BASF Plastics, D-6700 Ludwigshafen, FRG.
12. Product Literature, Hoechst–Celanese, 6230 Frankfurt am Main 80, FRG.
13. Product Literature, Dart Plastics, Amoco, PO Box 5867, Augusta, GA 30906, USA.
14. Product Literature, DuPont Composites Center, Chestnut Run 702, Wilmington, DE 19898, USA.
15. Product Literature, Mitsui–Toatsu, Speciality Products Business Development Division, 3-2-5 Kasumigaseki, Chiyoda-ku, Tokyo 100, Japan.
16. Product Literature, Rogers Corporation, Rogers, CT, USA.
17. Leach, D. C., Cogswell, F. N. & Nield, E., High temperature performance of thermoplastic aromatic polymer composites. In *Proc. 31st SAMPE Symposium, Las Vegas, April 1986*, pp. 434–48.
18. O'Connor, J. E., Lou, A. Y. & Brady, D. G., Polyarylene sulfide composites. In *Proc. 1st Technical Conference, American Composites Society, Dayton, OH, 7–9 Oct 1986*, pp. 21–35.
19. Jones, D. P., Leach, D. C. & Moore, D. R., Mechanical properties of poly(ether-ether-ketone) for engineering applications. *Polymer,* **26** (1985) 1385–93.
20. Augl, J. M., Moisture effects with mechanical properties of Hercules 3501-6 resin. Naval Surface Weapons Center Report, TR 79-41, March 1979.
21. Johnston, N. J., High performance thermoplastics: a review of neat resin and composite properties. In *Proc. 32nd SAMPE Symposium, Anaheim, 6–9 April 1987*, pp. 1400–12.
22. Johnston, N. J., Synthesis and toughness properties of resins and composites. *ACEE Composite Structures Technology Conference, NASA CP-2321, Seattle, August 1984*, pp. 76–95.
23. Hahn, H. T. & Williams, J. G., Compression failure mechanisms in unidirectional composites. NASA Technical Memorandum 85834, August 1984.
24. Boyce, R. J., Gannett, T. P., Gibbs, H. H. & Wedgewood, A. R., Processing, properties and applications of K-polymer composite materials based on Avimid K-III pre-pregs. In *Proc. 32nd SAMPE Symposium, Anaheim, 6–9 April 1987*, pp. 167–84.
25. Sheppard, C. H., House, E. E. & Stander, M., Advanced thermoplastic composite development. In *Proc. 36th Conf. Reinforced Plastics/Composites Industry, Society of Plastics Industry, 16–20 Feb 1981*, Paper 17-B, pp. 1–5.
26. Phillips, L. N., The properties of carbon fibre reinforced thermoplastics moulded by the film stacking method. Royal Aircraft Establishment, Farnborough, Report No 76140, 1976.
27. Gantt, B. W., Edie, D. D., Lickfield, G. C., Drews, M. J. & Ellison, M. S.,

Thermoplastic coating of carbon fibers. Paper presented at ASTM Symposium on Advances in Thermoplastic Matrix Composite Materials, Bal Harbour, 19–20 October 1987.

28. Hartness, T., Thermoplastic powder technology for advanced composite systems. In *Proc. 33rd SAMPE Symposium, Anaheim, 7–10 March 1988*, pp. 1458–72.
29. Silverman, E. M. & Jones, R. J., Property, processing and performance of graphite/PEEK prepreg tape and fabrics. *SAMPE J.*, (July/August 1988) pp. 33–40.
30. Clemans, S. R., Western, E. D., & Handermann, A. C., Hybrid yarns for high performance thermoplastic matrix composites. *Materials Engng*, (March 1988), pp. 27–30.
31. Product Literature, APC Group, ICI Fiberite, 28271 Verdugo Drive, Laguna Hills, CA 92653, USA.
32. Product Literature, Engineered Materials Department, American Cyanamid Company, Polymer Products Division, One Cyanamid Plaza, Wayne NJ 07470, USA.
33. Iaconis, J. M., Process variables evaluation of PEEK APC-2 thermoplastic matrix composite. In *Proc. 32nd SAMPE Symposium, Anaheim, 6–9 April 1987*, pp. 104–15.
34. Beever, W. H., Ryan, C. L., O'Connor, J. E. & Lou, A. Y., Ryton-PPS carbon fiber reinforced composites: the how, when and why of molding. *Toughened Composites*, ASTM STP 937, ed. N. J. Johnston. American Society for Testing and Materials, PA, 1987, pp. 319–27.
35. Ma, C.-C. M., Hu, J.-T., Liu, W.-L., Hsia, H.-C., Shieh, B.-Y. & Liu, R.-S., The thermal, rheological and morphological properties of polyphenylenesulfide and polyetheretherketone resins and composites. In *Proc. 31st SAMPE Symposium, Las Vegas, 7–10 April 1986*, pp. 420–33.
36. Cogswell, F. N., The processing science of thermoplastic structural composites. *Int. Polym. Proc.*, **1** (4) (1987) 157–65.
37. Blundell, D. J. & Osborn, B. N., Crystalline morphology of the matrix of PEEK-carbon fiber aromatic polymer composites II. Crystallisation behaviour. *SAMPE Quart.*, **17** (1) (October 1985) 1–6.
38. Berglund, L., Fracture toughness of carbon fiber/PEEK composites. Linkoping Studies in Science and Technology, Dissertation No. 159, Linkoping University, Sweden, 1987.
39. Velisaris, C. N. & Seferis, J. C., Crystallisation kinetics of polyetheretherketone (PEEK) matrices. *Polym. Engng Sci.*, **26** (22) (December 1986) 1574–81.
40. Curtis, P. T., Davies, P., Partridge, I. K. & Sainty, J. P., Cooling rate effects in PEEK and carbon fibre–PEEK composites. In *Proc. 6th International Conference on Composite Materials, London, 20–24 July 1987*, pp. 4.76–4.88.
41. Davies, P., Benzeggagh, M. L. & de Charentenay, F. X., The delamination behavior of carbon fibre reinforced PPS. In *Proc. 32nd SAMPE Symposium, Anaheim, 6–9 April 1987*, pp. 134–46.
42. Blundell, D. J. & Willmouth, F. M., Crystalline morphology of the matrix of PEEK–carbon fiber aromatic polymer composites. *SAMPE Quart.*, **17** (2) (January 1986) pp. 50–7.

43. Cole, B., Torlon-C graphite composites. In *Proc. 30th SAMPE Symposium, Anaheim, 19–21 March 1985*, pp. 799–808.
44. Hartness, J. T., An evaluation of a high temperature thermoplastic polyimide composite. In *Proc. 32nd SAMPE Symposium, Anaheim, 6–9 April 1987*, pp. 154–68.
45. Hartness, J. T., Polyetheretherketone matrix composites. In *Proc. 14th SAMPE Technical Conference, 1982*, p. 26.
46. Chang, I. K., Thermoplastic matrix continuous filament composites of Kevlar Aramid or graphite fiber. *Compos. Sci. Technol.*, **24** (1985) 61–79.
47. Carlile, D. R., Leach, D. C., Moore, D. R. & Zahlan, N., Mechanical properties of the carbon fibre/PEEK composite APC-2/AS4 for structural applications. Paper presented at ASTM Symposium on Advances in Thermoplastic Matrix Composite Materials, Bal Harbour, 19–20 October 1987 (accepted for publication in ASTM STP 1044).
48. O'Connor, J. E., Murtha, T. P. & South, A., Polyarylene sulfide high performance thermoplastic composites. In *Proc. 6th International/2nd European Conference on Composite Materials, Imperial College, 20–24 July 1987*, Vol. 1. Elsevier, London, 1987, pp. 433–42.
49. Chung, T.-S. & McMahon, P. E., Thermotropic polyester amide–carbon fiber composites. *J. Appl. Polym. Sci.,* **31** (1986) pp. 965–77.
50. Product Literature, BASF Structural Materials, Thermoplastic Composites, 13504-A South Point Blvd, Charlotte, NC 28217, USA.
51. Curtis, P. T., An initial evaluation of a high strain carbon fibre reinforced epoxy. Royal Aircraft Establishment, Report No. TR 84004, January, 1984.
52. Carlile, D. R. & Leach, D. C., Damage and notch sensitivity of a graphite/PEEK composite. In *Proc. 15th SAMPE Technical Conference, 4–6 October 1983*, p. 82
53. Fife, B., Peacock, J. A. & Barlow, C. Y., The role of fibre–matrix adhesion in continuous carbon fibre reinforced thermoplastic composites: a microstructural study. In *Proc. 6th International/2nd European Conference on Composite Materials, Imperial College, 20–24 July 1987*, Vol. 5. Elsevier, London, 1987, p. 439.
54. Measuria, U. & Cogswell, F. N., Thermoplastic composites in woven form. Institute of Mechanical Engineers Reprints C21/86, 1986.
55. Turner, R. M. & Cogswell, F. N., The effect of fibre characteristics on the morphology and performance of semi-crystalline thermoplastic composites. In *Proc. 18th SAMPE Technical Conference, 1986*, pp. 32–44.
56. O'Connor, J. E., Lou, A. Y. & Beever, W. H., Polyphenylenesulfide — a thermoplastic polymer for high performance composites. In *Proc. 5th International Conference on Composite Materials, San Diego, 1985*, pp. 963–70.
57. Krueger, W. H., Khan, S., Croman, R. B. & Chang, I. K., High performance J-2 thermoplastic matrix composite reinforced with Kevlar Aramid fiber. In *Proc. 33rd SAMPE Symposium, Anaheim, 7–10 March 1988*, pp. 181–93.
58. Smiley, A. J. & Pipes, R. B., Rate sensitivity of interlaminar fracture toughness in composite materials. In *Proc. 1st Technical Conference, American Society for Composites*, Technomic, Basel, 1986, pp. 434–49.
59. Prel, Y. J., Davies, P., Benzeggagh, M. L. & de Charentenay, F. X., Mode I and Mode II delamination of thermosetting and thermoplastic composites.

Paper presented at 2nd ASTM Symposium on Fatigue and Fracture, Cincinnati, 27–28 April 1987.

60. Russell, A. J. & Street, K. N., The effect of matrix toughness on delamination: static and fatigue fracture under Mode II shear loading of graphite fiber composites. *Toughened Composites*, ASTM STP 937, ed. N. J. Johnston, American Society for Testing and Materials, PA, 1987, pp. 275–94.
61. Leach, D. C., Curtis, D. C. & Tamblin, D. R., Delamination behavior of carbon fiber/poly(etheretherketone) (PEEK) composites. *Toughened Composites*, ASTM STP 937, ed. N. J. Johnston, American Society for Testing and Materials, PA, 1987, pp. 358–80.
62. Davies, P. & de Charentenay, F. X., The effect of temperature on the interlaminar toughness of tough composites. In *Proc. 6th International/2nd European Conference on Composite Materials, Imperial College, 20–24 July 1987*, Vol. 3. Elsevier, London, 1987, pp. 284–94.
63. Hine, P. J., Brew, B., Duckett, R. A. & Ward, I. M., Failure mechanisms in continuous fibre reinforced PEEK. In *Proc. 6th International/2nd European Conference on Composite Materials, Imperial College, 20–24 July 1987*, Vol. 3. Elsevier, London, 1987, pp. 397–404.
64. Crick, R. A., Leach, D. C., Meakin, P. J. & Moore, D. R., Interlaminar fracture morphology of carbon fibre/PEEK composites. *J. Mater. Sci.*, **22** (1987) 2094–104.
65. Purslow, D., Matrix fractography of fibre reinforced thermoplastics, part 1. Peel failures. *Composites*, **18** (5) (1987) 365–74.
66. Bishop, S. M., The mechanical performance and impact behaviour of carbon-fibre reinforced PEEK. *Compos. Struct.*, **3** (1985) 295–318.
67. O'Brien, T. K., Fatigue delamination behavior of PEEK thermoplastic composite laminates. In *Proc. 1st Technical Conference, American Society for Composites*, Technomic, Basel, 1986, pp. 297–312.
68. Griffin, C. F., Damage tolerance of toughened resin graphite composites. *Toughened Composites*, ASTM STP 937, ed. N. J. Johnston, American Society for Testing and Materials, PA, 1987, pp. 23–33.
69. Evans, D., Morgan, J. T., Robertson, S. J. & Zahlan, N., The physical properties of carbon fibre reinforced PEEK composites at low temperatures. In *Proc. Cryogenic Engineering/International Cryogenic Materials Conference, St Charles, USA, June 1987.*
70. Cogswell, F. N. & Hopprich, M., Environmental resistance of carbon-fibre reinforced polyetheretherketone. *Composites*, **14** (3) (July 1983) 251–3.
71. Peake, S. L. & Maranci, A., Thermoplastic matrix composites based on polyetherimides. In *Proc. 32nd SAMPE Symposium, Anaheim, 6–9 April 1987*, pp. 420–32.
72. Horn, W. J., Saikh, F. M. & Soeganto, A., The degradation of the mechanical properties of advanced composite materials exposed to aircraft in-service environments. In *Proc. 27th Conference on Structural Design and Materials, American Institute of Aeronautics and Astronautics, San Antonio, 19–21 May 1986.*
73. Pritchard, G. & Randles, S. J., Long fibre thermoplastics for aerospace

applications. In *Proc. British Plastics Federation Reinforced Plastics Group Congress, Blackpool, 7–10 November 1988.*

74. Stober, E. J., Seferis, J. C. & Keenan, J. D., Characterization and exposure of polyetheretherketone (PEEK) to fluid environments. *Polymer,* **25** (December 1984) 1845–52.
75. Grayson, M. A., Pao, P. S. & Wolf, C. J., Transport of methylene chloride in poly(aryl-ether-ether-ketone) (PEEK). *J. Polym. Sci., Part B, Polym. Phys.,* **25** (1987) 935–45.
76. Kim, R. Y. & Hartness, J. T., Time-dependent response of AS-4/PEEK composite. In *Proc. 19th SAMPE Technical Conference, 13–15 October 1987*, pp. 468–75.
77. Xinran, X., Viscoelastic response of thermoplastic matrix composites. Doctoral Thesis, Free University of Brussels, November 1987.
78. Kim, R. Y. & Hartness, J. T., An evaluation of fatigue behavior of polyetheretherketone/graphite composite fabricated from pre-preg tape. In *Proc. 29th SAMPE Symposium, 3–5 April 1984*, pp. 765–76.
79. Dickinson, R. F., Jones, C. J., Harris, B., Leach, D. C. & Moore, D. R., The environmental fatigue behaviour of carbon fibre reinforced polyetheretherketone. *J. Mater. Sci.,* **20** (1985) 60–70.
80. Curtis, P. T., An investigation of the mechanical properties of improved carbon fibre composite materials. Royal Aircraft Establishment Technical Report 86021, April 1986.
81. Curtis, D. C., Moore, D. R., Slater, B. & Zahlan, N., Fatigue testing of multi-angle laminates of carbon fibre/PEEK. *Composites,* **19** (6) (1988) 446–52.
82. Croman, R. B., Flex fatigue of AS-4 graphite reinforced thermoplastics. In *Proc. 6th International/2nd European Conference on Composite Materials, Imperial College, 20–24 July 1987*, Vol. 4. Elsevier, London, 1987, pp. 76–88.
83. Moore, D. R., Fatigue of thermoplastic composites. In *Thermoplastic Composite Materials*, ed. L. A. Carlsson. (in press).
84. Briggs, P. J., Leach, D. C. & Carlile, D. R., Mechanical and fire properties of aromatic polymer composites. In *Proc. 3rd European Symposium on Spacecraft Materials in Space Environment, Noordwijk, The Netherlands, 1–4 October 1985*, ESA-SP-232, November 1985.
85. Shue, R. S., Fire safety testing of PPS thermoplastic composites. In *Proc. 33rd SAMPE Symposium, Anaheim, 7–10 March 1988*, pp. 626–33.
86. Carlin, D., Private communication, Materials Laboratory, Wright-Patterson Air Force Base, OH 45433-6533, USA.
87. Barnes, A. J. & Cattanach, J. B., Advances in thermoplastic composite fabrication technology. In *Proc. Materials Engineering Conference, London, 5–7 November 1985.*
88. Cattanach, J. B. & Cogswell, F. N., Processing with aromatic polymer composites. In *Developments in Reinforced Plastics — 5*, ed. G. Pritchard. Elsevier Applied Science Publishers, London, 1987.
89. Griffiths, G. R., Hillier, W. D. & Whiting, J. A. S., Thermoplastic composite manufacturing technology for a flight standard tailplane. In *Proc. 33rd SAMPE Symposium, Anaheim, 7–10 March 1988*, pp. 308–16.
90. Cowen, G., Measuria, U. & Turner, R. M., Section pultrusions of

continuous fibre reinforced thermoplastics. Institute of Mechanical Engineers Reprints, C22/86, 1986.

91. Beever, W. H., Rhodes, V. H. & Wareham, J. H., Continuous length thermoplastic laminates. *SAMPE J.*, (January/February 1988) 8–11.
92. Mallon, P. J. & O'Bradaigh, C. M., Development of a pilot autoclave for polymeric diaphragm forming of continuous fibre-reinforced thermoplastics. *Composites*, **19** (1) (1988) 37–47.
93. Mallon, P. J. & O'Bradaigh, C. M., Polymeric diaphragm forming of continuous fibre reinforced thermoplastics. In *Proc. 33rd SAMPE Symposium, Anaheim, 7–10 March 1988*, pp. 47–61.
94. Egerton, M. W. & Gruber, M. B., Thermoplastic filament winding demonstration of economics and properties via in-situ consolidation. In *Proc. 33rd SAMPE Symposium, Anaheim, 7–10 March 1988*, pp. 35–46.
95. Witzler, S., Laying thermoplastic tape pre-pregs — a progress report. *Adv. Compos.*, (January/February 1988) 53–4.
96. Winkel, J. D. & Hurdle, J. R., Mechanical performance of filament wound polyphenylenesulfide thermoplastic composite tubular structures. In *Proc. 33rd SAMPE Symposium, Anaheim, 7–10 March 1988*, pp. 816–28.
97. Leeser, D. L., Adhesive bonding of APC-2 laminates. Fiberite Technical Report, ICI Composites, 2055 E Technology Circle, Tempe AZ 85284, USA.
98. Kone, J. R., Murtha, T. P., Stirling, J. A., Bonding and joining techniques for 'Ryton' polyphenylenesulfide composites. In *Proc. 33rd SAMPE Symposium, Anaheim, 7–10 March 1988*, pp. 829–43.
99. Benatar, A. & Guttowski, T. G., A review of the methods for fusion bonding thermoplastic composites. *SAMPE J.*, (January/February 1987) 33–9.
100. Stein, B. A., Hodges, W. T. & Tyeryar, J. R., Rapid adhesive bonding of advanced composites and titanium. Paper presented at AIAA/ASME/ASCE/AHS 26th Structures, Structural Dynamics and Materials Conference, Orlando, 15–17 April 1985, AIAA Paper No 85-0750-CP.
101. Barnes, J. A. & Cogswell, F. N., Flow processes in continuous fibre reinforced thermoplastic composites. In *Proc. 2nd Conference, American Society for Composites, Delaware, 23–25 September 1987*, p. 35.
102. Smiley, A. J., Diaphragm forming of carbon fiber reinforced thermoplastic materials. Doctoral Thesis, Center for Composite Materials, University of Delaware, Newark, DE, December 1987.
103. Guttowski, T. G., Cai, Z., Soll, W. & Bonhomme, L., The mechanics of composite deformation during manufacturing processes. In *Proc. 1st Conference, American Society for Composites, Dayton, OH, October 1986*.
104. Soll, W. & Guttowski, T. G., Forming thermoplastic composite parts. *SAMPE J.*, (May/June 1988) 15–19.
105. Harvey, M. T., Designing with thermoplastic composites — product form and process economics. Paper presented at CoGSME/SME Conference 'Fabricating Composites '86', Baltimore, September 1986.
106. Duthie, A. C., Engineering substantiation of fiber reinforced thermo-

plastics for aerospace primary structure. In *Proc. 33rd SAMPE Symposium, Anaheim, 7–10 March 1988*, pp. 296–307.

107. Christensen, S. & Clark, L. P., Thermoplastic composites for structural applications — an emerging technology. In *Proc. 31st SAMPE Symposium, Las Vegas, 7–10 April 1986*, pp. 1747–55.

Chapter 3

High Performance Fibres

W. Bishop

Aero Consultants (UK) Ltd, Stonehill, Huntingdon, Cambridgeshire, UK

1 INTRODUCTION

1.1 Historical Background to Fibre Development

Composite materials consisting of two or more clearly distinguishable constituent components working together to achieve a blend of mechanical properties superior to the properties of the individual constituents are familiar to us in their natural forms as wood, bone, muscle and other tissues. However, it is only within the last half-century that we have had fibrous materials that could be combined with synthetic resins or metallic matrices to form 'new' engineering materials with unique mechanical properties previously unavailable to us.

One of the earliest fibre reinforced materials produced was based on the flax fibre reinforcement of a phenolic resin matrix. This material, Gordon Aerolite, developed during the early part of the Second World War by Professor J. E. Gordon of RAE Farnborough and Dr N. E. DeBruyne of Aero Research Ltd, Cambridge, was designed to offer a similar structural performance to the grades of aluminium alloy in common use at that time as a strategic alternative to aluminium alloy if supplies of bauxite had become scarce (Fig. 1).[1]

The untwisted flax fibres were impregnated with phenolic resin and formed into ribbons 6 in (15 cm) wide which were then assembled into a cross-ply 0°/90° form to make up sheets which were subsequently hot

Fig. 1. The first aerospace primary structure component to be made in fibre reinforced composite — the Gordon Aerolite Spitfire fuselage 1940.[1]

pressed to cure the laminate fully. These pre-cured sheets were then assembled on the airframe structure with conventional rivetting techniques (Fig. 2). The pressures of wartime priorities accelerated the development to such a point that a complete Spitfire fuselage was manufactured in this material and structurally tested to show a satisfactory performance in comparison with a conventional aluminium alloy structure.

The specific tensile strength of Gordon Aerolite was approximately equal to that of the Duralumin in use at that time, whereas the specific tensile modulus of the material was approximately three-quarters of that of Duralumin.

The continuing free availability of bauxite supplies meant that this work never proceeded beyond the demonstration of a structural test fuselage but it established the fact that fibre reinforced composites could achieve comparable performance to the best aerospace material of the day and the past 20 years have seen the re-birth and rapid growth

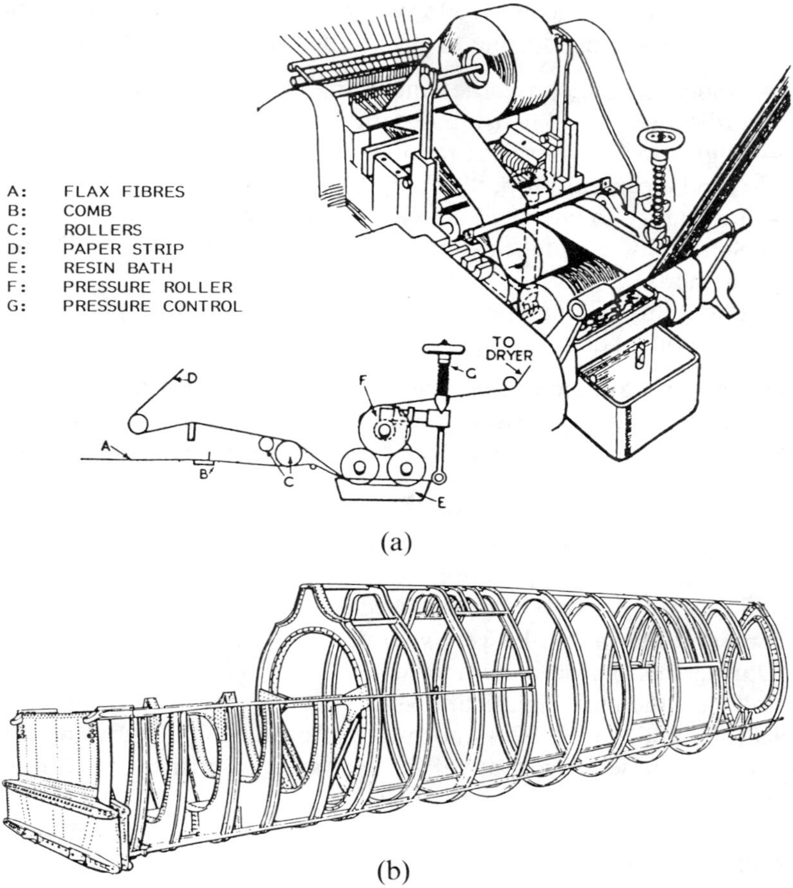

Fig. 2. (a) Pre-impregnation equipment for phenolic resin impregnation of flax fibre for the Gordon Aerolite Spitfire fuselage. (b) View of frames and longerons.

of fibre reinforced composite materials to a position where they now dominate the aerospace materials technology field.

The present range of high performance, fibre reinforced composites can be traced back to the wartime development of glass fibre reinforced materials predominantly for the manufacture of radar-transparent radome components by combining glass fibre with polyester resin.

From the end of the Second World War, glass fibre reinforced composites enjoyed a strong growth rate based on the following advantages:

- Good physical properties including strength, modulus, impact resistance and dimensional stability.
- High strength-to-weight ratio.
- Good electrical properties.
- Resistance to chemical attack and outdoor weathering.
- Suitability for fabrication by a wide range of production methods.
- Possibility of 'one-operation' moulding of large, complex shaped components.

These performance characteristics prompted the use of glass fibre reinforced materials in a wide range of industries including aerospace, marine, electrical equipment, surface transport, chemical tank and pipe industries.

This growth was further reinforced in the late 1960s when new, higher performance fibres were introduced into the market, initially for high cost, defence related applications. Composite materials based on these new fibres became known as advanced composite materials and in turn this market has enjoyed a very vigorous growth.

During this period a number of advanced composite fibres were introduced, the most significant being boron fibre (1966), carbon fibre (1968), aramid fibre (1972) and high performance polyethylene fibre (1987). It is these fibres that will be examined in further detail in this overview of high performance fibres.

1.2 Fibre Performance

The task of describing and characterising the wide range of advanced composite fibres currently available is a difficult one, since the only meaningful measure of a fibre's performance is assessed through the mechanical performance of the finished composite structural element consisting of a fibre reinforced matrix material. In turn, the performance of the composite structural element is influenced by many factors, some of which are tabulated in Fig. 3. These comments essentially relate to thermosetting resin matrix materials reinforced with fibrous filaments. Since different fibres, available in a range of different forms, are influenced to a differing degree by the choice of matrix resin, processing cycle and range of operating environments, the problem of describing

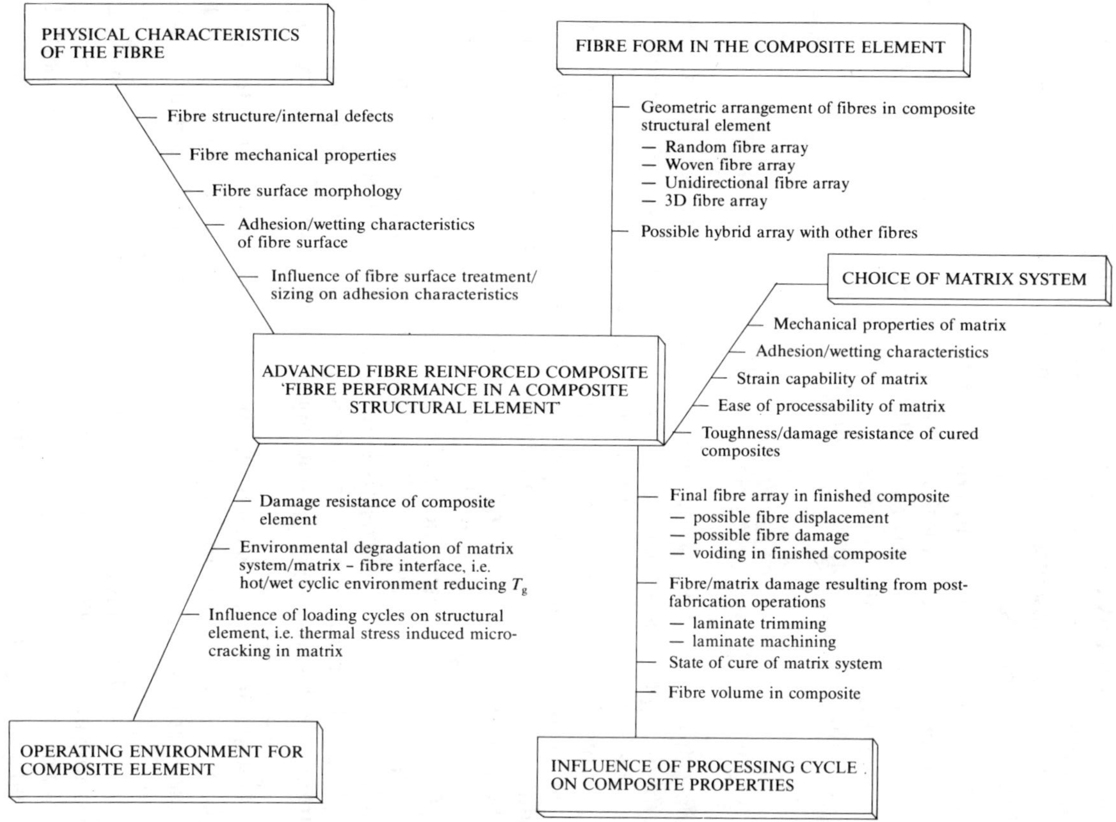

Fig. 3. Factors influencing the behaviour of a high performance fibre in a thermosetting resin matrix composite.

fully the fibre performance becomes rather involved. Because of these factors a fully comprehensive review of the range of fibres available in their various forms and matrix system combinations would involve much more space than is available here.[2,3]

This chapter presents a simplified overview of each of the various fibrous materials, summarising:

— Physical characteristics of the fibre.
— Fibre manufacturing methods.
— Range of fibre forms available.
— Approximate fibre costs.
— Future trends in fibre development.

2 GLASS FIBRES

Glass fibres are the most widely used fibrous reinforcement in the reinforced plastics industry and over 90% of the structural reinforced plastics applications involving thermosetting resins utilise glass fibre.

Many different compositions of mineral glasses have been used to produce fibres. The most common are based on silica (SiO_2) with additions of oxides of calcium, boron, sodium, iron and aluminium. These glasses have an amorphous structure with a lack of orientation in comparison with aramid and carbon fibres.

Glass does not burn and retains good mechanical properties up to about 50% of its strength to 370°C and 25% of its strength up to 540°C. It has excellent moisture resistance and glass fibres do not swell, stretch, disintegrate or undergo other chemical changes when wet. Except for strong alkalis and hydrofluoric acid, glass has excellent corrosion resistance (Chapter 5).

Glass fibre is produced by feeding the raw glass, in the form of marbles or batch, into an electrically heated fiberising element referred to as a 'bushing'. This vessel has several hundred small diameter holes in its base through which the molten glass flows (Fig. 4).

These molten glass filaments are mechanically drawn through the bushing at over 3000 m/min and as the rapidly cooling glass filaments leave the bushing they are sprayed with an aqueous or solvent dispersed organic sizing solution. This sizing operation fulfils four functions:

— Acting as a lubricant to allow relative movement between the filaments.

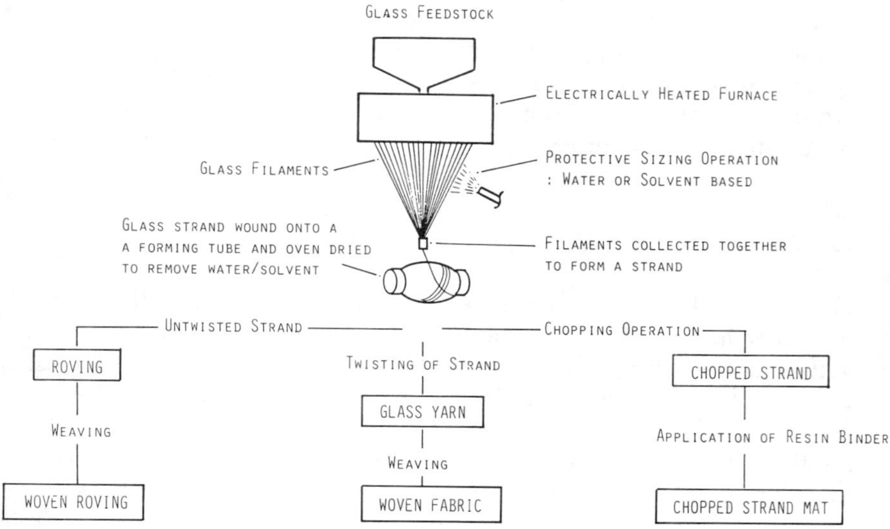

Fig. 4. Schematic diagram of glass fibre manufacturing process and the available forms of glass fibre products.

— Protecting filaments from abrasion damage by provision of a covering film over the filament surfaces.
— Providing a coating to the filaments that will be compatible with subsequent matrix resins and assist wetting and adhesion.
— Giving anti-static properties to the filaments.

The sized glass filaments are gathered together into a bundle, called a strand, which is then wound on to a forming tube. This wound array of glass strand is removed from the winder and dried in an oven to remove the water or solvent used in the size application.

The glass strand can then be converted into a number of alternative forms by subsequent operations as shown in Fig. 4.

The most frequently utilised glass type in glass fibre composites is E-glass (E designating electrical), which has good drawing characteristics and exhibits excellent strength, stiffness, electrical and weathering properties. For selected high strength applications a high tensile strength grade of glass is used, S-glass, with a tensile strength approximately 30% greater than E-glass and a tensile modulus approximately 20% higher than E-glass.

The typical mechanical properties of E- and S-glass are tabulated in Table 1 and shown in graphical form in Fig. 5. In common with all of the other high performance fibres, E- and S-glass fibres have a linear stress–strain curve when tested in tension and the typical stress–strain curves for unidirectional composites based on the various fibres are shown in Fig. 6.

2.1 Pre-preg forms

For structural applications we most frequently utilise glass fibres and the other high performance fibres in various forms pre-impregnated with a thermosetting resin which is then advanced in cure to such an extent that the resulting material can be cut, handled and assembled to form multi-ply laminates which are then cured to a finished form by the application of heat and pressure. These materials are generally referred to as 'pre-pregs'. The most common pre-preg types are as follows.

2.1.1 Unidirectional Tape (Unitape) Pre-preg

In this form the untwisted glass strand is aligned in a parallel array and thermosetting resin is applied, most frequently in the form of pre-coated thin resin film on a release paper backing which is hot-rolled into the parallel fibre array. This is referred to as pre-preg production by the film impregnation technique. Alternatively, the resin can be applied to the fibres by passing the fibres through a low viscosity resin/solvent solution bath and nip-rolling the excess resin from the fibres as they emerge from the bath. The solvent is then removed in a heated drying tower or tunnel and the state of cure of the resin is advanced to the required degree.

This is referred to as solvent impregnation of pre-preg. These alternative pre-preg production methods are shown schematically in Fig. 7.

2.1.2 Woven Fabric Pre-preg

In this form the twisted or untwisted fibres are woven into a range of dry fabric weave styles to achieve the required blend of mechanical properties from the finished laminate and practical formability of the pre-preg fabric to assist in the placement of the finished pre-preg fabric over double-curvature forms with the minimum of wrinkling and fabric distortion.

A wide variety of fabric weave styles have been developed and some

TABLE 1
Mechanical Properties of Typical High Performance Fibres

Fibre types	*Fibre diameter*		*Fibre density*		*Tensile strength*		*Tensile modulus*	
	(inch × 10^{-3})	*(μm)*	*(lb/in³)*	*(g/cm³)*	*(KSI)*	*(GPa)*	*(MSI)*	*(GPa)*
E-glass	0·3–0·6	8–14	0·092	2·54	500	3·45	10·5	72·4
S-glass	0·3–0·6	8–14	0·090	2·49	665	4·58	12·5	86·2
HP-polyethylene[a]	0·4–0·47	10–12	0·035	0·97	392	2·70	12·6	87·0
Aramid (Kevlar 49)[b]	0·47	12	0·052	1·44	525	3·62	19·0	130·0
Aramid (Kevlar 149)[b]	0·47	12	0·052	1·44	503	3·47	27·0	186·2
High strength carbon (T300)[c]	0·28	7·0	0·063	1·76	514	3·53	33·6	230·0
High modulus carbon (HM 400)[d]	0·26	6·5	0·067	1·87	450	3·10	58·7	405·0
Ultra-high modulus carbon (GY 80)[e]	0·33	8·4	0·071	1·96	270	1·86	83·0	572·0
Boron filament	2–8	50–203	0·094	2·60	500	3·44	59·0	406·7
Silicon carbide filament (SiC on carbon)	5·6	142	0·11	3·04	611	4·21	60·0	413·6

Manufacturers' trade names/references are as follows:

[a] Dyneema reference (Dyneema SK60).
[b] Du Pont registered trade mark.
[c] Toray Industries reference.
[d] Courtaulds Grafil reference.
[e] BASF reference.

The above values have been taken from manufacturer's literature as typical characteristic properties; a wide range of values has been published and this information should be taken as an approximate guide.

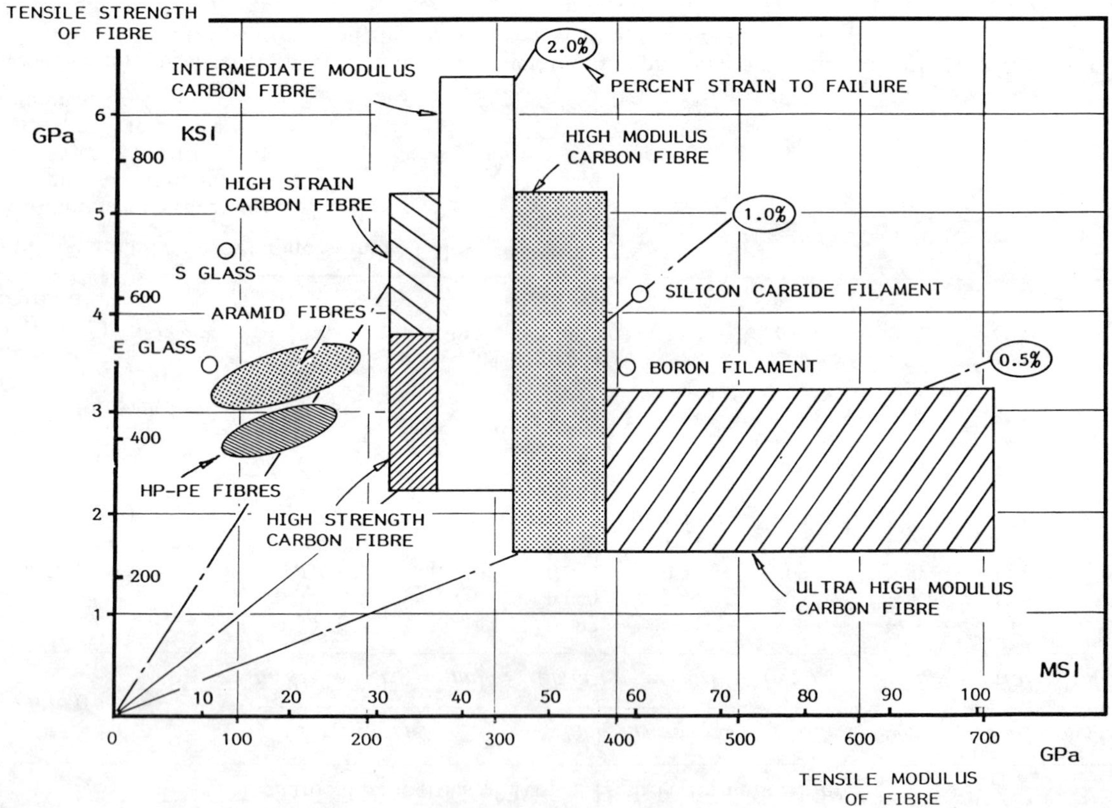

Fig. 5. Tensile strength and modulus of key high performance fibres.

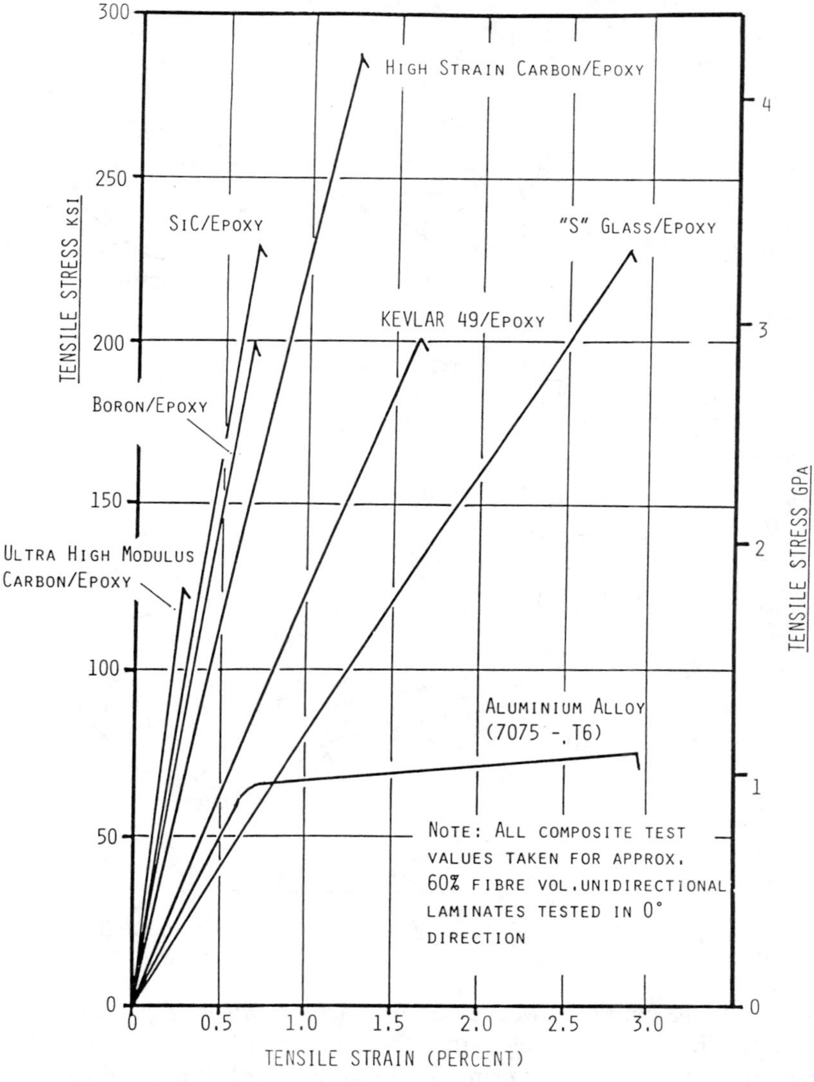

Fig. 6. Stress–strain behaviour of unidirectional composites.

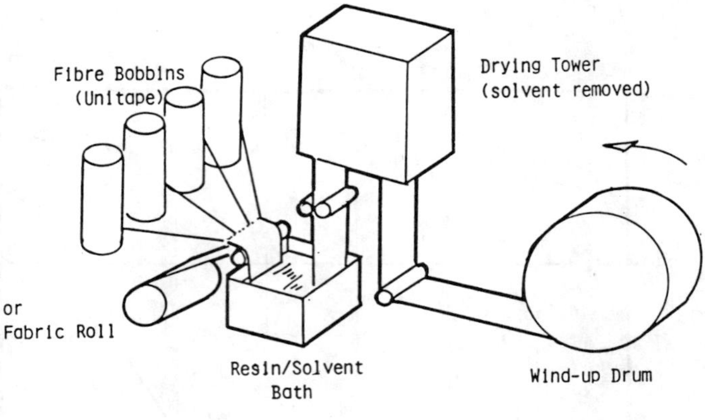

Solvent Impregnation - Unitape or Fabric

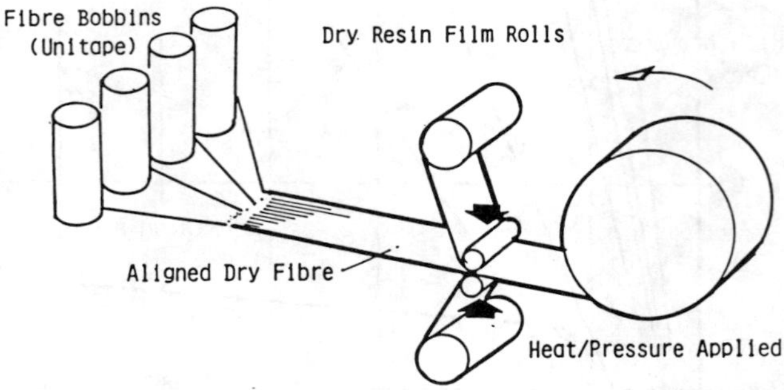

Film Impregnation - Unitape

Fig. 7. Schematic diagram of pre-preg production methods.

of the more common weave patterns are shown in Fig. 8. It will be evident from these sketches that the most simple plain weave fabric imparts a high degree of crimp to the constituent fibres and hence we can expect lower mechanical performance from laminates made with this fabric (Chapter 10). Also the square weave fabric allows very little slippage to occur between the fibres; hence this fabric style is difficult to

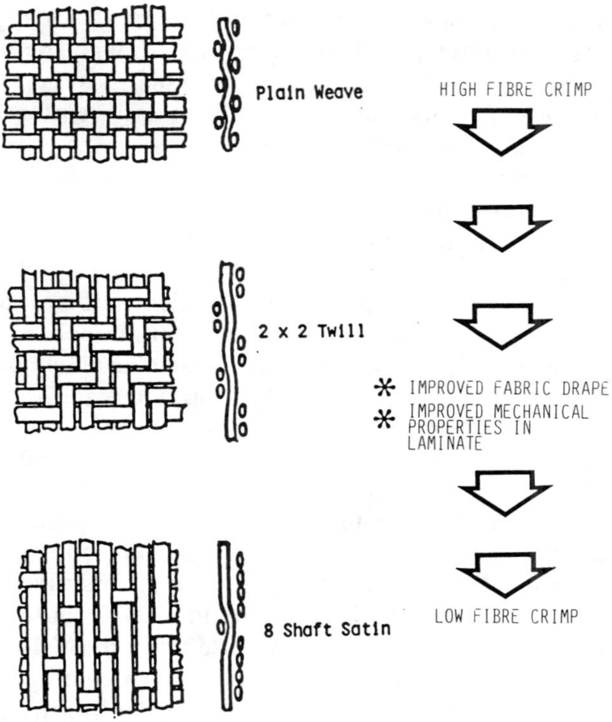

Fig. 8. Fabric weave styles used with high performance fibres.

form over double-curvature shapes. The relative ease with which these fabrics conform to complex shapes is known as the fabric 'drape'. As the fabric styles impart progressively less crimp to the fibres by moving to twill weave and satin weave styles, the mechanical properties of the resulting composite laminates and the degree of drape both improve.

These woven fabrics are most frequently impregnated with thermosetting resin by the solvent impregnation techniques.

3 BORON FILAMENT

Boron filament was the first of the so-called advanced composite fibres to be produced and was developed in the early 1960s in the USA with the support of heavy defence department funding, mainly for applications involving aerospace structural components. Although the first applications of this fibre involved its use with resin matrix systems, boron is now becoming more widely used as a filamentary reinforcement for metal matrix components (Chapter 6).

Boron filaments are produced by a vapour deposition process, involving the deposition of boron on a moving incandescent tungsten filament substrate and, although development materials have been made depositing boron on a carbon monofilament substrate, the majority of boron produced is produced by deposition on tungsten.

The most common method for producing continuous boron filament is a chemical vapour deposition process in which the reduction of boron trichloride by hydrogen gas takes place on a moving incandescent tungsten filament. A boron production line basically consists of an assembly of plating chambers (Fig. 9). The tungsten filament, 12.7 μm in diameter, is drawn continuously through the length of the reactor tube and at both ends of the tube it passes through liquid mercury, held in place by capillary action, serving as electrodes connecting the filament to a power supply. Current from the power supply resistively heats the section of the filament in the tube to about 1200°C and the gaseous mixture of hydrogen and boron trichloride flows through the reactor and reacts chemically with the heated filament. A coating of boron is deposited in the tungsten filament substrate, and hydrogen chloride gas is the by-product. As the filament passes through the plating chamber its diameter is enlarged by chemical deposition, and the boron filament is finally taken up on the spool. The hot tungsten substrate is subjected to an out-gassing treatment before the boron deposition stage, when it is heated to about 1200°C in a hydrogen atmosphere to remove any surface contamination such as oxide layers or die lubricant resulting from previous drawing operations.

Since only about 5–6% of the hydrogen–boron trichloride mixture reacts with the tungsten filament during this operation, a chemical recovery system is essential, where the hydrogen chloride is scrubbed out, unreacted hydrogen is vented to the atmosphere and unreacted boron trichloride is condensed, recovered, mixed with fresh hydrogen and recirculated through the reactor.

Boron filaments are available in continuous lengths, in a range of diameters from 50 to 203 μm.

Boron filaments in epoxy matrix systems were used for the first major structural applications of advanced composites in the aerospace

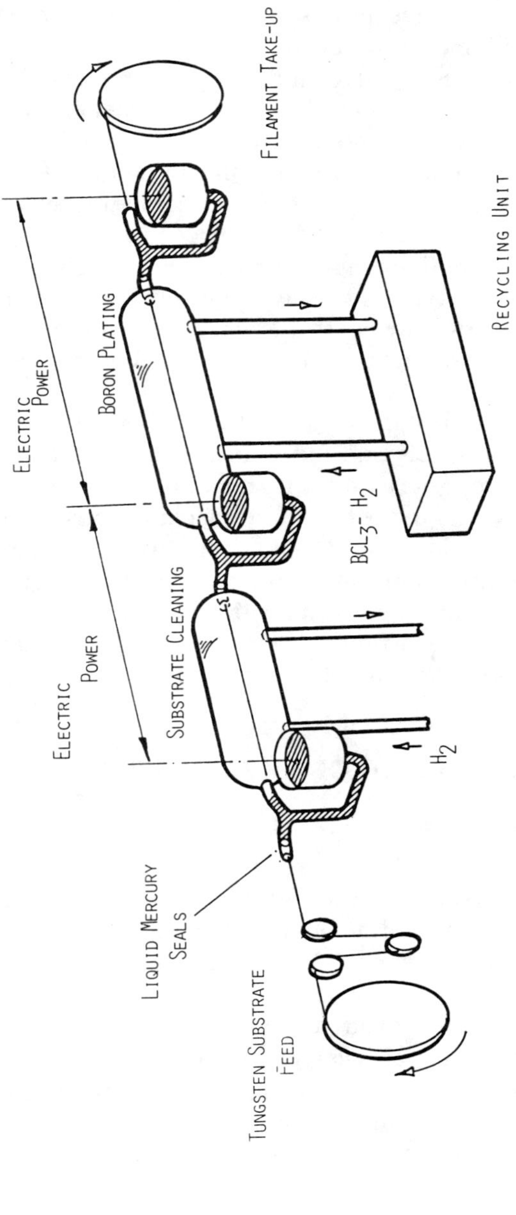

Fig. 9. Schematic diagram of boron deposition process.[2]

industry and the current generation of US military aircraft in service utilise these materials extensively. However, the relative cost of boron has made it an expensive fibrous reinforcement and since the relatively large diameter filaments are generally difficult to handle, the application of boron has been restricted to unidirectional pre-pregs. Whilst one can stabilise unidirectional ribbons of boron filament with a nylon weft thread to form a unidirectional dry tape, it is not possible to weave boron in the conventional sense since the fibres are so large in diameter and have such intrinsic stiffness that they could not be deformed into a crimped pattern for weaving without extensive damage.

Although boron filaments have excellent mechanical properties (Figs 5 and 6; Table 1) they also have relatively high fibre density; hence the specific properties are not so attractive when compared with current generation high strength and high modulus carbon fibres (see Fig. 14).

The use of boron filament with resin matrix systems has declined sharply over the last ten years and almost all new structural applications for advanced composites utilise either carbon or aramid fibres because of the lower cost and the better handling characteristics of these fibres in their various unidirectional and fabric forms. However, over this period metal matrix composites have grown to a significant market size and boron filament coated with a silicon carbide layer has become a standard material for fibrous reinforcement of aluminium alloy matrices in aerospace and defence related industries (Chapter 6).

4 SILICON CARBIDE FIBRE

With the growth of fibre reinforced metal matrix technology, silicon carbide filaments have been developed for these applications. Silicon carbide is more oxidation resistant than boron and less susceptible to attack by molten aluminium whilst retaining more of its strength and stiffness at elevated temperatures. Silicon carbide is produced, like boron, by a chemical vapour deposition process and both tungsten and carbon monofilament substrates have been used.

Silicon carbide filaments can be used with resin matrix systems and some curves for unidirectional composite performance are shown in Fig. 6. However, these fibres are predominantly used in metal matrix applications and only have limited use in applications involving resin matrix systems (Chapter 6).

5 CARBON FIBRES

In the USA in the early 1960s extensive experimental work was carried out in the thermal conversion of various organic precursor materials into carbon and graphite fibres and fabrics. This effort was mainly directed towards the production of carbon and graphite cloth for use with phenolic resins in ablative components for missile rocket motors.

These initial developments used rayon as the precursor material and resulted in relatively low performance fibres with a high weight loss during pyrolysis and a low carbon content in the finished fibre. However, during this period theoretical work indicated that a graphite single crystal should have an elastic modulus of about 965 GPa. Single graphite whiskers with a perfectly ordered basal plane crystal structure were produced and found to give tensile strength and modulus values of 20.7 GPa and 690 GPa respectively. This confirmed the tremendous potential for the development of a high performance fibrous material based on the carbon structure.

5.1 Internal Structure

Carbon fibres are characterised by a combination of low density, high strength and high stiffness. The fibre tensile modulus and to a lesser extent the tensile strength depend upon the degree to which the carbon-layer planes are oriented parallel to the fibre axis. All carbon fibres are made by the pyrolysis of organic precursor fibres in an inert atmosphere. Pyrolysis temperatures can range from 1000 to 3000°C with the higher process temperatures generally leading to higher modulus fibres. The high tensile strength and modulus of carbon fibres is a result of the intrinsic strength and stiffness of the basic graphite molecular structure. In a graphite single crystal the carbon atoms are arranged in hexagonal arrays and although these atoms in the layer or basal planes are held together by very strong covalent bonds they are only held by relatively weak van der Waal forces between the layers. This marked anisotropy is a feature of the basic crystal unit.

Carbon fibres are made up of small crystallites of 'turbostratic' graphite which resembles graphite single crystals in structure but with no regular packing, plane to plane between the layer planes. High strength and stiffness can be obtained from these fibres if layer planes of graphite are aligned parallel to the axis of the fibre. In reality the alignment of the discontinuous crystalline units is not perfect and the presence of voids within the parallel graphite crystallite planes is a source of stress concentration and fibre weakness. The carbon fibre manufacturers have devoted considerable effort towards eliminating these defects with a degree of success.

Different carbon fibres exhibit markedly different structures, where some polyacrylonitrile (PAN) based fibres have a thin circumferential skin of axially aligned layer plane sheets with an internal core of more randomly aligned, folded crystallites. In contrast, some pitch based fibres have a radially aligned crystallite layer structure. These differences in structure have a strong influence on the properties of the fibre.

The marked improvements made in fibre performance over the past few years by the carbon fibre producers have been achieved mainly by reducing the contamination and improving the purity of the basic precursor material and producing the basic precursor fibre under 'clean room' conditions. It can be expected that this improvement, which in one instance has yielded a 20% improvement in fibre strength over a two-year period,[4] will be a continuous process.

5.2 Precursor Types

Over the years three precursor materials have evolved as starting points for the commercial scale production of carbon fibres.

— Rayon
— Polyacrylonitrile (PAN)
— Pitch.

5.2.1 Rayon Precursor

The first generation of carbon fibres was based on a rayon precursor with the fibres being stretched to several times their original length at temperatures above 2800°C. This however was a costly and relatively wasteful process with a final carbon fibre yield of only around 15% by weight of the precursor rayon material. This technique involved high process costs, low yield rates and difficulties in obtaining the appropriate grades of rayon precursor. These factors have, in effect, made rayon based carbon fibres obsolete.

5.2.2 PAN Precursor

This process represents the most widely used carbon fibre manufacturing technique in current use and was developed by Rolls Royce and RAE

Farnborough in the late 1960s from original Japanese work. PAN is a polymer closely resembling polyethylene in molecular structure in which every hydrogen side group is replaced by a nitrile grouping.

The schematic diagram summarising PAN based carbon fibre production is shown in Fig. 10,[5] and consists of the following basic stages.

(1) The basic bulk PAN material is converted into a special acrylic fibre by a spinning process where a degree of hot stretching helps to align the polymer groups.
(2) The fibre is then heated under tension in an oxygen environment which leads to an exothermic chemical reaction and the formation of a crosslinked ladder molecule structure in the fibre.
(3) Carbonisation of the fibres takes place in an inert atmosphere furnace at temperatures up to 1000°C and this process generates quantities of off-gases which must be efficiently removed to prevent the condensation of tars on to the surface of the fibres passing through the furnace.
(4) A graphitisation process follows in a higher temperature inert atmosphere furnace to complete the final carbon fibre structure. The gas flow rates and precise temperatures influence the final structure of the fibre to a high degree. The typical variation of fibre strength and stiffness with final graphitisation temperature for PAN based fibre is shown in Fig. 11.[6]
(5) A surface treatment and surface finishing stage follows to modify the surface condition of the carbon fibre to achieve a balanced level of adhesion between the matrix resin system and the fibre. This is a very important stage and overtreatment of the fibre surface can result in very poor brittle failure performance characteristics in the finished laminate.
(6) PAN based carbon fibres are assembled into yarns containing from 1000 to 12 000 filaments. (In general the lower filament count yarns are more costly than the higher filament count yarns.)

5.2.3 Pitch Precursor

The use of conventional pitch materials as a precursor for carbon fibre production has been known for some considerable time; however, fibres made from these isotropic pitches are themselves isotropic. Although

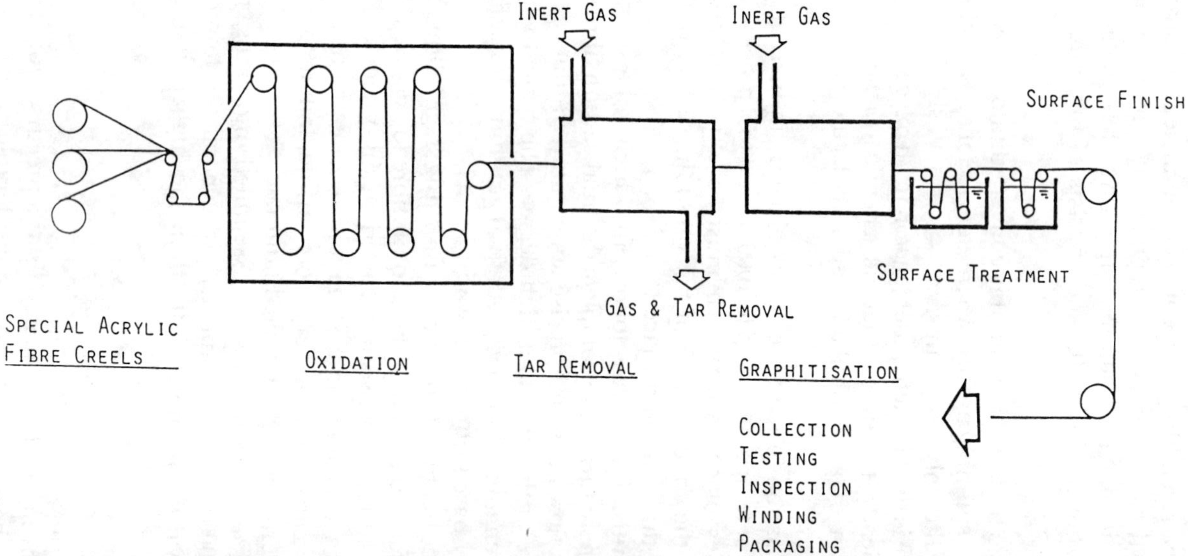

Fig. 10. Schematic diagram of PAN based carbon fibre production process.[5]

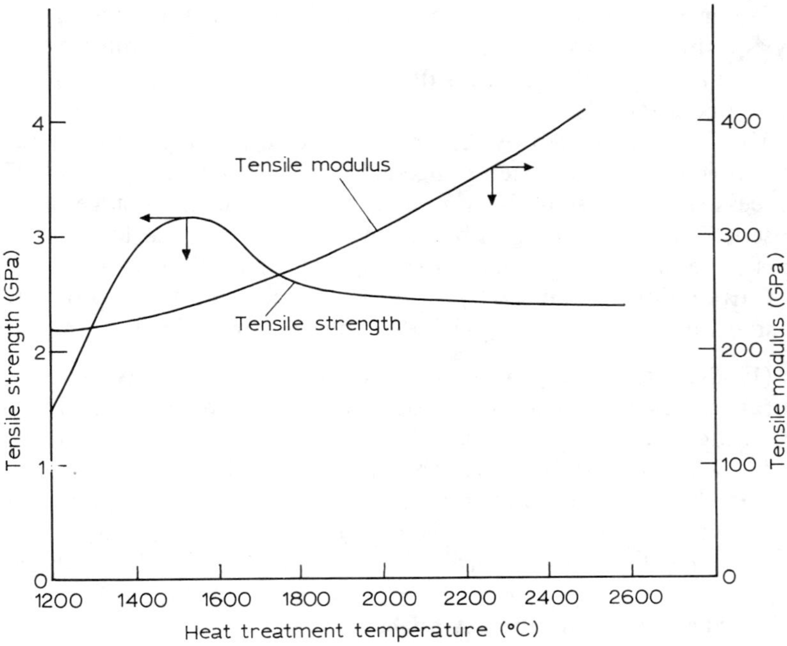

Fig. 11. Effect of heat treatment temperature on the strength and stiffness of carbon fibres produced from a PAN precursor.[6]

these relatively low strength and stiffness fibres can be substantially improved in performance by hot stretching at temperatures between 2700 and 3000°C, the process is intrinsically expensive and no fibres are currently produced by this route.

During the mid 1970s high performance, pitch based carbon fibres became commercially available manufactured by a technique where the ordinary pitch is first converted into a mesophase or liquid crystal phase. The mesophase pitch is obtained by heating the ordinary pitch to about 350°C or higher, when the pitch undergoes a dehydrogenative condensation reaction with the planar aromatic molecules aggregating into a liquid crystalline phase. This phase separates from the surrounding isotropic pitch in the form of small droplets which coalesce and grow in size as the reaction continues. When the mesophase content reaches about 40% a phase inversion occurs and the anisotropic material becomes the continuous phase.

The mesophase pitch can then be converted into a filamentary form by spinning or extrusion processes, which further improves the molecular orientation as a result of the high shear forces induced by these operations.

The fibres produced by this method are still thermoplastic and an oxidative crosslinking step is needed to thermoset the material. This process stage is essentially similar to the carbonisation stage of PAN based fibre production, with the important difference that no fibre tension is necessary to maintain preferred orientation.

Carbon fibres produced by this process enjoy significant production cost advantages compared with PAN based fibres, since

(1) the pitch precursor is much cheaper than the PAN material;
(2) the carbon yield in mesophase pitch fibre is about 75–80%, depending upon molecular weight and process conditions, whereas the yield for PAN based fibres is only around 50%;
(3) when thermosetting and carbonising the mesophase pitch-based fibre, no fibre tension is required and carbonisation can be carried out at a faster rate than is possible with PAN fibres.

5.3 Range of Carbon Fibres Available

Well over 60 different grades of carbon fibre are currently available from approximately 18 different carbon fibre manufacturers world-wide and an excellent guide to the range of materials available can be found in Ref. 7.

This situation is by no means a static one since fibre producers are constantly improving their fibre mechanical properties, particularly in the area of improved fibre strength and modulus, higher strain-to-failure and reduced coefficients of variation for mechanical properties. To give some impression of the range of carbon fibres available and their associated mechanical properties, the tensile strength and tensile modulus properties for 54 of the most commonly used carbon fibres are shown in graphical form in Fig. 12.

It is convenient to classify carbon fibres by their tensile strength and tensile modulus into one of five distinct performance groups.

Ultra-high modulus	Modulus >394 GPa
High modulus	Modulus 317–394 GPa
Intermediate modulus	Modulus 255–317 GPa
High strength	Strength <3·8 GPa
High strain	Strength >3·8 GPa

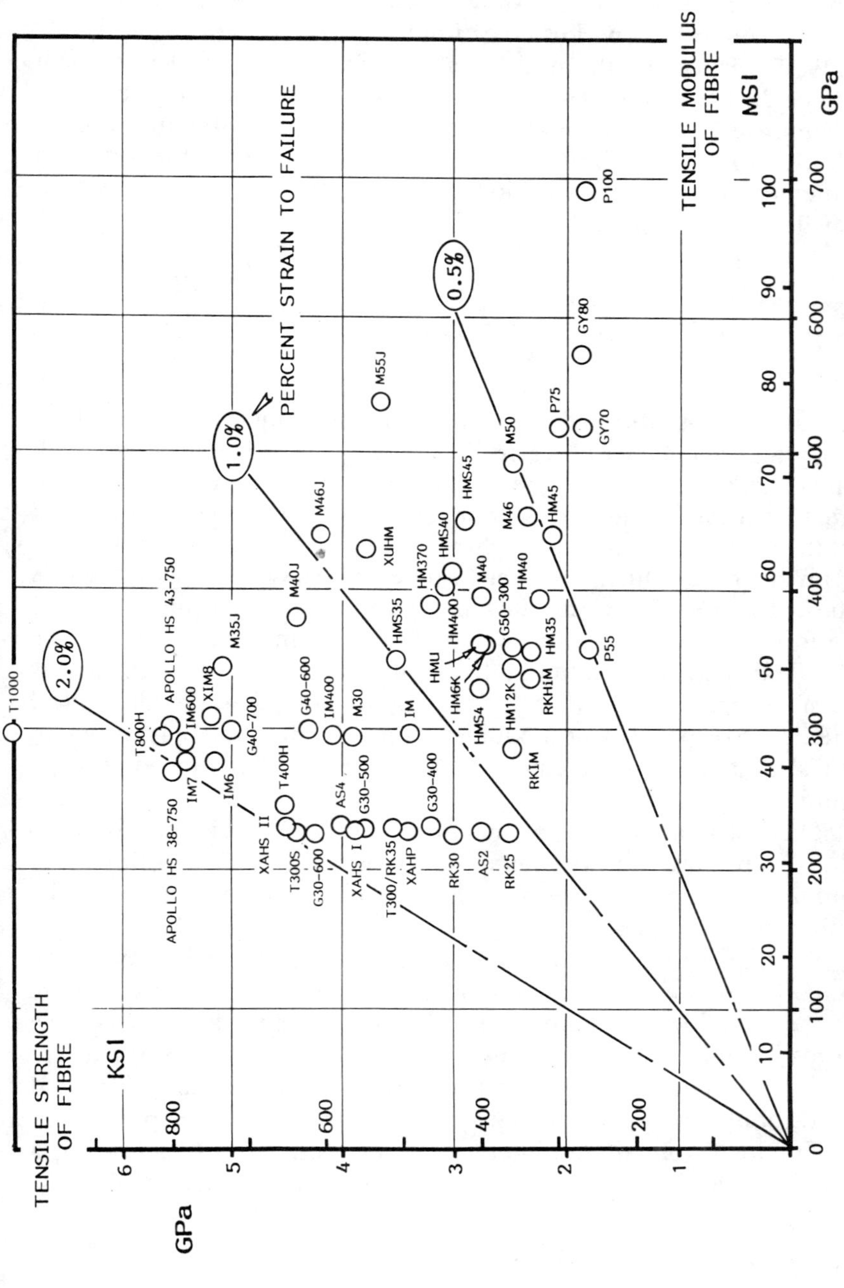

Fig. 12. Mechanical properties of currently available carbon fibres — August 1988.

The above classifications are shown graphically in Fig. 5 in comparison with the other high performance fibres.

We can expect that the main thrust of future carbon fibre development will be directed towards the improvement of high strain fibres in terms of improved ultimate tensile strength with corresponding increases in fibre strain to failure since the majority of carbon fibres used in structural composite applications are from this high strain, very high tensile strength group.

6 ARAMID FIBRES

It has been established for some time that highly aligned, high performance polymer fibres are capable of giving high tensile strength and modulus properties and recent work on experimental fibres produced from rigid-rod polymers of high molecular weight has confirmed this.[8] In this work rigid-rod poly(*p*-phenylene benzobisoxazole) (PBO) was spun into fibres which were then heat treated. These fibres gave a tensile strength of 5·8 GPa and tensile modulus of 345 GPa. The high modulus performance was the result of nearly perfect molecular orientation, excellent lateral packing and infrequent occurrence of chemical defects such as chain ends. It is interesting to note that the theoretical prediction of axial tensile modulus for the PBO rigid-rod polymer chain is 630 GPa, indicating the potential for high modulus aligned polymer fibres.

The first of the aligned polymer fibres to be commercially available to the composites industry were the aramid fibres. These fibres are produced by spinning long-chain polyamide polymers using special spinning techniques.

The most well known aramid fibre in current commercial production is Kevlar† which was introduced into the composites market by Du Pont in the early 1970s. The fibre was originally developed to replace steel cords in radial tyres for commercial vehicles where significant weight savings are possible with increased strength and durability. The first grade of fibre produced, Kevlar 29, was designed for these applications having high strength and intermediate modulus; shortly afterwards a higher modulus aramid fibre was introduced, Kevlar 49, and this is the

†Trade mark of E. I. du Pont de Nemourns & Co. Inc.

aramid fibre most widely used today in structural composite applications.

Kevlar 49 is thought to be an aromatic polyamide called poly(*p*-phenylene terephthalamide) and is produced by an extrusion and spinning process. A solution of the polymer in a suitable solvent is held at a temperature between −50 and −80°C before extrusion into a hot walled cylinder at 200°C whereupon the solvent evaporates. The extruded fibre at this stage has a tensile strength of 840 GPa and a tensile modulus of 3·6 GPa. The fibre is then given a stretching and drawing treatment to align further the polymer chains to achieve the properties given in Table 1 and shown graphically in Fig. 5.

The fibre manufacturing process results in the alignment of the stiff polymer molecules parallel to the fibre axis and the fibre modulus of 130 GPa confirms that a high degree of alignment has been achieved. The supramolecular structure of the Kevlar 49 fibre is interesting and has been investigated in some detail.[9]

A schematic representation of this structure is shown in Fig. 13. The

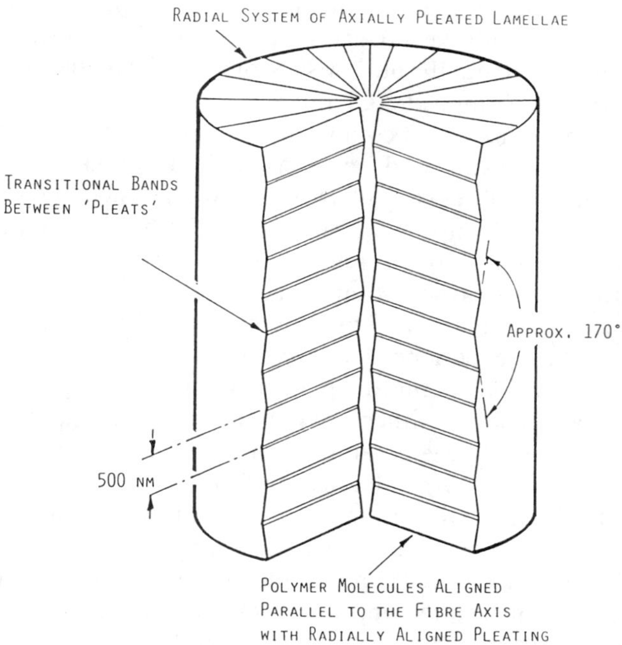

Fig. 13. Schematic representation of the supramolecular structure of aromatic polyamide fibres. Kevlar 49.[9]

molecules form rigid planar sheets with the chain-extended molecules hydrogen bonded together. The sheets are stacked to form a crystalline array but there is only weak bonding between these sheets which are arranged in a radial system of axially pleated lamellae. Each component of the pleat is about 500 nm long and they are separated by short transitional bands; the angle between neighbouring sections of the pleat is about 170°.

Fibres with this structure are likely to have low longitudinal shear modulus, poor transverse properties and low axial compression strength and it is the relatively poor compression performance of Kevlar 49 that is one of its major shortcomings. As an example, whilst glass and carbon fibres show approximately similar tensile and compression strengths when tested in the form of unidirectional composites, Kevlar 49 laminates show a compression strength of around 25% of their tensile strength. This is the result of the anisotropic structure of the fibre and its low shear stiffness.

However, this unique structure of Kevlar 49 enables it to offer an outstanding performance advantage not enjoyed by most other high performance fibres — namely impact resistance and ballistic energy absorption. As a result, Kevlar 49 is frequently used in hybrid composite applications to enhance the otherwise poor impact damage performance of a carbon fibre composite (Chapter 11).

Kevlar 49 unidirectional composite laminates have a linear stress–strain performance and good tensile strength in comparison with other high performance fibres (Fig. 6). However, when tested in compression or flexure, the low compression strength of the basic fibre produces a non-linear, ductile stress–strain behaviour with ultimate strengths lower than glass or carbon composites.

The interfacial bond strength between aramid fibres and resins is much lower than experienced with carbon fibres and work has been carried out on various sizing and plasma treatments to improve this. An interesting approach to interfacial bond strength improvements is based on the controlled generation of small fibrils on the surface of the fibre which considerably enhance the fibre–resin bond strength and improve the delamination resistance of the aramid laminate.[10]

A new aramid fibre, Kevlar 149, has recently been introduced by Du Pont with a tensile modulus approximately 40% higher than Kevlar 49 with a similar tensile strength. Kevlar 149 gives a tensile strength and modulus performance very close to typical high strength carbon fibres (Table 1).

A further recent development is the commercial availability of an alternative range of aramid fibres from AKZO Fibres and Polymers, marketed under the trade name Twaron.

A factor that must be appreciated is that aramid fibres have a much lower density than glass or carbon fibre (Table 1). This lower density means that when aramid fibres are compared with other high performance fibres in specific strength and modulus terms they show significant advantages (Fig. 14).

The combination of excellent specific strength and modulus performance combined with 'toughness' and impact resistance makes aramid fibre composites an attractive choice where minimum weight, 'toughness' and damage resistance are key requirements.

7 HIGH PERFORMANCE POLYETHYLENE FIBRES

Recent improvements in the mechanical properties of man-made fibres have resulted in the commercial availability of high performance polyethylene (HP-PE) fibres.

These fibres are made from ultra-high molecular weight polyethylene and are made by a gel spinning process where the very long molcules are dissolved in a volatile solvent and spun through a spinnerette. In the solution the long molecules become disentangled and remain so after cooling in the fibre. The fibre is then drawn and a very high level of macromolecular orientation is achieved.

The resulting HP-PE fibre has a high degree of parallel orientation and crystallinity which gives the fibre its excellent mechanical properties.

HP-PE fibres are now becoming commercially available in Europe from DSM in the Netherlands under the trade name Dyneema and in the USA from Allied-Signal Inc. under the trade name Spectra.

In terms of tensile strength and modulus HP-PE fibres have a performance close to aramid fibres (Table 1 and Fig. 5); however, since HP-PE fibres have a density about 30% lower than aramid fibres, their specific mechanical properties are significantly higher than those of aramid fibres.

HP-PE fibres have excellent chemical and abrasion resistance as well as low moisture absorption and good resistance to degradation from exposure to ultraviolet light. However, one serious difficulty is that

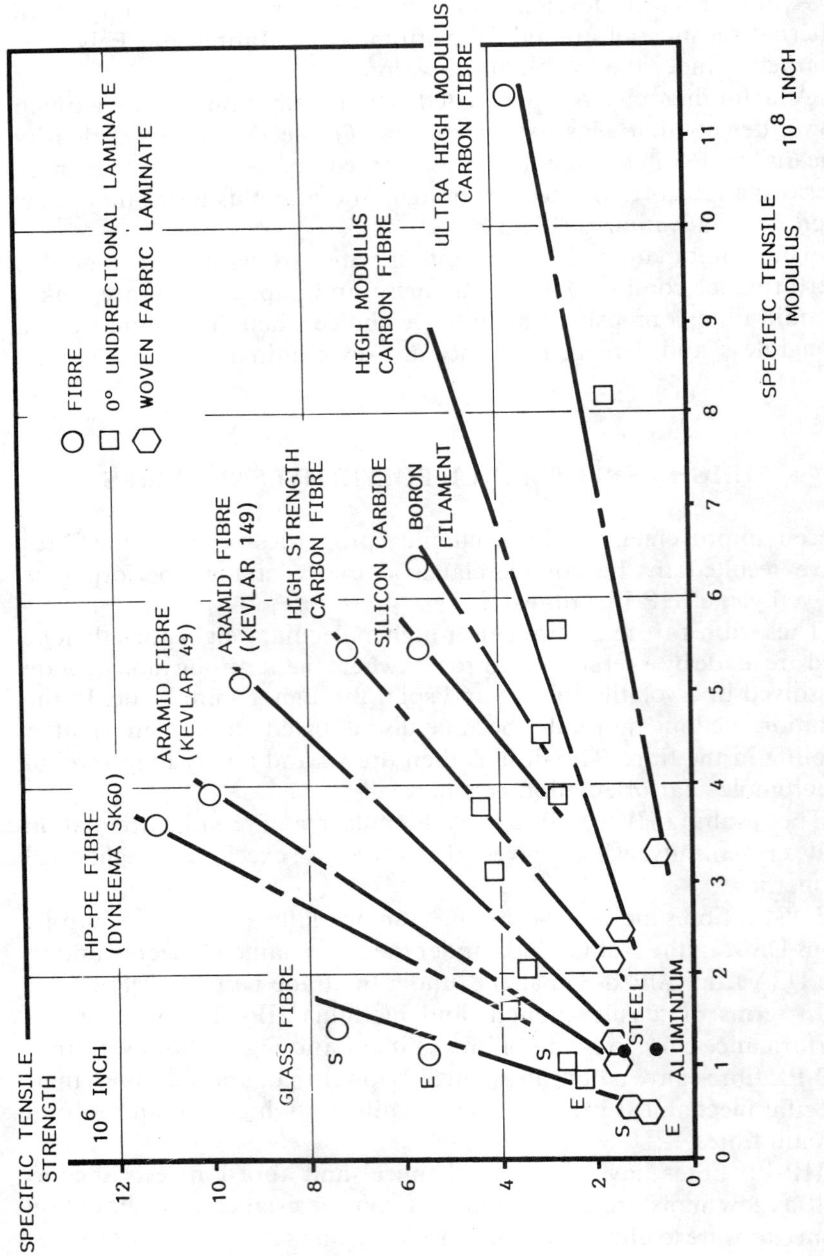

Fig. 14. Comparison of advanced composite fibres and laminates in terms of specific strength and stiffness.

HP-PE fibres melt around 150°C and suffer significant loss of mechanical properties above 130°C.

Although HP-PE fibres show linear stress–strain characteristics with a tensile modulus close to that of high strength carbon fibre, the compression and interlaminar shear performance of the HP-PE based composite laminates is relatively poor.[11] These characteristics are predominantly a result of the poor interfacial bond between HP-PE fibres and the resin matrix system. A range of fibre treatment techniques have been examined with the aim of improving the interfacial bond, including chemical etching, corona, flame, mechanical and cold gas plasma treatment. Current indications are that the most successful technique is cold gas plasma treatment and some significant improvements in interfacial bond strengths have been achieved.[12–14]

In terms of impact resistance and damage tolerance, composites based on HP-PE fibres demonstrate excellent characteristics, similar to aramid laminates, where the HP-PE fibres show no brittle fracture, absorbing energy by widespread delamination and splitting. This makes HP-PE fibres a good choice for ballistic applications such as helmets and composite armour. HP-PE fabrics could also be used in a hybrid application to protect carbon fibre structures from impact damage. The impact resistance, electrical insulation and radar transparency characteristics of HP-PE fibres make them attractive materials for composite radome applications.[12]

We can expect continuing improvements in the mechanical properties of HP-PE fibres and this promises to be an exciting area for future high performance fibre development.

8 HIGH PERFORMANCE FIBRE PRICES

The relative prices of the various high performance fibres are shown in Fig. 15 for the period 1970–1988. These figures are intended to give an indication of the order of magnitude of fibre prices over this period and should not be interpreted as a precise statement of fibre price, since this will depend upon the specific fibre type and filament count.

Glass fibres are the lowest priced of all the fibre types reviewed, but even here the influence of fibre filament count and the particular fibre surface treatment required can result in significant differences in fibre price. Commercial grade E glass is the lowest priced material whereas aerospace grade E and S glass fibres are considerably higher in price.

The price trend for carbon fibre over the period 1970–1988 represents a typical high strength fibre and it will be seen that significant price reductions have taken place since 1970. Over this period the world demand for carbon fibre has risen from a few tonnes a year to approximately 5000 tonnes a year in 1988. We now have a situation where a considerable overcapacity exists world-wide for carbon fibre and this will probably contribute to a continuing decline in the price of carbon fibre. It will be noted that a very wide range of carbon fibre types are currently available with prices ranging from $12/lb up to $850/lb. The typical current price ranges for the various carbon fibre types are shown in Fig. 15.

Aramid fibres now overlap in price with the lower priced carbon fibres and with the entry into the market of a second aramid fibre supplier, AKZO Fibres and Polymers, it is anticipated that the aramid fibre prices will probably stabilise at around their current price levels.

High performance polyethylene fibres are only just becoming commercially available and it is expected that they will be priced at approximately the same level as aramid fibres when volume production commences.

Boron prices have been increasing since the late 1970s when the market for boron fibre started to decline. The competitive situation with boron fibre is also a factor since only one boron fibre manufacturer currently exists producing commercial quantities of material. It is anticipated that the price trend for boron fibre will continue to increase, influenced by reducing demand and the single source of supply.

The pattern of prices for high performance fibres established over the past 20 years seems to be stable and it is unlikely that any major changes to these price trends will take place in the short term.

9 FIBRE COMPARISONS

It is difficult to compare fibres in an overall sense, since in addition to a simple comparison of fibre mechanical properties, each fibre has very specific advantages and disadvantages often associated with practical aspects of the fibre utilisation in composite structures, fibre handle-ability, and ease of processing finished composite laminates.

Table 1 tabulates the relative mechanical properties for the various

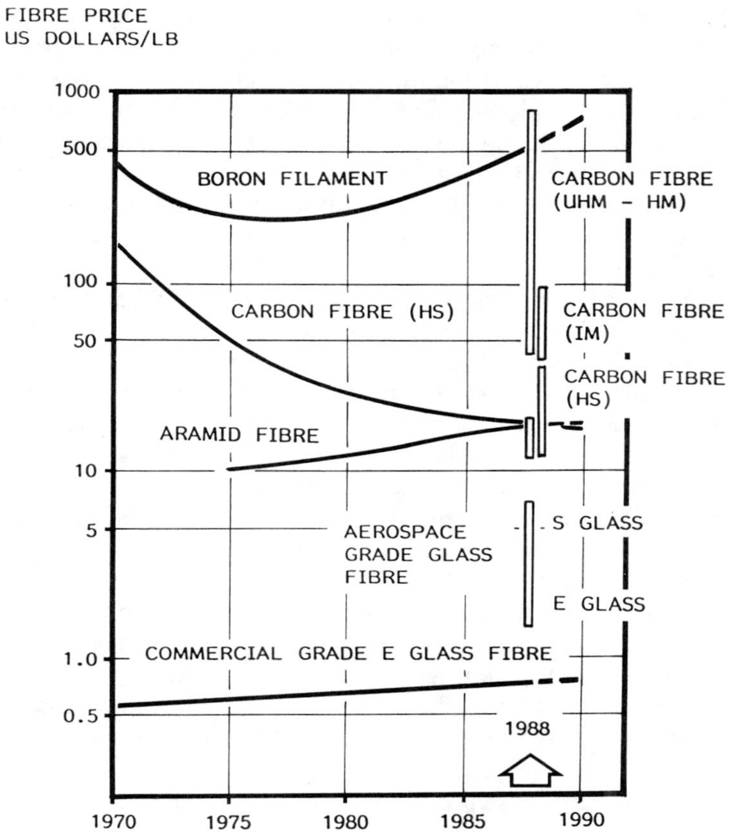

Fig. 15. High performance fibre price trends, 1970–1988. These figures are intended only as an approximate indication of relative prices since the exact fibre price will depend on the fibre grade/tex/filament count.

fibres and these are shown graphically in Fig. 5. Additionally, the stress–strain characteristics of each of the major fibre groups are shown in Fig. 6 and it can be seen that all fibres when processed into unidirectional composites exhibit linear stress–strain characteristics. A comparison of the fibre tensile strength and modulus is frequently used as a means of demonstrating the relative performance of these fibres, and a comparison is frequently made in this form between unidirectional fibre properties on a specific basis and the mechanical properties of metallic materials (Fig. 14).

This, however, is a rather unfair comparison since the tensile properties of a fibre cannot reasonably be compared with the properties of metallic materials and a more reasonable comparison would be that of unidirectional laminates manufactured from these fibres and woven fabric composites based on the various fibres. This comparison is shown in Fig. 14 and it can be seen that the specific strength and stiffness performance of a particular fibre is considerably reduced when we consider its performance in unidirectional composite and woven fabric composite form.

When we consider the practical aspects of fibre and composite handling we find that each fibre has some very particular characteristics which may influence fibre selection for a particular structure. The handling, laminating and trimming of cured glass resin composites is relatively straightforward, and presents very little difficulty in a production situation. Similarly, it has been found that carbon fibre based composites can also be readily handled and processed without any particular problems. Aramid fibres, however, because of their basic structure and inherent 'toughness', have presented some problems in terms of trimming and cutting of the aramid fabric pre-preg and the finished aramid composite. The initial difficulties with aramid fibres have now been overcome with the development of special cutting techniques and equipment. However, one must expect to spend more time and effort developing appropriate cutting and trimming operations for aramid based composites than would be necessary with glass or carbon fibre based materials.

Boron filament based pre-pregs and cured laminates are also relatively difficult materials in terms of mechanical handling, since the very large diameter filaments are difficult to cut in the pre-preg form and the finished laminate must be handled carefully to avoid damage or displacement of these large filaments in the cured matrix system.

10 SUMMARY

- A comprehensive range of high performance fibres is currently available offering a spectrum of mechanical properties when incorporated into a composite structural element.
- The performance of the fibre in the structural element is influenced by a number of factors and an understanding of these

factors and their effect on the finished composite structure is essential if one is to utilise these materials efficiently.

— As a first step the overview information in this chapter should enable one to make an initial selection of one or two fibre types for further detailed investigation.

— A useful source of information on the finished composite performance of the various materials is to be found from the fibre and pre-preg suppliers, who make this information readily available.

— The composite materials industry has grown enormously over a relatively short period and we can anticipate significant improvements in fibre and matrix resin technology from the major companies involved in this market over the next few years.

REFERENCES

1. A fighter fuselage in synthetic material, *Aero Research Technical Notes, Bulletin,* **34**, Aero Research Ltd, Duxford, Cambridge, October 1945. (Now Ciba-Geigy Plastics).
2. Schwartz, M. M., *Composite Materials Handbook*. McGraw-Hill, New York, 1984.
3. Hull, D., *An Introduction to Composite Materials*. Cambridge University Press, Cambridge, 1981.
4. Holleyman, C. F., High performance fibres: the development of improved properties in the Hysol–Grafil range of products. Paper presented at Carbon Fibre III Conference, London, October 1985.
5. Trewin, E. M., The development of high strain carbon fibre. Paper presented at SAMPE Conference, Engineering with Composites, London, March 1983.
6. Watt, W., Phillips, L. N. & Johnson, W., High strength, high modulus carbon fibres. *Engineer*, **221** (1966) 815–16.
7. Lovell, D. R., A comparison of available carbon fibres. Paper presented at Carbon Fibre III Conference, London, October 1985.
8. Adams, W. W., Wierschke, S. G., Lenhert, P. G., Stewart, J. J. P., Klei, H. E., Eby, R. K. & Jiang, H., Theoretical and experimental studies of the axial tensile modulus of high-performance rigid-rod polymer fibres. Paper presented at The Plastics and Rubber Institute 7th International Conference: Deformation, Yield and Fracture of Polymers, Cambridge, April 1988.
9. Dobb, M. G., Johnson, D. J. & Saville, B. P., Structural aspects of high modulus aromatic polyamide fibres. *Phil. Trans. R. Soc. London A,* **294** (1980) 483–5.

10. Roebroeks, G. & van Dreumel, W. H. M., Mechanical interfacing of aramid fibre. Paper presented at 7th International SAMPE Conference, Munich, 1986.
11. Scholle, K. F. M. G. J., Jacobs, M. J. N., Omloo, H. A. A., Loete, G. J. H. & Janse, G. H. A., Performance of Dyneema SK600 polyethylene fibre/epoxy composites. Paper presented at 9th International SAMPE Conference, Milan, 1988.
12. Cordova, D. S. & Donnelly, D. S., Spectra® reinforced composite systems for high impact/microwave transparent radar domes. Paper presented at 33rd International SAMPE Symposium, Anaheim, 1988.
13. Kaplan, S. L. & Rose, P. W., Gas plasma treatment of Spectra® fiber. Paper presented at 33rd International SAMPE Symposium, Anaheim, 1988.
14. Nguyen, H. X., Riahi, G., Wood, G. & Poursartip, A., Optimisation of polyethylene fibre reinforced composites using a plasma surface treatment. Paper presented at 33rd International SAMPE Symposium, Anaheim, 1988.

Chapter 4

Approaches to Toughness Enhancement

C. B. Bucknall

School of Industrial Science, Cranfield Institute of Technology, Bedford, UK

1 INTRODUCTION

Engineers have long been aware of a need for enhanced toughness in advanced composites. The problem became the focus of attention in the early 1970s, when Rolls-Royce developed designs based on carbon fibre reinforced epoxy resins for the compressor blades of the RB211 jet engine. The blades were required to survive collisions with birds, and in trials it was found that the composites available at that time could not meet the design specification for impact resistance. The intervening period has seen an expanding effort to overcome these problems, which has resulted in some significant improvements in performance, and a growing number of applications in the aerospace industry. Current developments offer considerable promise for future expansion.

Materials suppliers have followed two contrasting routes to improved composites. The more radical is to abandon thermosetting resin matrices altogether, and to replace them with ductile thermoplastics; this approach is discussed in Chapter 2. The second route involves retaining the thermosetting matrix resin, but modifying the composite by adding a rubber or thermoplastic. Four separate techniques for adding a second polymer to the composite have been explored actively by the industry, and are reviewed in some detail here. These are:

(a) Rubber-toughening the matrix resin.
(b) Mixing a thermoplastic with the thermosetting resin.

(c) Coating the fibres with an elastomer or ductile plastic.
(d) Interleafing the laminae of fibre reinforced resin with thin films of a ductile thermoplastic or toughened thermosetting polymer.

In each of these technologies, the emphasis is on improving the balance of properties of the final composite. There is little point in increasing the toughness of the product if this is achieved by sacrificing other essential properties. The properties of most concern in the present context are stiffness and strength, both of which could be reduced as a result of adding a second polymer.

Stiffness and strength tend to be dominated by the fibres in areas of the composite where the fibres are in tension, but under shear or compressive loading the mechanical properties of the matrix and of the fibre–matrix interface must be taken into account. Composites show a reduction in compression strength as the resin approaches its glass transition temperature T_g, which in epoxy resins generally lies between 80 and 220°C. Water acts as a plasticiser for epoxies, so that T_g is reduced when the resin equilibrates with atmospheric moisture.

Addition of a polymer with a low T_g, whether rubber or thermoplastic, inevitably causes some reduction in the modulus of the matrix at temperatures between the T_g of the additive and that of the epoxy resin. It is important to ensure that the polymeric modifier does not impair the ability of the composite to operate at elevated temperatures. A standard method for studying these effects is the 'hot–wet compression test',[1] in which a quasi-isotropic $[+45°/90°/-45°/0°]_S$ plate specimen is immersed in water at 71°C (160°F) for 14 days, then tested in compression at 93°C (200°F), with edge restraints to prevent buckling.

It is not easy to define exactly what is meant by toughness in a composite. The practical goal is a material that can withstand the impact of the proverbial dropped spanner without suffering loss of function. By contrast with the response of most materials, impact damage in composites may be completely undetectable to the eye, and become apparent only when the material is loaded in compression. Test procedures have therefore been developed to simulate service conditions. For example, in the CSAI (Compression Strength After Impact) test[1] a sheet of quasi-isotropic laminate is subjected to a specified impact at its centre from a ball-ended striker, and the impacted sheet is then tested in in-plane compression, again with the aid of anti-buckling plates. An alternative procedure is to measure the area of impact damage within the plate using ultrasonic crack detection. Because loading conditions and failure modes in the plate impact and CSAI tests are complex and difficult to analyse, it is helpful to correlate the data with results obtained from the simpler and more controlled delamination tests discussed in Chapter 9. However, it must be recognised that higher values of, for example, G_{IC} in delamination do not necessarily guarantee improved impact resistance under service conditions. Other properties, notably those involving interlaminar shear, may well be more important.

2 RUBBER TOUGHENING

The technology of rubber toughening was first developed for thermoplastics, and was extended to epoxy resins during the 1960s.[2–5] The basic principle is that brittle polymers can be made tough and ductile by addition of 5–20% of a suitable rubber. The rubber is dispersed as small spheres of the order of 1 μm in diameter, which act as stress concentrators, and initiate deformation in the surrounding matrix.[6] A recent conference reviewed developments in the subject.[7] Several different deformation mechanisms are known to contribute to toughening, the extent of each contribution depending upon a number of factors, including the type of matrix, test temperature, and applied stress. In rubber toughened epoxy resins, the principal mechanism appears to be shear yielding in the resin, which is accompanied by cavitation within the rubber particles.[8–10] Advances in rubber toughening of thermosetting resins up to 1984 are reviewed in a recent book.[11]

In order to produce the required dispersion of rubber spheres, it is necessary to dissolve the rubber in the liquid resin and to precipitate it during the curing reaction. This requirement restricts the range of suitable rubbers. A second desirable property of the rubber is an ability to react chemically with the resin. In the case of epoxy resins, by far the most widely used class of rubber toughening agents is based on the liquid, low molecular weight, carboxyl terminated butadiene–acrylonitrile (CTBN) copolymers, represented by the formula:

$$HOOC{-}[{-}(CH_2CH{=}CHCH_2)_x{-}(CH_2CHCN)_y{-}]_z{-}COOH$$

where x, y and z typically have values of 5, 1 and 10 respectively. The solubility of the rubber in epoxy resin depends upon the ratio of x to y. The function of the carboxyl end groups is to react with the epoxy group of the resin to form a block copolymer. This reaction is best carried out before adding hardener; otherwise the hardener will compete against the carboxyls for the unreacted epoxy units.[12, 13]

Liquid CTBN copolymers have also been used to toughen polyester,[14] vinyl ester,[14] and bismaleimide[15] resins. Butadiene–acrylonitrile copolymers with other types of end groups, including hydroxyl —OH, amine $—NH_2$, and vinyl $—CH_2CH(CH{=}CH_2)$, have also been studied as toughening agents for epoxy resins.[16] Functionally terminated polybutadienes,[17] fluoroelastomers,[18] and siloxanes[19] have also been used to toughen epoxy and other thermosetting resins.

In order to achieve toughening, the solution of rubber in resin must undergo phase separation during the curing reaction. If the rubber remains in solution it will act basically as a plasticiser, reducing the glass transition temperature of the resin, but not necessarily increasing its fracture resistance. There appear to be two major factors that contribute to phase separation in blends of epoxy resin with CTBN: (a) the increasing molecular weight of the resin, which results in a decrease in the entropy of mixing per unit mass of material;[20] and (b) the increasing polarity of the resin due to formation of hydroxyl groups from epoxy rings, which causes an increase in the enthalpy of mixing per unit mass.[21] The increase in density of the resin during curing, the formation of block copolymer, and the temperature of the system also affect miscibility.

The changes occurring in a rubber modified thermoset during cure can be summarised in a time–temperature–transformation diagram, as illustrated in Fig. 1.[22, 23] At any given cure temperature, the resin–rubber mixture illustrated passes successively through phase separation, gelation of the resin-rich phase, and vitrification of the resin. Both gelation and vitrification prevent further phase separation in the resin, and it is therefore essential to ensure that the miscibility characteristics of the resin and rubber lead to a high degree of phase separation before either of these stages is reached. Phase separation can of course continue in the rubber-rich phase, which remains uncrosslinked and above its own T_g. Resin precipitated from this phase may produce sub-inclusions within the rubber particle or form a rubber-free shell around the particle. Sub-inclusions increase the effective volume fraction ϕ of the rubber particles, and thereby contribute to toughness.[24] Epoxy resin that has reacted on the molecular scale with the rubber also contributes to ϕ.

As the concentration of rubber is increased above about 20 vol.%, the morphology changes. The rubber-rich phase now forms the matrix, and

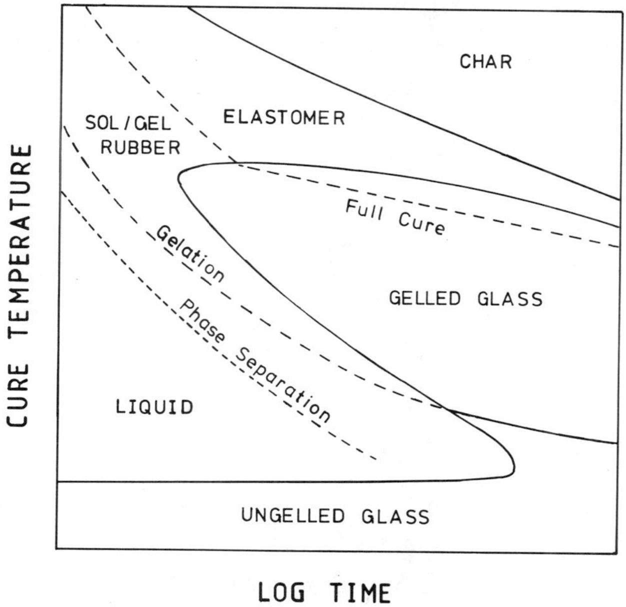

Fig. 1. Schematic time–temperature–transformation diagram for a solution of rubber in thermosetting resin which undergoes phase separation before gelation. After Gillham & Chan.[22]

the resin is dispersed as particles within it.[25,26] The modulus, which has been decreasing slowly with increasing rubber content, falls sharply, and the fracture surface energy G_{IC} reaches a peak and begins to decrease, as shown in Figs 2 and 3.[27] In an unstirred polymerising system, the structure of the final product is determined by the volume fractions of the two phases at the onset of phase separation. At rubber concentrations above about 20%, the rubber-rich solution occupies a larger volume than the polymerised resin-rich solution, and therefore forms the continuous phase. Further polymerisation increases the proportion of resin-rich phase, which in a stirred system would result in a phase inversion. In the absence of stirring, the rapid increase in solution viscosity and reducing molecular mobility prevent the inversion. However, the volume fraction of the resin phase may be so high that neighbouring particles are actually in contact, so that the product is an interpenetrating network (IPN) rather than simply a dispersion of resin particles in a rubber matrix.

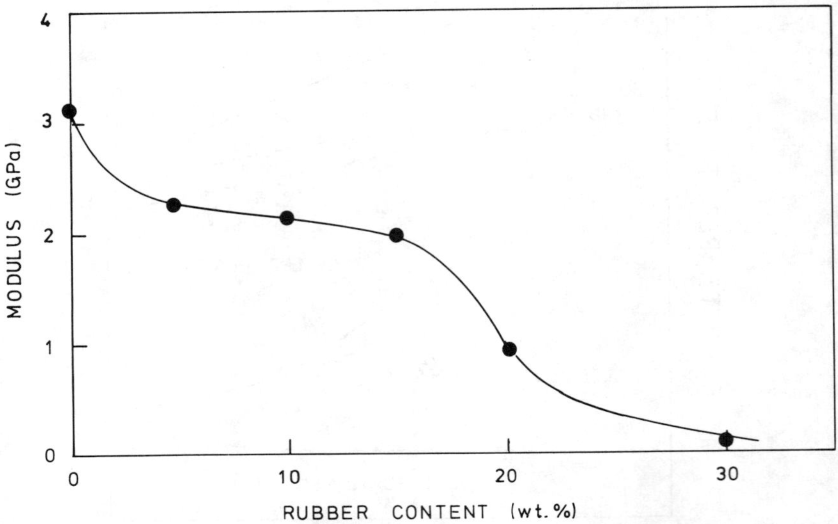

Fig. 2. Relationship between Young's modulus and CTBN rubber content in blends based on DGEBA/piperidine. After Bascom *et al.*[27]

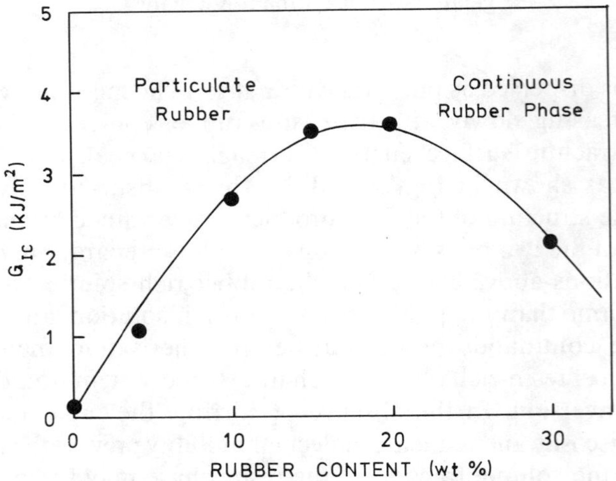

Fig. 3. Relationship between G_{IC} and CTBN rubber content in bulk specimens of DGEBA/piperidine. After Bascom *et al.*[27]

3 ADDITION OF THERMOPLASTICS

Addition of thermoplastics to thermosetting resins is a more recent development than rubber modification, but the two processes have much in common. The thermoplastic is dissolved in the uncured resin and phase-separates during curing, forming the disperse phase at low concentrations, and the matrix phase at moderate or high concentrations.[28] Further phase separation within the thermoplastic-rich phase produces sub-inclusions of the thermosetting resin. The morphologies are illustrated in Fig. 4. As in the case of rubbers, high concentrations of added thermoplastics tend to produce interpenetrating networks rather than simple matrix–particle structure.

The advantage of thermoplastics over rubbers as additives for high performance resins lies in their higher glass transition temperatures.

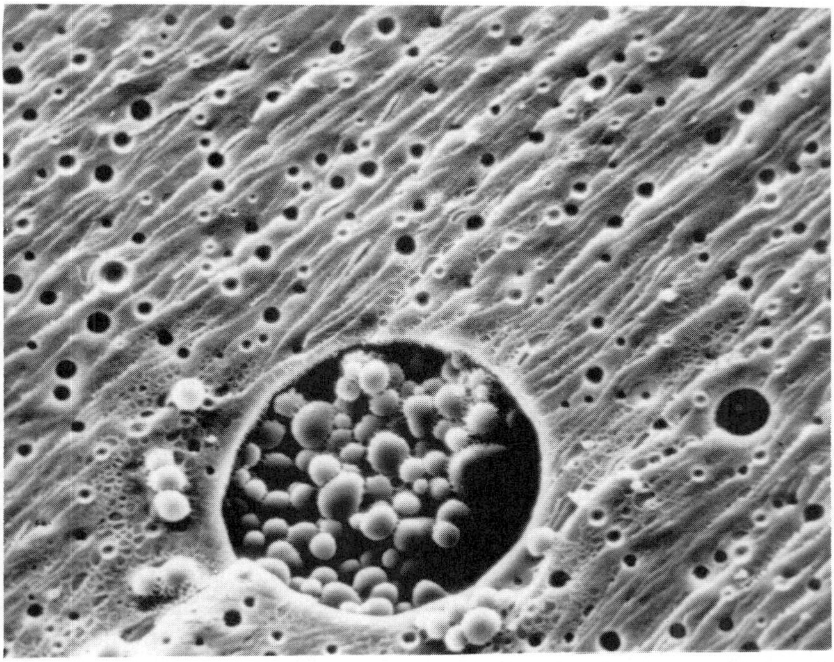

Fig. 4. Scanning electron micrograph showing phase separation in a blend of epoxy resin with 5 phr of poly(methyl methacrylate). Fracture surface etched with methyl ethyl ketone (MEK) and coated with Au/Pd (10 mm = 2 μm). Micrograph by A. Gilbert.

The preference is therefore for the higher-softening ductile plastics, but the choice is restricted to polymers that are soluble in the uncured resin. Fortunately, there are several commercially available thermoplastics which meet this specification, including polysulphones, poly(ether sulphone)s, polycarbonates, poly(phenylene oxide)s and poly(ether imide)s. The slowness with which these polymers dissolve in the resin and the high viscosity of the resulting solutions present significant practical problems in manufacturing the blends, and it is usually necessary to employ organic solvents which must be removed later.[29,30] The processability of these systems can be improved by lowering the molecular weight of the thermoplastic, following the example of the liquid rubbers discussed earlier. However, the supply of short chain thermoplastics is limited, and there is as yet little information on the degree to which molecular weight can be reduced before the strength of the blends begins to fall. The increased viscosity of the solutions is a positive advantage during curing of composites, because it enables pressure to be exerted on the resin to expel air and eliminate voids (see Chapter 1).

4 COATED FIBRES

Coating the fibres with a thin layer of a rubber or thermoplastic prior to impregnation with resin offers another possible approach to toughening of composites. The obvious method is to coat from solution. One limitation of this method is that coating tends to be uneven. Another is that the coating can subsequently dissolve in the resin, so that the final product is simply a fibre reinforced, rubber toughened or thermoplastic modified, resin. To overcome this problem, it is necessary either to crosslink the coating or to select a coating that is basically insoluble in the resin.

Electrodeposition appears to produce rather more stable coatings. Work by Wilson has shown that CTBN and VTBN (vinyl terminated butadiene–acrylonitrile) can be electrocoated on to carbon fibres from solution in dimethylformamide containing $NaNO_3$ as electrolyte.[31] Best results were obtained by making the carbon fibres the cathode, with a platinum anode, at a voltage of about 10–25 V. Little deposition occurred when the carbon was the anode. As in solution coating, it is difficult to coat all of the fibres uniformly. However, electrocoating does appear to give a better distribution of rubber, probably because coating takes place preferentially on previously uncoated regions of the fibre, where the conductivity is highest.

5 INTERLEAFING

Interleafing is the newest of the approaches to toughening considered in this review. Preferred interleafing materials are rubber toughened thermosetting resins[32] or high-softening ductile thermoplastics which form adhesive bonds with the fibre reinforced layers of thermosetting resin.[33] The list of suitable thermoplastics includes poly(ether imide), polyimide, poly(amide imide), polyaramid, polyamide, polycarbonate, and poly(ester carbonate). Layers of thermoplastic having thicknesses between 10 and 50 μm are placed between the laminae of fibre reinforced resin, which are typically about 200 μm thick. A convenient method is to bond the plastic film to one side of the pre-preg using rollers. The composite is then laid-up in the usual way.

6 TOUGHNESS OF POLYMER MODIFIED RESINS

Figure 5 shows the dramatic increase in G_{IC} that can be obtained on adding CTBN rubber to a standard epoxy resin.[24] The material illustrated is Ciba–Geigy MY750, which consists basically of the diglycidyl ether of bisphenol A (DGEBA). The untoughened resin has a G_{IC} of about 0·5 kJ/m^2. The fracture surface energy can be increased to almost 10 kJ/m^2 on addition of 8·7 wt% of CTBN rubber, which after allowing for density differences corresponds to 11 vol.% of rubber for a fully phase-separated system. Except for the pure epoxy resin, at $\phi = 0$, all of the materials plotted in Fig. 5 contained the same amount of CTBN rubber, and the variations in toughness are due to variations in structure resulting from the use of different amounts and types of curing agent. Epoxy resin sub-inclusions, together with copolymerised resin molecules within the rubber particle, increase the effective rubber particle volume fraction ϕ by a factor of between 1 and 2, depending upon the formulation, and G_{IC} clearly increases linearly with ϕ. Similar increases in G_{IC} have been demonstrated in bismaleimide resins, but from a lower baseline: the untoughened resin has a fracture surface energy of approximately 10 J/m^2, which is increased to 130 J/m^2 on adding 30 vol.% of CTBN rubber.[15]

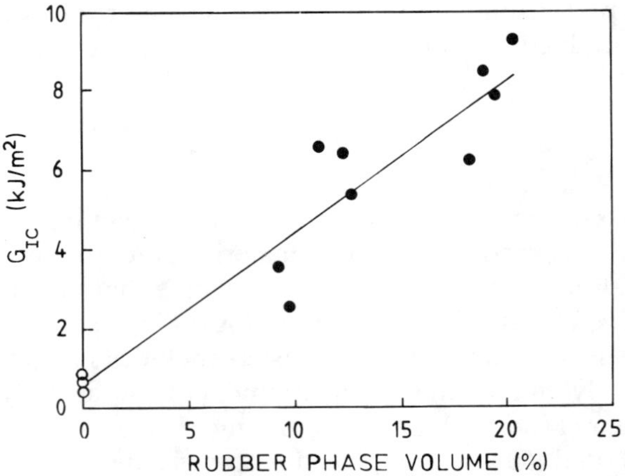

Fig. 5. Relationship between G_{IC} and rubber phase volume in DGEBA-based epoxy resins: (○) untoughened resin; (●) resin containing 8·7 wt.% CTBN rubber, with various hardener systems. After Bucknall & Yoshii.[24]

This comparison emphasises one of the main problems in toughening thermosetting resins. The materials that can be most effectively toughened are the less tightly crosslinked resins such as the DGEBA family, in which large strains can develop, at least locally, in the presence of rubber particles. Under tensile stress, yielding takes place in the resin matrix, accompanied by void formation within the rubber particles, and large amounts of energy are absorbed in the yield zone.[9, 10, 34–36] Tighter crosslinking restricts the ability of the matrix to deform in this way, and therefore limits the level of toughening that can be achieved even under the most favourable conditions. This point is well illustrated in Fig. 6, which shows the relationship between G_{IC} and the molecular weight between crosslinks for two sets of epoxy resins, one without added modifier, and the other containing 10 phr of CTBN.[9]

Relatively little has been published on the mechanical properties of thermoplastic–thermoset blends. Bucknall & Partridge measured K_{IC} in a range of epoxy resins containing up to 25 phr of a low molecular weight poly(ether sulphone), and found only marginal increases in toughness.[37] On the other hand, a recent patent application reports an increase in notched Charpy impact strength from 1·7 to 5·5 kJ/m² on adding 10 phr of Ultem® poly(ether imide) (PEI) to a tetrafunctional epoxy resin.[30] Increasing the PEI content to 20 phr raised the impact

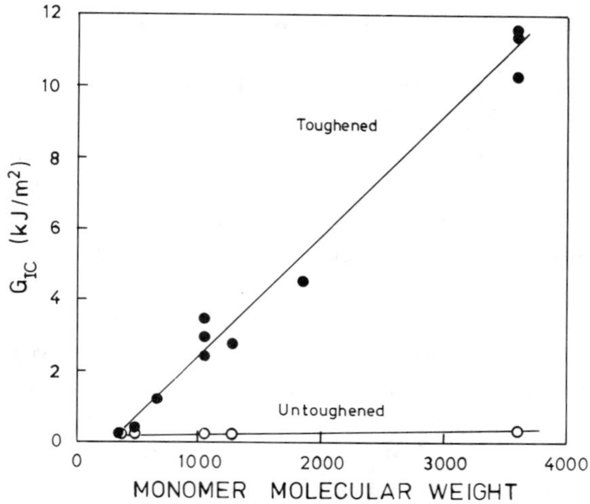

Fig. 6. Relationship between G_{IC} and resin monomer molecular weight in epoxy resins cured with DDS: (○) without rubber; (●) with 10 phr of CTBN rubber. After Yee & Pearson.[10]

strength to 6·4 kJ/m². The heat distortion temperatures of the three materials were 215°, 200°, and 161°C respectively. Neither PEI nor PES significantly affects room temperature modulus.[30, 38]

Bucknall & Gilbert have shown that PEI is precipitated from solution in tetrafunctional epoxy resin during curing,[39] producing a complex two-phase structure. Below about 20 phr of PEI, the epoxy resin forms the matrix, with the PEI dispersed into particles; spherical sub-inclusions of epoxy resin constitute a large fraction of the total particle volume, as illustrated in Fig. 4. At higher concentrations, PEI forms the matrix. Over the whole of this composition range, there is a continuous increase in G_{IC} and K_{IC} with PEI concentration.

Scanning electron microscope studies of fracture surfaces from these materials show that ductile tearing of the thermoplastic additive is largely responsible for the observed increases in toughness. The PEI particles act as ductile bridges spanning the growing crack, absorbing energy in a similar manner to that proposed for rubber particles by Merz *et al.*[40] and later by Kunz-Douglass *et al.*[41]

This mechanism could obviously occur in other phase-separated blends of thermosetting resins. There is ample scope for extending the principle of using ductile thermoplastic additives in thermosets.

7 TOUGHNESS OF POLYMER MODIFIED COMPOSITES

Increasing the toughness of the resin produces a corresponding increase in the toughness of the composite, as shown in Fig. 7, which is based upon a survey by Hunston.[42] Increasing G_{IC} of the resin over the range 10 to about 250 J/m^2 results in a very rapid increase in the interlaminar G_{IC} of the composite. This initial part of the curve has a slope of about 2, perhaps reflecting the greater fracture surface area in the composite as compared with the unreinforced resin. These effects relate to materials in which energy absorption occurs close to the fracture surface itself, and does not involve a large volume of sub-surface polymer: the variations in G_{IC} are due to differences in the degree of crosslinking of the resin and in the resulting local mobility of the resin molecules. Provided that the fibre–matrix interface is sufficiently strong, the presence of the fibres increases the true area of crack surface rather than limiting the depth at which yielding occurs.

Further increases in G_{IC} of the matrix are achieved through yielding at the crack tip involving material which might be up to a millimetre away from the immediate fracture surface. The presence of 60 vol.% of 7-μm diameter carbon fibres will obviously restrict the energy that is

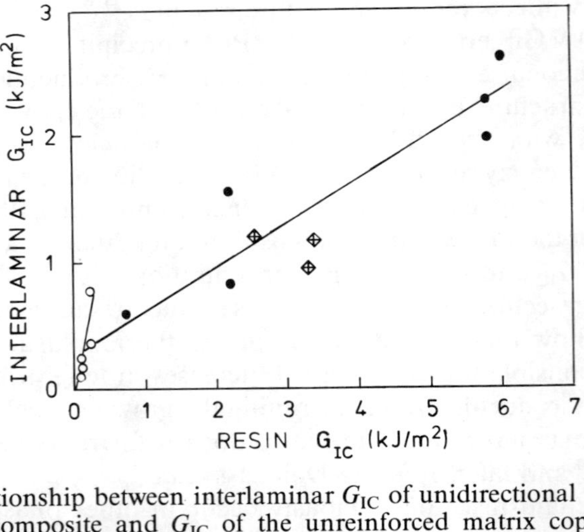

Fig. 7. Relationship between interlaminar G_{IC} of unidirectional carbon fibre reinforced composite and G_{IC} of the unreinforced matrix copolymer: (○) unmodified resin; (●) rubber-toughened resin; (⊕) thermoplastic. After Hunston.[42]

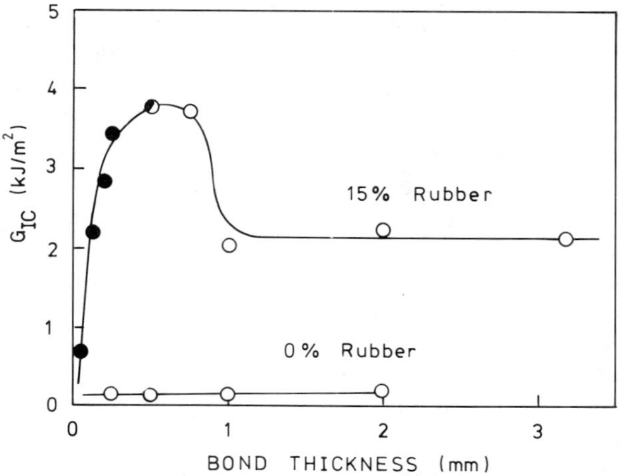

Fig. 8. Relationship between G_{IC} and bond thickness in adhesive joints based on DGEBA/piperidine epoxy resin with and without CTBN rubber: (○) unstable crack propagation; (●) stable propagation. After Bascom *et al.*[27]

absorbed in the yielding process. A similar effect is seen in rubber toughened epoxy adhesives bonded to metal substrates. Figure 8 shows the way in which G_{IC} for the adhesive falls when the layer thickness is reduced below 1 mm because metal occupies the place previously taken by yielding polymer.[27] Interaction between the crack tip and the metal substrate causes a maximum in G_{IC} of the rubber toughened resin at a bond thickness of about 1 mm.

Whilst an increase in interlaminar G_{IC} is clearly welcome, it must not be assumed that it guarantees improved practical performance. There is evidence that compression strength after impact, for example, is related to in-plane shear rather than to crack opening properties, and that G_{IIC} would therefore be a better indicator than G_{IC} of toughness in the composite. The difficulties of measuring G_{IIC} are discussed by Bascom *et al.* in connection with adhesive layers.[43] However, after making due allowance for these difficulties, it appears that increases in G_{IC} on adding rubber are not reflected in a corresponding increase in G_{IIC}. Another property of relevance to the impact resistance of laminates is the interlaminar shear strength (ILSS), although it must be noted that strength and toughness are quite distinct properties.

Further doubts about the relevance of G_{IC} measurements on fibre composites arise when the mechanisms responsible for high G_{IC} values are examined. In many cases, increased energy absorption occurs as a result of more extensive fibre bridging, as illustrated in Fig. 9.[44] The materials are epoxy resins containing 60 vol.% of carbon fibres, one without modification, and the other having a thin coating of CTBN on the fibres. A serious drawback of composites containing rubber coated fibres is that ILSS decreases as the amount of rubber used increases.[31] This reduction offsets the increase in G_{IC} obtained by the process.

Interleafing does not suffer from these problems. The action of the interleaved polymer appears to depend upon using a polymer that deforms in shear just below the ILSS of the adjacent composite layers, so that shear fracture is delayed or prevented in the reinforced material.[1,45] Shear strains as high as 50% have been reported in the interleaf material.[46] A laminated plate containing interleafed ductile layers can therefore survive an impact much better than a standard laminate. The stiffness of the composite is reduced as a result of adding unreinforced layers, but the total thickness of interleafed polymer is only of the order of 5%, so that the reduction is not a major consideration.

8 COMPARISONS AND OUTLOOK

The four approaches outlined above offer considerable scope for the materials scientist and engineer to develop high performance composites which combine toughness with stiffness and creep resistance at high operating temperatures. Because of their low shear moduli, rubbers in the form of particles or fibre coatings reduce the stiffness of the matrix resin. The extent of this reduction depends on the effective rubber volume fraction, and is usually less than 30%. By contrast, thermoplastics below their glass transition temperatures have little effect on modulus. Neither rubbers nor plastics have any significant effect upon the T_g of the resin provided that they separate from solution during curing. However, it is possible for the resin to crosslink or vitrify before phase separation is complete, in which case the T_g of the resulting resin will lie between those of the pure resin and the additive. The product may be satisfactory when the additive is a high-softening thermoplastic, but will suffer a serious reduction in maximum operating temperature if the additive is a rubber or low-softening plastic. Rubber modified systems in which the additive forms the matrix have low moduli down to the T_g of the rubber.

The technology of using rubbers as toughening additives for epoxy

Fig. 9. Intralaminar fracture in unidirectionally reinforced epoxy resin containing rubber coated carbon fibres, showing extensive fibre bridging (15 mm = 200 μm). From Preud'homme.[44]

resins and other thermosets is well established, and several systems are available commercially, mainly for use as adhesives, where there is less emphasis than in composites upon very tightly crosslinked resins. Progress in extending the technology to composites has been slow by comparison.

The use of thermoplastics additives as modifiers for epoxy resins is also well established. Here the technology is directed specifically towards composites. Poly(ether sulphone) (PES) is added to a commercially successful grade of epoxy pre-pregs for aerospace applications, principally as a means of raising viscosity at cure temperatures, but also with a view to optimising toughness. Interest in similar systems based on other thermoplastics is high.

Fibre coating and interleafing are currently at the development stage, and the extent to which they will eventually be used in engineering applications is difficult to predict. Preliminary results suggest that interleafing has a promising future. A commercial interleafed system has been developed by American Cyanamid.[46]

There is of course no reason why the various technologies should not be combined to optimise the balance of properties. Each has its own advantages. The high performance composites of the future are likely to exploit two or more of the approaches discussed in this review to provide materials with the required stiffness, strength, and toughness characteristics for the demanding applications of aerospace. In the short term, much of this technology will be based on epoxy resins, because there is an established body of knowledge and experience relating to these resins, and manufacturing capability is in place. In the longer term, resins offering T_g values equal to or higher than those of epoxies, without the attendant problems of water absorption, might well have a bright future.

REFERENCES

1. Krieger, R. B. Paper presented at National SAMPE Symposium, Reno, NV, 3–5 April 1984.
2. Rowe, E. H., Siebert, A. R. & Drake, D. S., *Mod. Plast.*, **47** (1970) 110.
3. McGarry, F. J., *Proc. Roy. Soc., London,* **A319** (1970) 59.
4. Sultan, J. N., Liable, R. C. & McGarry, F. J., *J. Appl. Polym. Sci.*, **16** (1971) 127.
5. Sultan, J. N. & McGarry, F. J., *Polym. Engng Sci.*, **13** (1973) 29.
6. Bucknall, C. B., *Toughened Plastics*. Applied Science Publishers, London, 1977.
7. *Second International Conference on Toughening of Plastics*, Plastics and Rubber Institute, London, 2–4 July 1985.
8. Yee, A. F. & Pearson, R. A., NASA Contractor Rept. 3718, 1983; 3852, 1984.
9. Garg, A. C. & Mai, Y.-W., *Compos. Sci. Technol.*, **31** (1988) 179, 225.
10. Yee, A. F. & Pearson, R. A. In Ref. 7, Paper 2.
11. Riew, C. K. & Gillham, J. K., eds, *Rubber Modified Thermoset Resins*, Advances in Chemistry Series No. 208. American Chemical Society, Washington, DC, 1984.
12. Siebert, A. R. & Riew, C. K., *ACS Div. Org. Coat. Plast. Chem.*, **31**(1) (1971) 552.
13. Siebert, A. R. In Ref. 11, p. 179.
14. Siebert, A. R. In Ref. 7, Paper 7.
15. Kinloch, A. J., Shaw, S. J. & Tod, D. A. In Ref. 11, p. 101.
16. Riew, C. K., Rowe, E. H. & Siebert, A. J., *ACS Div. Org. Coat. Chem.*, **34**(2) (1974) 353
17. Dusek, K., Lednicky, F., Lunak, S., Mach, M. & Duskova, D. In Ref. 11, p. 27.
18. Mijovic, J., Pearce, E. M. and Nir, Z. In Ref. 11, p. 293.
19. Yorkgitis, E. M., Tran, C., Eiss, N. S., Hu, T. Y., Yilgor, I., Wilkes, G. L. & McGrath, J. E. In Ref. 11, p. 137.
20. Williams, R. J. J., Borrajo, J., Adabbo, H. E. & Rojas, A. J. In Ref. 11, p. 195.
21. Bucknall, C. B. & Partridge, I. K., *Polym. Engng Sci.*, **26** (1986) 54.
22. Gillham, J. K. & Chan, J. In Ref. 7, Paper 1.

23. Gillham, J. K. In *Developments in Polymer Characterisation—3*, ed. J. V. Dawkins. Applied Science Publishers, London, 1982.
24. Bucknall, C. B. and Yoshii, T., *Brit. Polym. J.,* **10** (1978) 53.
25. Soldatos, A. C. & Burhans, A. S., *ACS Adv. Chem. Ser.,* **99** (1971) 531.
26. Visconti, S. & Marchessault, R. H., *Macromolecules,* **7** (1974) 913.
27. Bascom, W. D., Cottington, R. L., Jones, R. L. & Peyser, P., *J. Appl. Polym. Sci.,* **19** (1975) 2545.
28. Bucknall, C. B., Davies, P. & Partridge, I. K., *Polymer,* **26** (1985) 109.
29. Biermann, T. F. & Hopper, L. C., NASA Contractor Rept. 172459, 1985.
30. European Patent Application No. 0 108 476. BP Chemicals, 1983.
31. Wilson, J. E. & Bucknall, C. B., *SERC Polymer Engineering Directorate Conference*, Loughborough, 11–13 April 1983.
32. US Patent 3 472 730. American Cyanamid.
33. US Patent 4 604 319. American Cyanamid, 1986.
34. Kinloch, A. J., Shaw, S. J., Tod, D. A. & Hunston, D. L., *Polymer,* **24** (1983) 1341.
35. Kinloch, A. J., Shaw, S. J. & Hunston, D. L., *Polymer,* **24** (1983) 1355.
36. Kinloch, A. J. & Hunston, D. L., *J. Mater. Sci. Lett.,* **5** (1986) 1207.
37. Bucknall, C. B. & Partridge, I. K., *Polymer,* **24** (1983) 639.
38. Bucknall, C. B. & Partridge, I. K., *Brit. Polym. J.,* **15** (1983) 71.
39. Bucknall, C. B. & Gilbert, A. H., *Polymer,* **30** (1989) 213.
40. Mefl, E. H., Claver, C. G. & Baer, M., *J. Polym. Sci.,* **22** (1986) 325.
41. Kunz-Douglass, S., Beaumont, P. & Ashby, M. F., *J. Mater. Sci.,* **15** (1980) 1109.
42. Hunston, D. L., *ASTM Composites Technology Review,* **6** (1984) 176.
43. Bascom, W. D., Jones, R. L. & Timmons, C. O. In *Adhesion Science and Technology*, ed. L.-H. Lee, Plenum, New York, 1975.
44. Preud'homme, T. A. R. MSc thesis, Cranfield 1983.
45. Krieger, R. B. In *Progress in Advanced Materials and Processes*, Materials Science Monographs 29, ed. G. Bartelds & R. J. Schliekelmann, Elsevier, Amsterdam, 1985, p. 189.
46. Georgiadis, K., *Insights,* **1** (1), American Cyanamid, Wayne, NJ, 1984.

Chapter 5

Reinforced Plastics in Anti-Corrosion Applications

Geoffrey Pritchard

School of Applied Chemistry, Kingston Polytechnic, Kingston upon Thames, Surrey, UK

1 INTRODUCTION

Few constructional materials can safely be used in prolonged contact with aggressive chemicals, especially at elevated temperatures, without corroding, swelling, cracking or dissolving. Fortunately for those who design and build chemical process equipment, fibre reinforced plastics (FRP) generally perform well in this respect and material selection has therefore been made easier. It is not true that FRP materials are inert to chemicals, but they tend to perform well against dilute acids and alkalis and certain other, more harmful chemicals which pose severe problems for steel, brass, zinc and aluminium. Because of their different weaknesses and complementary strengths, metals, glass and FRP together constitute a useful range of options, enabling the designer to choose a suitable material for most kinds of fluid environment. This chapter discusses some of the principles involved in selecting and using reinforced plastics for structural applications where chemical or physical attack by the environment is a major concern.

The discussion has particular and obvious relevance to chemical engineering plant, but there are many other applications where the interaction between FRP materials and the chemical environment has to be taken into account. For example, the principles are relevant to marine applications, hot water and fuel storage tanks, sewer pipes, aerospace and automotive parts, vehicle components and storage containers for food and drink.

Corrosion can be induced by a wide variety of chemical environments and their influence can be complicated by considerations of stress, temperature, etc. Published research has not yet covered all the possibilities, and has concentrated heavily on the mechanisms, kinetics and effects of degradation of glass fibre composites by water, acids and organic solvents.

The kind of chemicals the author has in mind during this discussion are water, steam, brine, acids, alkalis, hydrocarbon solvents, chlorinated solvents, powerful oxidising agents and oxygenated solvents, halogens and bleaching agents.

2 TYPICAL ANTI-CORROSION APPLICATIONS

2.1 Chemical Plant

First consider tanks, including process tanks and vats. FRP tanks have several advantages besides corrosion resistance; they are comparatively light in weight and therefore easier to transport and install than steel ones; they also need less maintenance. There are unique fabrication opportunities, with scope for essentially monolithic construction.

There are also serious disadvantages. The mechanical properties of all but the most advanced fibre composites are inferior to those of steel and other metals (this comment refers especially to stiffness, toughness, hot strength and creep resistance). High temperature performance is poor by comparison with metals and some other traditional materials, and FRP is hardly an option yet (apart from special varieties: see Chapter 1) above 160°C.

The majority of organic polymers burn too easily, often producing dark smoke, with serious hazards to personnel. Reinforced thermosetting plastics are very dependent on fabrication quality and, once made, the components are not easy subjects for quality control. Finally, standards and codes of practice are not widely established, and in some cases have still not been pioneered. Nevertheless, outstanding corrosion resistance means longer product life and reduced down-time, and so reduced production losses. Combined with initial cost-effectiveness, this is sufficient reason to outweigh a great many minus factors in the chemical process industry.

Other chemical process industry applications include reaction vessels, exhaust ducting and hoods, scrubbers and towers, chimney stacks, valves and pumps, agitators, gas accumulators, electrostatic precipitators and cryogenic pressure vessels. Pipework is a major end use for FRP. Pipe can be overhead within a process plant, buried, under-sea, or part of an oil well. Again, temperature is a limitation; current materials for pipe are limited to about 120°C. FRP is also more restricted than steel in terms of pressure.

Certain processes within or allied to the chemical industry are especially dependent on FRP. Examples are activities involving chlorine and hydrochloric acid, dilute sulphuric acid, alkalis, bleach chemicals and those concerned with wood pulp and paper, metal refining, plating, etc., fertilisers, and food and drink storage.

2.2 Vehicles

Road tankers made of FRP offer attractive weight saving and therefore fuel saving opportunities.

Fuel tanks for motorbikes, cars and other vehicles are made in FRP in some countries, and vehicle bodies are made of composites in countries where labour costs are low.

Under-bonnet components such as valve covers are increasingly made of FRP. Biddle & Graham[1] describe the evaluation of a thermosetting moulding compound for a diesel engine valve cover. The active chemical concerned was an acidic oil, and the continuous use temperature was about 160°C. Creep and potential losses of modulus at this temperature are causes of concern in such products.

2.3 Underground Tanks and Containers

Petrol tanks at filling stations constitute a potentially advantageous use of FRP. The low corrosion rate could prove a boon because corrosion of steel tanks results in safety hazards. Unfortunately, the lower mechanical strength of early FRP tanks led to some of them being damaged on sharp stony beds during installation. New varieties of fuel, such as unleaded or alcohol based petrol have necessitated re-thinking of the FRP formulations used.[2]

FRP containers are also used for applications such as transformer housings, where moisture barrier properties are required.

2.4 Aircraft

FRP parts for aircraft are not chosen primarily for chemical resistance, but the materials have to be assessed for resistance to jet fuel, hydraulic fluid, anti-icing fluid, brake fluid, paint stripper, etc. Some paint strippers are particularly aggressive and can cause environmental stress cracking.[3] The future is likely to see increased caution in the use of dichloromethane based strippers, as most polymers are affected by them. One solution is to apply a highly coloured primer capable of acting as a permanent barrier coating on the structure. Table 1 shows some of the effects of aircraft fluids on two epoxy systems.[4]

The effects of moisture should not be underestimated. It can cause reductions in strength, modulus and glass transition temperatures. In supersonic aircraft, absorbed moisture can suddenly be lost through heating during a supersonic dash; repeated 'thermal spikes' of this kind can induce physical damage. In the case of carbon fibre reinforced composites joined by metal bolts, electrolytic corrosion can occur.[5] Fuel tanks for helicopters and military aircraft have been developed with the aim of crash survivability.

2.5 Civil Engineering

Overhead and buried pipes are commonly made of FRP. Buried pipe is in demand for sewerage, water and wastewater. The chemical action of warm sewage is considerable, and has been likened to that of dilute sulphuric acid, so this is used for material assessments. Any trials must be carried out with the test specimens under load, as an imposed stress greatly reduces glass reinforced composite lifetimes in certain corrosive environments.

2.6 Other Applications

FRP is a very successful material in the marine industry, where water resistance has to be maintained over periods of as long as 30 years. Osmosis is a problem, however, in connection with gelcoat blistering in boat hulls.[6] Prompt treatment can prevent blisters from advancing from the cosmetic irritant to the structural problem category.

TABLE 1
Properties of Two Epoxy Resins and Their Carbon–Epoxy Laminates (60 vol.% Fibre) after Immersion for 91 Days in Aircraft Fluids[4]

Fluid	*Epoxy system A*			*Epoxy system B*		
	Cast resin		*Laminate*	*Cast resin*		*Laminate*
	Flexural strength (MPa)	*Flexural modulus (GPa)*	*ILSS (MPa)*	*Flexural strength (MPa)*	*Flexural modulus (GPa)*	*ILSS (MPa)*
Control	61·8	1·62	77·0	45·1	1·85	109·0
Skydrol 500B at 50°C	56·8	1·51	70·0	45·4	1·87	112·4
Skydrol 500B at 100°C	47·8	0·98		50·9	1·79	116·2
Hydraulic Oil MIL/H-5605 at 50°C	45·1	1·75	74·7	44·6	1·81	105·2
Hydraulic Oil MIL/H-5605 at 100°C	56·6	1·55	73·5	63·5	1·88	115·5
Aerosafe 2300W at 50°C	51·4	1·54	74·3	56·7	1·84	107·1
Aerosafe 2300W at 100°C	73·8	1·44	80·8	58·3	1·92	
Jet Fuel at ambient	65·3	1·63	73·1	62·0	1·84	103·9
Lubricating Fluid at 50°C	56·8	1·75	80·7	49·3	1·85	111·9
Anti-icing Fluid[a] at ambient	47·9	1·74	76·9	42·9	1·82	107·5

[a]Isopropanol.

The mining industry also provides a market for FRP. Food and drink storage too can present serious corrosion problems. Composites have proved better than metals in some instances. Naturally containers for food and drink must not contaminate the contents, and special grades are required.

3 FAILURE

The chemical resistance of composites becomes important where interaction with the environment threatens to result in impairment of function. A distinction should be drawn here between the total failure as the stress analyst may envisage it, e.g. creep rupture, and simple failure to perform to expected standards. The mere roughening of a pipe surface could constitute failure if it leads to contamination or loss of transport efficiency. The slightest leakage is usually unacceptable. Changes in appearance, and especially loss of translucency, can sometimes constitute failure. The designer should not assume constancy of material properties.

4 MECHANICAL PROPERTIES

Mechanical and other properties are just as important as chemical ones, and FRP materials must show a balance of properties. For example, resistance to thermal or mechanical shock has to be evaluated. Steam inlets make severe demands by causing simultaneous heating and mechanical vibration.

Mechanical properties can sometimes be the crucial ones in selecting anti-corrosion materials. Filament wound epoxy glass pipe is widely used in oilfields where it has to resist crude oil or natural gas, sea and salt water, and hot soil. Polyester and vinyl ester resins also have the necessary corrosion resistance, but epoxy systems have the necessary superior mechanical properties for withstanding high internal pressures and rough handling, so they are still preferred.

Oswald & Moore[7] report that epoxy pipe sections, dug up after eight years service in a Texas oilfield, showed no loss of burst strength, but after 20 years there was a substantial fall-off in the cyclic pressure strength (measured by repeated tests at 8·3 MPa). Failure was generally by wall weeping.

5 RESINS

5.1 Classification

At first sight the number of candidate resins for use in anti-corrosion applications might seem bewildering. However, this chapter deals mainly with the commonest ones. They can be divided into (a) thermosetting resins, which are hardened irreversibly by heat and/or addition of catalysts, and (b) thermoplastics, which can be repeatedly re-shaped by heating.

5.2 Thermosetting Resins

All thermosetting resins have certain fabrication advantages which are particularly useful in the chemical process equipment field, notably their suitability for unique and/or very large mouldings, and for pipe. Figure 1 shows a cooler/filter vessel designed to withstand the extremely corrosive atmosphere of a wet chlorine gas. The diameter at the top is 3·35 m. The resin used was a bisphenol fumarate polyester (see next section) and is designed to stand temperatures up to 80°C. Figure 2 shows part of a network of 4000 m of polyester process pipework in a chlor-alkali plant at Wilton, UK. However, thermosets have serious disadvantages too. They are not well suited to mass production, except in respect of filament wound pipe (see Chapter 7) and pultrusion. Their chemical structures are almost invariably hydrophilic, which makes them inferior to semi-crystalline thermoplastics in resistance to fluid uptake and in toughness.

The most important thermosets in the anti-corrosion market are:

(i) Unsaturated polyesters. These subdivide into three groups: orthophthalic, isophthalic and bisphenol varieties.

Orthophthalic resins cost less than the other polyesters, and are favoured for general purpose laminates but are considered inferior to other polyesters in chemical resistance. They are widely used for boat building, but not as gelcoats in direct contact with water. They are also used in situations where corrosion is possible, but not severe, such as in oilfield storage tanks.

The isophthalics have improved resistance to aqueous reagents, and are more expensive; finally the bisphenols are in some respects better still, and have been marketed for many years for anti-corrosion

Fig. 1. Bisphenol polyester glass laminate was used to make this cooler/filter vessel at an ICI Mond Division chlorine plant. The vessel has withstood an extremely corrosive atmosphere of wet chlorine gas since 1974. It is 19·3 m high and the top filter section is 3·35 m in diameter. There are 35 branch outlets. (Courtesy of ICI Speciality Chemicals.)

purposes, especially the chemical process equipment and pipe markets.

The simple view that orthophthalic resins are inferior to isophthalic ones in water resistance and mechanical property retention is contested in some quarters. Short-term comparisons can give one

Fig. 2. A bisphenol polyester and terephthalic acid polyester were used to make the glass reinforced plastic pipe shown here at ICI's Wilton site. A total of 4000 m of pipework was installed, ranging from 3·75 to 105 cm in diameter. The pipe was for a chlor-alkali plant. (Courtesy of ICI Speciality Chemicals.)

order of priority, whilst long-term tests over a year or more can give different results.

Many polyesters with impressive heat distortion temperatures are too brittle to withstand rough handling, and are best used in modified form, blended with a more flexible grade. The resin elongation and modulus are affected by blending (see Chapter 4).

(ii) Epoxides. These too have several varieties. The differences here lie not only in the base resins themselves, but in the greater likelihood of modification by additives and in the contribution to the final structure of the curing agents, which can impart various properties, as discussed at length by Scola.[8] Epoxides have some overlap with polyesters in their applications, competing (where economic considerations allow) as laminating resins. Their cure shrinkage is much lower than that of the unsaturated polyesters, they bond better to steel substrates and inserts, and they adhere well to glass reinforcement; this in turn means improved mechanical properties. It is because of their outstanding mechanical properties that they are used for such critical applications as double-walled storage tanks for liquid petroleum gas. The ceiling temperatures of conventional difunctional epoxy resins are not quite good enough for the demands of the aerospace industry, and newer resins based on tetraglycidyl diaminodiphenylmethane (TGDDM) have been promoted. Further improvements in high temperature performance will lead to resins such as bismaleimides[9] rather than epoxies.

Epoxy resin chemical resistance is better than that of polyesters towards hot alkalis, and is comparable with that of the best polyesters in many other respects. Coatings are outside the scope of this discussion, but the use of epoxy resins in flooring compositions for industrial chemical installations and the use of filled epoxy resins in chemical laboratory benchtops testify to the chemical resistance of epoxy resins.

(iii) Phenolic laminating resins. Phenolics have been known for many years as moulding powders, and they were the first resins used as protective liners. They are still very satisfactory in this respect, but the chemical environment must be carefully chosen as they are vulnerable to alkalis. Very recently, phenol–formaldehyde laminating resins have been marketed for hand lay-up. Their great advantage for laminating is that they burn only with difficulty, and with low smoke evolution (Table 2). Hart & Pritchard[10] have reported on the chemical resistance of carbon fibre reinforced phenolic and aralkyl resin laminates. Their study shows that there are differences between the chemical resistance of various fibre grades, but the nature of the resin is more important in most environments than the fibre grade.

(iv) Vinyl esters. These resins were specially developed for the anti-corrosion market. They are a compromise between epoxy and poly-

TABLE 2
Properties of Phenolic Laminates

(a) Flammability[27, 29]

Reinforcement	*Property*	*Test method*	*Value*
30 wt.% CSM	Oxygen index[a]	ASTM 2863-74	70%
30–32 wt.% CSM	Oxygen index[a]	NFT 51071	47%
20 wt.% Continuous filament	Smoke density: max. value	NBS Chamber	22
30–32 wt.% CSM	Smoke density: time to maximum	NBS Chamber, NFX 10701	20 min
30 wt.% CSM	Flame spread	BS 476: Part 7: 1971	Class 1

[a]The oxygen index is affected by the catalyst.

(b) Chemical Resistance[27, 29]

Fibre	*Fibre content (wt.%)*	*Chemical*	*Conditions*	*Performance*
Glass[a]	30	Caustic	Immersion for 1 year	Very poor
Glass[a]	30	Acetone	Immersion for 1 year	Very poor
Glass[a]	30	Hydrocarbon fuels	Immersion for 1 year	Very good
Glass[a]	30	Chlorinated solvents	Immersion for 1 year	Very good
Carbon[b]	50	Superheated steam	223°C for 50 days	Flex. strength halved
Carbon[b]	50	Caustic alkalis (30 wt.% NaOH)	50°C for 100 days	Severe swelling
Carbon[b]	50	Sodium hypochlorite	23°C for 100 days	Slight loss of weight and shear strength
Carbon[b]	50	Glacial acetic acid	23°C (also 50°C) for 100 days	No change in weight or volume

[a]After immersion for one year. CdF Chemie Norsphen resin.
[b]BXL novolak resin, VW 61933.

ester resins. The base resins are essentially epoxides of various kinds, which have been reacted with unsaturated acids such as acrylic or methacrylic acid. This reaction removes the epoxide chemical groups and replaces them with crosslinking sites to facilitate curing by the same general method as that used for polyesters. They are dissolved in styrene and a multicomponent free radical catalyst system is added. Figure 3 shows a simplified representation of the chemical structures of some vinyl ester resins. (Commercial resins can have very different structures from those shown, but the principles are the same.) The advocates of vinyl esters put forward three reasons for their superior corrosion resistance. Consideration of these reasons can be instructive.

(a) Vinyl esters, despite their name, possess very few ester groups. Ester groups are undesirable because they are hydrophilic and, more importantly, are easily hydrolysed by aqueous solutions at high temperatures. In acids the hydrolysis is reversible, but in alkalis it is irreversible and more severe.
(b) The few ester groups they possess — two per molecule — can be protected sterically, by selecting the right unsaturated acid for reaction with the original epoxide. The protection is not absolute, but it helps.
(c) The reactive unsaturated groups which take part in the cure reaction are all at the ends of the vinyl ester molecules. This makes them more accessible and therefore they react rapidly, as well as being more likely to be completely utilised. It is undesirable to have unused crosslinking sites, because it reduces the chemical resistance to solvents, halogens and heat ageing. The same structural feature, i.e. the absence of mid-chain crosslinking sites, means that the resin is not too highly crosslinked, and brittleness is avoided. Thus embrittlement in aggressive media is less likely, and fatigue performance is improved.

In addition to the chemical arguments listed above, it is claimed that the low molecular weight of vinyl esters, and consequent low viscosity, allow rapid and thorough wetting of the reinforcement.

Vinyl esters have show themselves particularly well suited to the handling of chlorine and caustic alkali.

(v) Furan resins. These are made from furfuryl alcohol rather than

Standard Derakane type

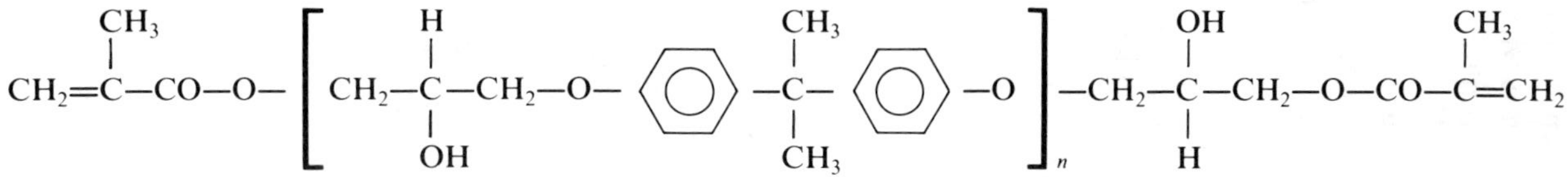

Schematic of Derakane 470 series

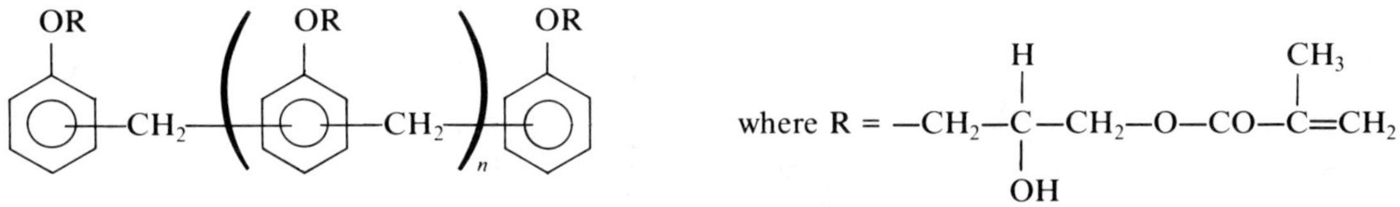

Diacryl 101

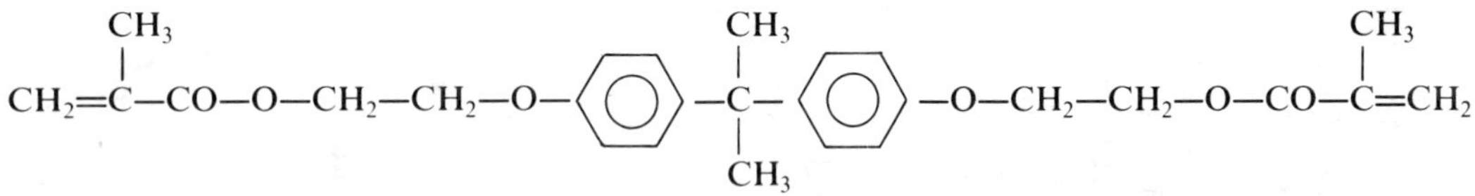

Fig. 3. Vinyl ester resin structures. Derakane is the trade name of Dow Chemical Co. These resins are marketed in the UK by Freeman. Diacryl is manufactured by AKZO Chimie.

petrochemical feedstocks. They are better known as the basis of chemically resistant cements, made by mixing them with chopped fibres or flakes, than as resins for load bearing laminates. But cold-curing structural laminates are avoidable, and have much to commend them. The resins themselves have excellent resistance to acids, caustic alkalis, oxidising agents and a wide range of organic solvents. They are relatively expensive, and dark in colour, making void detection difficult.

(vi) Novel thermosetting developments. Several novel crosslinkable resins have been produced in recent years. Examples are the addition polyimides, the bismaleimides, PSP (polystyrylpyridine), the Friedel–Crafts resins, the acrylic modified urethanes[11] (see Table 3) and acetylene terminated resins such as Thermid 600. Many of the newer resins have been designed mainly for improved high temperature durability for advanced aerospace applications (see Chapter 1).

5.3 Thermoplastics

Thermoplastics can be roughly divided into two kinds:

(a) the glassy, amorphous kind such as polystyrene and the acrylics [i.e. poly(methyl methacrylate), PMMA] and
(b) the semi-crystalline variety, such as polyamides and polyethylene.

The first kind have very poor solvent resistance indeed, and are prone to

TABLE 3
Chemical Resistance of Acrylic Modified Urethanes[11]

Chemical[a]	*Flexural strength (MPa)*
Control	138
25% H_2SO_4	104
15% HCl	33
5% NaOH	99
Sat[a]. NaCl	105
Dist. H_2O	92
Kerosene	130
Toluene (18 h)	Failed
Perchloroethylene (4 days)	Became flexible

[a]One month at 100°C.

solvent crazing. They are inferior under non-static loading; intermittent loading changes their behaviour from ductile to brittle, with a reduction in strength. Nevertheless rigid poly(vinyl chloride), i.e. unreinforced UPVC, which is largely but not entirely amorphous, has a dominant position in the pipe industry. Many semi-crystalline thermoplastics materials have very good chemical resistance indeed, notably polypropylene. Figure 4 illustrates schematically the difference between a crosslinked, amorphous polymer and a semi-crystalline one. In the latter case, the close-packed arrangement of the ordered macromolecular structure facilitates high intermolecular forces, which in turn restrict the penetration of solvent molecules. The chemical resistance of semi-crystalline thermoplastics such as polyethylene, polypropylene, the polyamides and the saturated polyesters is much better than that of amorphous ones such as polystyrene.

Thermoplastics are widely used in the chemical process industry in their unreinforced form as protective coatings for metal. Poly(vinylidene fluoride) (PVDF) can be sprayed electrostatically or by fluidised bed or plasma, and resists most acids, other than oleum, as well as many organic liquids.

Nevertheless, despite the good chemical resistance of classical semi-crystalline thermoplastics such as polypropylene, their application as fibrous composites remains severely limited by the following factors.

(1) Low maximum use temperatures (below 120°C).
(2) Difficulties in obtaining high volume fractions of fibres because

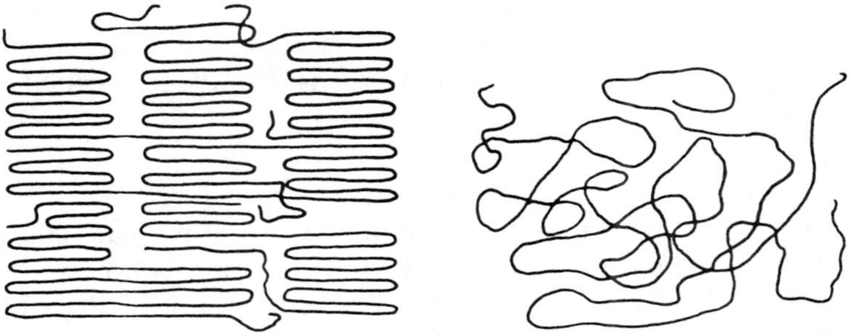

Fig. 4. On the left is the schematic representation of a regular folded chain structure in semi-crystalline thermoplastics. The structure is associated with close packing, molecular cohesion, and insolubility, contrasted with the loose, disordered arrangement of macromolecules in amorphous glassy polymers shown on the right.

of their high melt viscosities. A loading of 30 wt% short glass fibres in polypropylene does not compare well with 60 wt% glass in epoxy resins in such mechanical properties as stiffness and creep resistance.

(3) Considerable unsolved problems in the fabrication of large structures.

Much effort has been necessary to improve the fibre–resin interfacial bond in thermoplastics such as polypropylene, so as to optimise the mechanical properties (see Chapter 2).

Some chemical resistance data for thermoplastics are given in Table 4. Brief mention of individual thermoplastics is appropriate.

5.3.1 Polyethylene

This is a good chemically resistant lining material for use at moderate temperatures, but it is not used much in the fibre reinforced form. There are several varieties of polyethylene, but only the high density variety (HDPE) and crosslinked polyethylene have any claim to be structural materials.

5.3.2 Polypropylene

The chemical structure differs from that of polyethylene in an important respect. The tertiary carbon atom means that polypropylene is much more readily oxidised at high temperatures. The chemical resistance is otherwise good, and it has found favour as a lining for glass reinforced plastics laminates. The thermoplastic has to be bonded to the FRP by incorporating glass fabric into it and allowing the thermosetting matrix resin to impregnate the glass fibre ends protruding from the polypropylene sheet. Such liners have many advantages — including the ability to withstand steam treatment.

5.3.3 Polyamides

Glass reinforced polyamides are common engineering materials, and are used in automotive applications where they must withstand oils, petrol, brake fluid, etc. Their outdoor weathering is not good, and they do not have such a high degree of chemical inertness as the previous types. The polyamides have been successfully used as petrol tanks in vehicles. Recently, glass reinforced polyamide bolts have been marketed, with corrosion resistance as a major selling point. Other advantages claimed for the bolts include their electrical insulation, non-magnetic

TABLE 4
Chemical Resistance of Thermoplastics (Ambient Temperature)[a]

Thermoplastic	*Dilute acid*	*Weak alkali*	*Strong alkali*	*Aromatic hydrocarbons*	*Ketones*
UPVC	No effect	No effect	No effect	Attacked	Attacked
Polycarbonate	No effect	Slight attack	Attacked	Attacked	Attacked
Polyamides	Slight attack	Little effect	Little effect	Resistant	Resistant
Polypropylene	No effect	No effect	No effect	Swollen	Resistant
Acetal	Resistant	No effect	No effect	Resistant	Resistant
Polyethylene, high density	No effect	No effect	No effect	Attacked	Resistant
PTFE[b]	No effect	No effect	No effect	No effect	No effect
PBT	No effect	Attacked	Attacked	Little effect[c]	Attacked

[a]The effects at higher temperatures may be very different. All the data in this table are generalised and should be confirmed for specific grades and conditions.
[b]PTFE is swollen by fluorinated hydrocarbons at ambient temperature, by fused or dissolved alkali metals, and by certain halogens at high temperatures.
[c]Benzene is more harmful than most others in this category.

qualities, and low density. Chemical resistance tests indicate that the bolts eventually lose about 40% of their strength during prolonged immersion in water, and are not recommended for acidic or alkaline solutions because of hydrolysis of the matrix, but they can withstand organic liquids such as oils, acetone and toluene, and have good stress corrosion performance in sea water.

Polyamides absorb moisture readily because of their numerous hydrogen bonding sites, and their dielectric strength falls off sharply after conditioning in a moist atmosphere.

5.3.4 Saturated Polyesters

The family name of these polymers is of course derived from the fact that they contain many ester groups, and they therefore have the same vulnerability to hydrolysis by acids and more especially by alkalis, but there the resemblance to thermosetting, unsaturated polyesters ends. The processing, properties and applications of the thermoplastic variety are much closer to those of polypropylene than of crosslinked polyesters. Important types of short fibre reinforced thermoplastic polyester for engineering components are (a) poly(ethylene terephthalate) (PET) and (b) poly(butylene terephthalate) (PBT). Both are used in under-bonnet automotive applications, where a degree of chemical resistance is required, and any absorption of moisture, oil, petrol, etc., would be undesirable. Moisture absorption is much lower than that of the polyamides, and the dielectric properties of unreinforced thermoplastics, such as dielectric strength, dielectric constant and dissipation factor, are hardly changed by conditioning in moist atmospheres. The glass reinforced varieties do undergo some changes in electrical properties when wet.

5.3.5 PEEK

Poly(aryl ether ether ketone) is one of an increasing number of aromatic, high performance, high cost thermoplastics. It can be reinforced by long carbon or glass fibres (see Chapter 2). Preliminary evidence suggests that carbon reinforced PEEK has excellent chemical resistance (see Table 5). Strong oxidising acids such as sulphuric acid dissolve it, and bromine also destroys it. There is slight swelling by a few solvents such as dichloromethane and tricresyl phosphate. PEEK appears to be a versatile material for anti-corrosion use, at least in the absence of high mechanical stresses. It also has other assets — a T_g of about 143°C, and outstanding fire performance (see Table 6). The availability of a matrix able to resist hot water, acids, alkalis and many hydrocarbon solvents means that when carbon fibres are employed as the reinforcement their corrosion resistance to these environments is not wasted by resin failure.

TABLE 5
Percentage Weight Increase of Carbon Fibre Reinforced PEEK

Environment[a]	*Wt increase (%)*
1M-Sodium hydroxide	0·18
0·1M-Ammonium hydroxide	0·18
50% Sulphuric acid	0·19
Conc. Nitric acid	4·45
30 vol.% Hydrogen peroxide	0·17
Bromine	Sample disintegrated
Isopropanol	0·53
Acetone	1·31
Dichloromethane	3·99
Kerosene	0·49
Skydrol 500B	0·59

[a]Sample size 45 mm × 25 mm × 1 mm. One month at ambient temperature, 16–24°C.

TABLE 6
Some Thermal and Fire Performance Characteristics of PEEK[a]

Material designation	*Property*	*Value*
Victrex PEEK 450 G	Oxygen index	35%
Victrex PEEK 450 G	Flammability (UL 94)	V–0
Victrex PEEK 450 G	Continuous use temp.	245°C
Victrex PEEK 450 G	Heat distortion temp.	153°C
Victrex PEEK 450 GL30[b]	Heat distortion temp. at 1·82 MPa	310°C
APC (PEEK/65 wt.% carbon fibre)	Smoke generation rating (specific optical density)[c]	1
APC (PEEK/65 wt.% carbon fibre)	Time to penetration of $\frac{1}{8}$-in sheet by 1000°C flame	>15 min

[a]Poly(ether ether ketone), ex ICI.
[b]Contains 30% short glass fibres.
[c]This compares with 800 for ABS, 500 for PVC and 200 for polysulphones.

5.3.6 PVC

Unreinforced PVC is used in the chemical process industry for pipe manufacture. Its main interest as a constituent of a composite material is probably as a pipe lining when used in conjunction with glass reinforced polyester or epoxy. It has good resistance to many chemicals and solvents, and is not prone to environmental stress cracking, but suffers from a narrow usable temperature range, losing strength and stiffness above 60°C, and it becomes decidedly brittle below 0°C. It also has a low impact strength.

6 FIBRE REINFORCEMENTS

Glass fibres are unaffected by organic solvents, but they are attacked and weakened by prolonged immersion in aqueous solutions, particularly acidic and alkaline ones. The mechanism is believed to involve hydroxyl or hydrogen ion penetration into the glass. Cations such as sodium in the fibre surfaces are removed and progressively replaced by smaller ions from the liquid environment. The size change is thought to enhance tensile stresses in the glass surface, and to aid the subsequent development of tiny surface cracks. The crack tips can be penetrated in turn by water molecules, which hydrolyse the siloxane groups of the glass and thus promote flaw growth, a process accelerated by an applied tensile stress. Consequently the stress corrosion performance of unidirectional glass reinforced resins in dilute hydrochloric acid or sodium hydroxide leaves something to be desired. Scrimshaw[12] has shown that sized E-glass strands, without resin protection, can be completely destroyed in dilute hydrochloric acid after 6 h at 88°C. He also showed that the time to failure of E-glass strands in dilute sulphuric acid was reduced 200-fold when the strain in the glass was increased from 0·10 to 0·48%. The pioneering work of Charles[13] on the time, t, to failure of glass rods in water under a stress σ resulted in the expression

$$\log 1/\sigma = 1/n \log t + 1/n \log K$$

where n and K are constants.

Many studies of the failure of unidirectionally reinforced and tensile-loaded glass–polyester and glass–epoxy laminates, in water and acid, have shown that at low strains, there is a characteristic smooth fracture surface, with almost complete absence of fibre pull-out (sometimes only part of the fracture surface shows this effect, and the remainder has the more typical brush type appearance). The smoother surface is associated with environmental stress cracking rather than purely mechanical failure, and has been studied in some detail.[14]

Pritchard & Speake[15] have considered the effect of temperature on the failure time in water of glass–polyester laminates. Increasing the water temperature accelerates the rate of diffusion of the water to the glass fibres, but if the change in temperature extends over an appropriate range, there can also be an increase in the apparent critical stress intensity factor of the matrix, and this reduces its tendency to undergo surface microcracking under a tensile load. As a result the overall rate of transport of water to the glass, and the overall effect on rupture time in water, can be unpredictable. However, the diffusion rate of hydrochloric acid into polyester resins has been found to be negligible.[16]

Marshall & Harrison[17] studied the processes leading to attack on the glass in marine laminates. They suggested that gelcoat cracks, whether caused by rough handling or as a long-term consequence of creep of the resin-rich layer under load, lead on to complete failure of the coating at low strains, as a result of stress concentrations. In turn this results in environmental stress cracking, and in degradation of the fibres at an advanced stage. Carswell & Roberts[18] have discussed in similar terms the marked adverse effect of acidic environments on the fatigue life of chopped strand mat laminates.

In view of the sensitivity of glass to acid and alkaline attack, it is noteworthy that osmotic blister fluids are generally acidic, and that some resin hardeners can produce acidic or alkaline products. Phenolic laminating resins are usually hardened with acid catalysts, and any residual catalyst should therefore be kept to a minimum. The epoxy hardener, dicyandiamide, reacts with hot water to produce ammonium hydroxide, and should not be allowed to remain in the resin as an unreacted excess.

Various types of glass with different compositions are available (see Table 7). The commonest type of reinforcing glass fibre is made of E-glass, which is much lower in sodium than common window-glass. S-glass is superior in moisture resistance but price and availability have restricted its applications. C-glass is so called because it is chemically resistant, and monofilaments of it have been used to make surface veils for incorporating in the protective, resin-rich layer of chemically resistant laminates. Without such a veil, a thick resin layer would crack

TABLE 7
Glass Fibre Compositions (% by weight)

	Glass type				
Constituent	*A*	*C*	*E*	*S*	*R*
SiO_2	72	65	55	65	60
Al_2O_3	2·5	4	15	25	25
B_2O_3	0·5	5	7	—	—
MgO	1	3	3·3	10	6
CaO	9	14	18·7	—	9
Na_2O	12·5	8·5	0·3	—	—
K_2O	1·5	—	0·2	—	—
Fe_2O_3	0·5	0·5	0·3	—	—

easily. New types of glass with very low boron content have also been developed, and are said to have much improved acid resistance.

Surfacing veils are also made from synthetic organic polymers such as thermoplastic polyester or acrylic. They are preferred for strongly alkaline solutions or strong oxidising agents.

It cannot be emphasised too strongly that the overwhelming majority of reinforced plastics composites are based on glass fibres, and moreover that in anti-corrosion markets the dominant matrix is still polyester. Nevertheless carbon fibres have some very attractive anti-corrosion properties, provided that a suitable matrix is used. The chemical resistance of carbon fibres has been reviewed at length by this author.[19] Carbon fibres are highly resistant to non-oxidising acids such as hydrochloric, and to hot water and steam up to near 1000°C, as well as to caustic alkalis. These properties are very desirable, but they can only be utilised if the resin matrix can also withstand the same environments, which is usually doubtful. For example, 30% sodium hydroxide causes gross swelling of the matrix in carbon fibre reinforced phenolic laminates, but the fibres are not affected.[10] PEEK is one matrix with the potential to solve this problem. Several reports show that carbon reinforced PEEK has good resistance to water (Fig. 5) including prolonged immersion in boiling water, and to immersion in ketones and hydrocarbon solvents. However, before any new fibre reinforced composite can be proved reliable in anti-corrosion applications, it needs to be evaluated over rather long periods, using several fibre orientations and under demanding combinations of temperature and

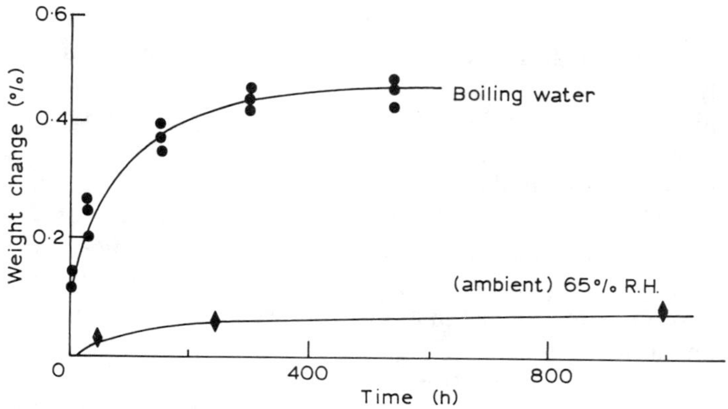

Fig. 5. Water absorption of a [0°/90°] carbon fibre reinforced PEEK laminate in boiling water (●) and in ambient conditions of 65% RH (♦).[26]

stress. Isolated carbon filaments can be attacked readily by powerful oxidising agents, such as nitric acid, dichromate solution and hypochlorite. They are also attacked by very hot air. The rate of attack on the fibres in a composite material depends on the degree of protection afforded by the matrix; for example, carbon fibres embedded in a metal matrix show gradual oxidation at high temperatures in air, starting from exposed fibre ends. Carbon fibres are also attacked, rather more slowly, by intercalating agents, such as bromine and ferric chloride, which insert themselves between the layer planes of the turbostratic (semi-graphitic) structure. The vulnerability to oxidising agents declines slightly with increasing fibre heat treatment temperature, as the surface becomes dominated by basal rather than edge planes, whereas the sensitivity to intercalation increases slightly because of the increasing degree of order.

Early carbon fibres contained appreciable sodium derived from the manufacture of the precursor fibre. This was found to be undesirable because it could be slowly removed by prolonged water action, thus setting up osmotic cells and promoting accelerated debonding from the matrix.

Having considered glass and carbon fibres separately, it is useful to speculate about the corrosion performance of unidirectional hybrid fibre composites. The inclusion of carbon fibres should increase the failure time, compared with all-glass equivalents, in stress corrosion conditions. This has been confirmed by French & Pritchard[20] for the case where the surface plies of the hybrid laminate, i.e. the plies exposed to the aggressive medium, contain only carbon, and there are no exposed glass ply edges.

Aramid fibres have excellent resistance to solvents, fuels, lubricants and salt water. Their resistance to acids and alkalis is good, but not complete. They absorb about 5% water at saturation, and according to Roylance[21] there are small increases in strain-to-failure and tensile strength of unidirectional aramid–epoxy laminates after water immersion. This is attributed to fibre re-alignment, facilitated by matrix plasticisation. Aramid fibres consist of aligned molecules with extensive use of hydrogen bonding to retain their orientation, and such bonding is vulnerable to water penetration.

Asbestos fibres have many excellent technical qualities, including their exceptional mechanical properties. Historically they have been prized for their resistance to high temperatures and to severe corrosive chemicals, with the notable exceptions of acids and salt solutions. The use of asbestos has declined in recent years, however, and it will not be considered further.

7 CORROSION MECHANISMS IN COMPOSITES

7.1 Introduction

Corrosion in reinforced plastics proceeds by several mechanisms. Electrochemical processes do not play a significant role, except when carbon fibres or other conducting materials are present and in close proximity with metals. The processes can be divided into (a) chemical ones, which in this context means those involving breakage of primary (usually covalent) bonds and the formation of new chemical entities, and (b) physical mechanisms, which do not involve chemical bond breakage. Degradation mechanisms such as fibre–resin debonding generally involve both physical and chemical mechanisms. The most important and common reactions involving the breaking of covalent bonds are:

(a) Hydrolysis by hot water, acids, and alkalis. This process accelerates greatly with increased temperature.
(b) Oxidation by air, ozone, liquid oxidising agents such as nitric acid, and acidic salts.

7.2 Stress Cracking

It has already been mentioned that an organic polymeric matrix absorbs liquids causing swelling and plasticisation. Swelling introduces changes in the internal stress fields within the composite. The resulting stresses can be computed from a knowledge of fibre content and orientation.[22] A certain amount of fibre–resin debonding often results. In the case of aqueous liquids, the drying-out process also causes problems, leaving the resin surface in tension and liable to undergo crazing and surface cracking.

7.3 Polymer Swelling and Solubility

Theories are available to predict whether a solvent will swell and/or eventually dissolve a polymer. These theories strictly apply only for linear, non-polar amorphous polymers above their T_g and are not so useful for the majority of matrix materials of interest in the present context, which are either crosslinked or crystalline, and if crosslinked, then below their T_g. However, the principles can be adapted to predict the extent of swelling of thermosetting resins (thermosetting resins do not dissolve completely in solvents, because of their covalent crosslinks, but they swell to an extent dependent on the chemical affinity of the solvent for the matrix structure). Usually swelling of thermosetting resins by ketones and chlorinated solvents is considerable, but hydrocarbon solvents are less effective and aqueous liquids such as water and brine have only a slight effect.

Consider the case of amorphous thermoplastics above their T_g. The solubility depends on the relative affinities of solvent molecules for other solvent molecules rather than for polymer molecules, i.e. on the thermodynamic heat of mixing. For non-polar polymers and solvents, the cohesive energy densities, E, of polymer and solvent must be similar. Hence if the values of all cohesive energy densities, or the closely related solubility parameters, δ, were to be known for all polymers and solvents, the problem would be one of finding matching pairs of similar E or δ values. The relationship between E and δ is given by

$$\delta = (E/V)^{0.5}$$

where V is the molar volume. Values of E for volatile solvents can be derived from the latent heat of vaporisation, but this property cannot be obtained for polymers, so instead solubility parameters are obtained indirectly by various means.

For example, Small[23] gave a method of prediction based on the summation of all the molar attraction constants for the various chemical groups in a typical repeat unit of the polymer. The constants themselves were estimated from literature values or vapour pressure and heat of vaporisation data for compounds not containing hydrogen bonding. Some of these constants are given in Table 8. For hydrocarbons, the principle of additivity appears to be soundly based, and can be used to calculate solubility parameters.

The principle is not quite so satisfactory for other organic solvents. However, there is in many cases reasonable agreement between values of solubility parameters calculated in this way, and those obtained by the longer method of equilibrium swelling studies, as Table 9 shows. Differences between calculated and observed values are not large, even for polar polymers.

Crystalline polymers do not readily dissolve in any cold solvents, because the entropy of mixing is insufficient to provide the necessary free energy for overcoming the existing inter- and intra-molecular cohesion. Semi-crystalline polymers containing hydrogen bonds therefore tend to dissolve at room temperature only in solvents with which they can also form hydrogen bonds, thus providing a positive heat of mixing. Examples are polyamides dissolving in formic acid, and PET in phenol. Polyethylene does not possess hydrogen bonding capability

TABLE 8
Molar Attraction Constants[30]

Group	*F* $(J\,cm^3)^{1/2}\,mol^{-1}$
$—CH_3$	303
$—CH_2—$	269
$>CH—$	176
$CH_2=$	259
$—CH=$	249
$—CH=$Aromatic	239
—O—	235
—O— (epoxide)	360
—CO—O—	668
—OH	462
OH (aromatic)	350

TABLE 9
Calculated Values of Polymer Solubility Parameters Compared with Experimental Values[23]

Polymer	*δ (J/cm³)^1/2^*	
	Calculated	*Experimental*
Polystyrene	18·6	18·6; 17·6–19·8
Poly(methyl methacrylate)	18·9	18·4–19·4
Polyethylene	16·6	16·2
Natural rubber	16·7	16·2; 16·3, 17·1
Cellulose diacetate	23·20	22·3
PTFE	12·7	—

and does not dissolve until heated to near its melting point. PEEK also resists most solvents.

7.4 Leaching

Solvents can remove low molecular weight substances from polymers if these substances are not bound by chemical bonds. Crosslinked polymers often contain a small quantity of uncrosslinked matter which can be leached out. When a polyester resin is exposed to alkaline solutions, chemical attack occurs, increasing the quantity of leachable material. It follows that when a polyester or similar polymer has been immersed for a long period in a solvent, and then re-dried, it is not the same material as before. The modulus will often be increased because of the loss of plasticising diluent.

7.5 Osmotic Cracks

Phase-separated water soluble substances in resins can set up osmotic pressure cells at which cracking eventually occurs, often at a characteristic level of fluid uptake. These cracks are small, numerous, and detrimental to resin strength, but do not have much effect on modulus.

7.6 Fibre–Resin Debonding

Adhesion between fibres and matrix has been variously explained in terms of chemical bonds, secondary bonds and mechanical locking.

Chemical bonds are promoted by surface treatment of the fibres; in the case of carbon fibres, an oxidation treatment is carried out, which increases the adhesion between reinforcement and matrix. Glass fibres are surface-treated with silanes, which react chemically with both fibre and matrix. The chemical and physical bonds between the silane coating and the glass fibre can be broken by water. Osmotic pressure pockets at the interface, induced by the leaching of ions from the glass fibres, can facilitate debonding. Debonding can often be partly reversed on careful drying, and shear strengths can be completely recovered in many instances.[24]

8 LAMINATE CONSTRUCTION

This section is concerned only with design to minimise chemical attack. The author is not qualified to discuss engineering design but this is obviously of vital importance, especially when strength and stiffness are lowered by high temperatures and by absorption of chemicals.

In simple FRP laminates such as those made from stacked plies of chopped strand mat or woven roving, the chemical environment can induce damage in three places:

(a) the resin;
(b) the reinforcement;
(c) the interface between them.

Progressive changes in the resin occur, and it is important to choose a suitable one but, as with fibre–resin debonding, the processes are slow and often do not have dramatic consequences. It has already been said that glass fibre is more seriously affected. Sudden failure can occur under stress in wet or especially in acid conditions. The prime need is to avoid catastrophic failure by chemical attack on the reinforcement. Therefore a strategy which takes this possibility into account is adopted in the design of corrosion resistant laminates for the chemical industry or for pipework. An example is the design of glass fibre reinforced laminates for use in aqueous and acidic environments. Figure 6 shows a chemically resistant glass laminate, made using unsaturated polyester resin. It contains several layers:

(1) A resin-rich surface, containing no reinforcing fibres, to minimise exposure of glass fibre to aqueous media such as hot water, acids

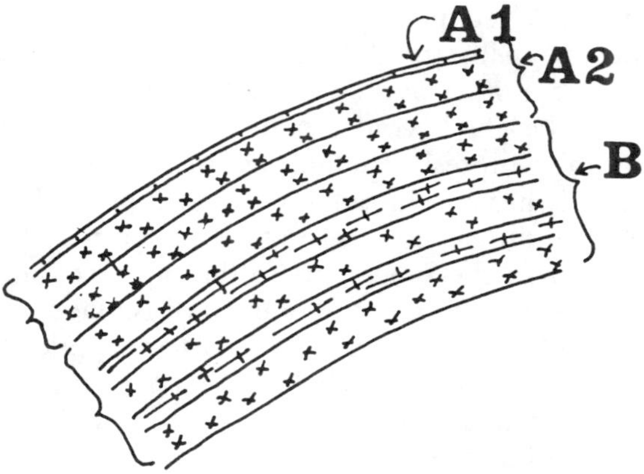

Fig. 6. Chemically resistant laminate structure. A, chemically resistant layer; B, structural layer. Region A1: resin-rich surface (90% resin) containing one ply of 10-mil C-glass surfacing veil. Region A2: Two plies of $1\frac{1}{2}$ oz (45 g) chopped strand mat (CSM) with soluble binder. Glass content 25–35 wt.%. Four plies are needed for chlorine. Region B: Alternating chopped strand mat and woving roving (50 wt.% glass) layers. Note: Two plies of CSM in the chemically resistant layer (regions 1 and 2) produces a 2-mm barrier.

or alkalis (carbon is unaffected by water up to about 1000°C, or by alkalis, or non-oxidising acids such as hydrochloric acid, and so it does not have to be protected, but it is vulnerable to oxidising acids).

(2) A veneer, or surfacing veil. This consists of one ply of a special chemically resistant glass fibre or thermoplastic mat. The mat provides enough reinforcement to prevent crazing as the resin shrinks slightly during cure.

(3) A few layers of resin impregnated reinforcing mat (chopped strand mat). The resin content is still high at this stage, typically 70% by weight.

(4) The reinforcing layer — some layers of glass cloth, i.e. woven roving. The resin content falls at this stage to about 50% by weight.

The essential feature of the laminate structure is the resin-rich protective region. A chemical fluid can diffuse into this region and it is vital that the main glass reinforcement should not be exposed to such liquids as hydrochloric acid, especially while under external load.

The corrosion resistant layer (from the resin coating to, and including, the chopped strand mat layer) should be thicker when chlorine and certain other particularly active chemicals are used.

Variations on the above construction are possible. Pipe made by centrifugal casting with resin-containing sand or a similar particulate filler automatically has a resin-rich layer on the inside. This is because the high-speed spinning action causes the heavier particles to move to the outer rather than the inner part of the pipe section. In another approach, the outer (structural) part of the pipe is made of steel. Resin-rich layers containing glass or graphite flakes are applied to the steel.

This can result in a low-permeability barrier and, with properly chosen filler, the expansion coefficient of the protective layer can be brought near that of a steel substrate. Oilfield pipe systems have been developed in which pre-tested epoxy pipes are inserted into high-pressure steel pipes and the annular space filled with cement or a suitable heat-cured adhesive.

9 DESIGN PROBLEMS

9.1 Thermal Mismatch

The thermal expansion coefficients of polyester resins and of many similar organic matrix materials are comparatively high; this results in high values for FRP, especially at low fibre loadings. Certainly the thermal expansion of FRP is much greater than that of steel, and this can result in stresses at junctions between an FRP tank and steel piping or lining. The result can be a cracked FRP tank, unless it is strengthened or the steel is allowed to move, e.g. by spring loading.

9.2 Joints

Pipework demands frequent joints and it is also necessary to join isolated components of process equipment, or to fit inlets and outlets to tanks, etc. Field experience has shown that FRP requires rather careful and skilled installation and assembly. It is not as resistant to mis-use as some other materials. Adhesive joints fail if untrained personnel do not mix adhesive components properly, or if they apply the adhesive to contaminated surfaces, or move a fabricated part before the adhesive is properly cured. The result is a high incidence of leaking joints.

9.3 Laminate Cure

Complete cure cannot be achieved with unsaturated polyesters and vinyl esters if the on-site temperature is too low, or the catalyst quantities are wrong, or an inhibitor has been inadvertently added. Inadequate cure means lowered chemical resistance. A great deal of effort has gone into detecting under-cure of an article, and other fabrication faults, by non-destructive means.

It is now recognised that an elevated temperature post-cure is advisable, even though it is difficult to achieve with large objects. Material properties can change noticeably even after the finished article has been left in strong sunshine for a few days.

9.4 Static Charge

Static electricity build-up can present an explosion hazard in non-conductive containers holding dust or hydrogen. The electrical conductivity of a composite can be increased by adding graphite filler or carbon fibres, or by using metal probes. Galvanic consequences should be considered.

9.5 Operating Conditions

The conditions under which chemical process equipment is used can be unexpectedly severe. The heat of mixing of chemicals (to say nothing of the heat of reaction) can cause severe thermal cycling. The temperature of a chemical delivered to a tank can be much higher than officially specified, or a liquid might be pumped into a tank from one side with excessive force. This can be countered with baffles, or by suitable positioning of the inlet pipe.

9.6 Fire

Fire is obviously a high risk in chemical plant, and also to some extent in aircraft and land vehicles using flammable fuel. Chemical plant containing ducting and fume extraction equipment is vulnerable to the build-up of combustible deposits, unless frequent cleaning is carried out.

There are some matrix materials with inherent advantages in terms of combustibility and low smoke evolution, notably PEEK and the phenolic resins. The commonest anti-corrosion resins, polyesters, the vinyl esters and epoxies, are all rather poor in these respects, whether assessed by oxygen index, flame spread rating or smoke density measurements. They can be much improved by suitable modification of their formulations, and thus enabled to meet demanding fire specifications, but the implications of these modifications for mechanical properties and anti-corrosion performance should be ascertained. Chemical plant has to be designed with the most suitable materials, but this is not enough; there must also be efficient automatic smoke warning and sprinkler systems, and damage limitation mechanisms such as cut-offs for the fuel and air supplies.

10 FURTHER INFORMATION

The major materials suppliers offer comprehensive information about the chemical resistance of their materials. Their contribution is essential because their knowledge of the proprietary formulations involved is required before an opinion can be offered about whether a particular grade will behave as expected. Independent data banks are being built up by the larger consultancy services and some of the companies concerned with chemical plant design and fabrication. The general reader can consult a large and increasing number of books and articles of various levels of technical detail for information about specific materials. A few examples are cited in the reference section following.[19, 22, 25, 26, 28, 31–35]

Standards organisations supply detailed information on test methods for chemical process equipment, pressure vessels, pipes, etc.

The scientific literature contains a great deal of information on the behaviour of specific composite materials. However, there is a shortage of information on stress corrosion performance because of the cost of collecting the data.

Key references are recommended below for further information on the subject.[22–25, 28]

REFERENCES

1. Biddle, M. B. & Graham, K. T. Paper 1-B presented at 38th Annual Conference, Reinforced Plastics/Composites Inst., SPI Inc., Houston, TX, Feb. 1983.

2. Craige, L. J., White, M. N. & Svatek, M. J. Paper 20-A presented at 41st Annual Conference, Reinforced Plastics/Composites Inst., SPI Inc., Atlanta, GA, Feb. 1986.
3. Yee, A. F., Critical issues in the use of thermoplastic matrices in composites. Paper 1 presented at Conference 'Polymers for Composites', Plastics and Rubber Institute, Solihull, UK, December 1987.
4. Pritchard, G., Mathews, B. J. & Stokes, F. C., The effect of aircraft fluids on the properties of carbon fibre laminates. MOD Contract AT/2055/06 MAT.
5. Brown, A. R. G. Paper 35 presented at Second International Conference on Carbon Fibres. The Plastics Institute, London, 1974.
6. Abeysinghe, H. P., Ghotra, J. S. & Pritchard, G., *Composites,* **14** (1) (1983) 57.
7. Oswald, K. J. & Moore, R. H. Paper 7-B presented at 38th Annual Conference, Reinforced Plastics/Composites Inst., SPI Inc., Houston, TX, Feb. 1983.
8. Scola, D. A. In *Developments in Reinforced Plastics — 4*, ed. G. Pritchard. Elsevier Applied Science Publishers, London, 1984, Chapter 5.
9. Landman, D., Advances in chemistry and applications of bismaleimides. In *Developments in Reinforced Plastics — 5*, ed. G. Pritchard. Elsevier Applied Science Publishers, London, 1986, Chapter 2.
10. Hart, G. L. & Pritchard, G., Mechanisms of corrosion in carbon fibre reinforced phenolic resins. In *Developments in Composite Materials — 1*, ed. G. Holister. Elsevier Applied Science Publishers, London, 1977.
11. Leitheiser, R. H., Szwarc, J. J. & Ryer, D. A. Paper 11-E presented at 29th Annual Conference, Reinforced Plastics/Composites Inst., SPI Inc., Washington, DC, Feb. 1974.
12. Scrimshaw, G. Paper 5 presented at Pipe Con Conference on Large Diameter Glass Reinforced Plastic Pipes, Fibreglass Ltd/Amoco Chemicals SA, London, June 1980.
13. Charles, R. J., *J. Appl. Phys.,* **29** (1958) 1529.
14. Price, J. N., & Hull, D., *Proc. ICCM-VI*, Vol. 4, ed. F. L. Matthews, N. C. R. Buskell, J. M. Hodgkins & J. Morton. Elsevier Applied Science Publishers, London, 1987, p. 365.
15. Pritchard, G. & Speake, S. D., *Composites,* **19**(1) (1988) 29.
16. Caddock, B. D., Evans, K. E. & Hull, D., The role of diffusion in the micromechanisms of stress corrosion cracking of E-glass polyester composites. Paper presented at Second Int. Conf., Fibre Reinforced Composites, Institute of Mech. Engineers/Plastics and Rubber Institute, Liverpool, UK, 1986.
17. Marshall, G. P. & Harrison, D., *Plastics Rubb. Proc. Appl.,* **2** (1982) 269.
18. Carswell, W. S. & Roberts, R. C., *Composites,* **11** (1980) 95.
19. Pritchard, G., *Polym. Plast. Technol. Engng,* **5**(1) (1975) 55.
20. French, M. A. & Pritchard, G., unpublished work, Kingston Polytechnic, Surrey, UK.
21. Roylance, Margaret E., *Polym. Engng Sci.,* **22**(15) (1982) 988.
22. Menges, G. & Lutterbeck, K., Stress corrosion in fibre-reinforced plastics in aqueous media. In *Developments in Reinforced Plastics — 3*, ed. G. Pritchard, Elsevier Applied Science, London, 1984, Chapter 4.

23. Small, P. A., *J. Appl. Chem.*, **3** (1953) 71.
24. Kasturiarachchi, K. A. & Pritchard, G., *J. Mater. Sci.*, **20** (1985) 1901.
25. Pritchard, G., Environmental stability of hybrid fibre composites. In *Fibre Composite Hybrid Materials*, ed. N. Hancox. Applied Science Publishers, London 1981.
26. Dickson, R. J., Jones, C. J., Harris, B., Leach, D. C. & Moore, D. R. Paper presented at AMAC Conference, Paris, September 1984; also published in *J. Mater. Sci.*, **20** (1985) 60.
27. CdF Chimie technical literature on Norsophen phenolic laminating resins. CdF Chimie SA, Tour Aurore, Place des Reflets, Cedex 5, Paris, France.
28. Butt, L. T. & Wright, D. C., *Use of Polymers in Chemical Plant Construction*. Applied Science, London, 1980.
29. Ciba-Geigy technical literature, Duxford, UK, 1981.
30. Hoy, K. L., *J. Paint Technol.*, **42** (1970) 76.
31. Hart, G. L., The chemical stability of carbon fibres and their composites. PhD thesis, Kingston Polytechnic, 1975.
32. *Chemical Resistance*, Crystic Monograph No. 1. Scott Bader Co. Ltd., Wollaston, Wellingborough, UK, 1971.
33. Fowle, D. J., *Chem. Ind.*, (11) (3 June 1978) 361.
34. Aveston, J. & Sillwood, J. M., Prediction of long term strength of glass reinforced plastics in wet environments. National Physical Laboratory Report DMA(A) 80, September 1984.
35. Menges, G. & Gitschner, H.-W. In *Proc. ICCM-3, Paris* Vol. 1, ed. A. R. Bunsell, G. Bathias, A. Martrenchar, D. Menkes & G. Verchery. Pergamon, Oxford, 1980, p. 25.

Chapter 6

Inorganic Matrix Composites

J. M. Quenisset

Laboratoire de Chimie du Solide du CNRS, Laboratoire de Genie Mecanique de L'IUT-A, Université de Bordeaux I, 33405 Talence Cédex, France

1 INTRODUCTION

The continual demand on high performance materials frequently leads to the same conclusions: the lightest elements combine with strong covalent or ionovalent atomic bonding giving rise to numerous types of ceramics (carbon, boron, oxides, carbides, borides, silicides, etc.). The main character of these materials is related to the stability of their chemical bonding: they are refractory and exhibit low expansion coefficients, high stiffness and good resistance to chemical aggression. In addition, the mechanical strength would also be very high if ceramics were not so sensitive to notch effects; the room temperature yield stress is too high to enable the material to relax the stress concentration corresponding to inevitable pre-existing flaws. As a result the mechanical behaviour of ceramics is generally controlled by crack propagation and they cannot exhibit any plastic deformation except at high temperature. It also means that a better use of the large potential of ceramics can be achieved by reducing the size of the largest flaws. One way aims to improve the processing techniques. Another way consists of producing a suitable coupling between multiple elements so small that the probability of encountering critical flaws is strongly decreased. In fact, composites cumulate both advantages since they associate small elements in a fibrous shape so that high quality ceramic materials and good load transfer between elements are easier to achieve.

Thus, a composite can be considered as offering better conditions to take advantage of the ceramic potential. It is noteworthy that apart from some polyamide fibres (Kevlar) which cannot be used at high temperature and metallic filaments (steel, tungsten) unsuitable for high specific performance, most fibrous reinforcements are ceramic fibres or whiskers. The main types of these reinforcements are reported in Table 1[1] (see also Chapter 3).

With such a view of composites, the matrix is nothing but a cement which controls load transfer between fibres depending on the nature of the matrix and fibre–matrix bonding. That is the case for continuously reinforced organic matrix composites for which the load is mainly supported by the fibres. However, the properties of the matrix may contribute to those of the whole composite. Such complementary functions of the matrix are particularly required when composites are used in severe environments or for specific applications. Thus, metal or ceramic must be considered as a possible matrix when the conditions of use of the composite reveal the following inadequacies of the organic matrices: instability at high temperatures (above 300 °C), low mechanical properties giving the composite poor transverse performance, weak surface resistance, sensitivity to moisture and UV, porosity and degassing effects, low electrical and thermal conductivity as well as problems in ensuring good fibre–matrix bonding.

TABLE 1
General Characteristics of the Main Fibrous Ceramic Reinforcements

Reinforcement	*Diameter (μm)*	*Density ($g\ cm^3$)*	*Young's modulus (GPa)*	*Ultimate tensile strength (MPa)*
Carbon HT (T 300)	7·8	1·75	230	3 450
Carbon HM (M 140)	8–9	2·02	390	2 450
Carbon UHM (P 100)	10	2·05	700	2 400
Al_2O_3-α(FP)	20	3·95	390	2 000
Al_2O_3 (Saffil)	3	3·4	250–320	1 000–2 000
Al_2O_3-SiO_2 (Nextel)	10–12	3·1	220	1 750
SiC (Nicalon)	10–15	2·5–2·6	180–200	2 500–3 000
B (CVD)	100–200	2·2–2·6	370–400	3 300–3 650
B (B_4C) (CVD)	140	2·5	420	4 200
SiC (CVD)	140	3	425	3 500–4 500
SiC { whiskers Tokamax	0·1–1	3·2	400–700	3 000–14 000

1.1 Expected Properties

When composites are to be used at high temperatures, various kinds of metal alloys or ceramics can be suggested, depending on the temperature range and environment related to the conditions of use: magnesium or aluminium alloys for 300–400 °C, titanium for 500–600 °C, nickel-cobalt based superalloys for 1000–1150 °C, silica glass or glass-ceramic for 1000–1200 °C, silicon carbide or oxides for 1300–1600 °C and carbon for above 2500 °C provided the atmosphere is non-oxidizing. Thus, as illustrated in Fig. 1, high temperature composites can be obtained provided the fibrous reinforcement remains stable in the related range of temperature. For example, grain growth phenomena occurring at high temperature in polycrystalline fibres such as Al_2O_3 fibres significantly reduce the reinforcement reliability.

The mechanical stiffness and resistance of metals and ceramics are able to give composites better transverse and shearing characteristics as well as a good resistance to brinelling effects. In addition, strong fibre-matrix bonding due to chemical interaction can be achieved with metal matrices.

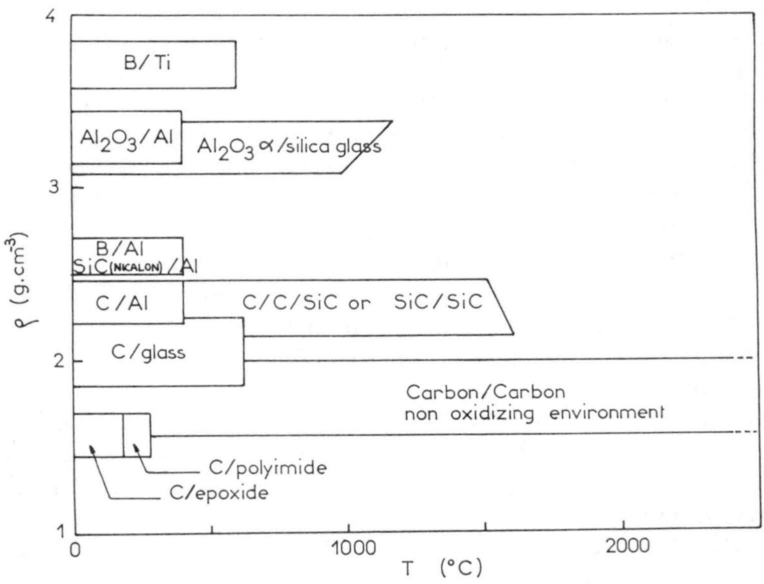

Fig. 1. Temperature range for the use of various non-organic matrix composites.

Furthermore, metals and ceramics are insensitive to moisture and UV, and exhibit good chemical inertness in various environments. That is particularly the case of carbides such as SiC and oxides at high temperature in oxidizing atmospheres. Metals and ceramics are not liable to degas; in addition, metals can give composites a level of imperviousness.

If desired, metals give the composites electrical and thermal conductivities and other specific properties related for example to tribological or electrochemical applications.

Finally, quite promising performance in terms of toughness, work of rupture, impact and thermal fatigue resistance can be expected from ceramic matrix composites (CMC), since, according to the remarks previously noted, the best conditions are combined to take advantage of all the ceramic qualities; the ceramic subdivision is organized to ensure an efficient load transfer and concerns not only the fibres but also the matrix localized in small interfibre spaces. Additionally, the function of the fibre–matrix interface (FMI) is so important that the FMI must be regarded as a third adjustable component of composites.

1.2 Limitations

Particular attention must be paid to the FMI since, in fact, this third component gathers all the problems related to the interactions between the two members of the coupling.

First of all, depending on the conditions of processing and use, fibres and matrix must be chosen in order to prevent any degradation due to a lack of compatibility. The main parameters to be considered are summarized in Table 2. It is often difficult to satisfy all of the conditions required to achieve the desired composite through this choice. For example, the association of high modulus and low thermal expansion fibres with a weak and compliant high thermal expansion matrix, inevitably results in matrix microcracking during cooling after processing.

Broadening the range of possible choices can be achieved by additional operations before or after processing. For instance, some fibre coatings can be applied to reduce the FMI chemical interaction, others allow the control of the deformation and fracture mechanism in the vicinity of the FMI. However, such treatments increase processing complexity and cost.

The compatibility criteria are more severe as the processing methods

TABLE 2
Physical Characteristics Influencing the Fibre–Matrix Compatibility and the Composite Damage

Physical characteristic	*Rupture strain,* α^R	*Poisson's ratio,* ν	*Expansion coefficient,* α	*Chemical reactivity,* c
Compatibility problem	Equal rupture strain enhances catastrophic failure	Mismatch leads to tensile stress at the fibre–matrix interface	Mismatch modifies the distribution of stress between fibre and matrix	Chemical interaction between fibre and matrix giving rise to interphases
Activation	Mechanical loading in the fibre direction	Mechanical fatigue	Significant temperature deviation and thermal fatigue	Thermal exposure and thermal fatigue
Damage	Brittleness	Fibre–matrix debonding, decrease in rigidity and strength	Microcracking or plasticity of the matrix and fibre–matrix debonding	Fibre degradation and fibre–matrix debonding

which are presently available to manufacture metal matrix composites (MMC) and CMC generally require high pressures and/or high temperatures. Such consitions give rise to many additional difficulties with respect to the processing techniques used for organic matrix composites (OMC). This results in a significant delay in the development of inorganic matrix composites and an insufficient reduction in costs. On the other hand, applications are increasingly becoming candidates for these kinds of materials.

2 PROCESSING

Processing of both MMC and CMC is based on three types of techniques:

— the liquid route, mainly the impregnation of yarns, fabric layers or fibrous pre-forms by a liquid alloy or for instance by an organometallic precursor;
— the solid route related to powder metallurgy or hot compression;
— the vapour route, namely the deposition of metal on fibres or the chemical vapour infiltration (CVI) of pre-forms.

Additionally, the different techniques can be combined to manufacture MMC, CMC and if desired hybrid matrix composites.

In any case, the processing method has to satisfy the following requirements:

— the incorporation of the fibres within the matrix or the densification of randomly oriented, unidirectional (1D) or multidirectional (2D, 3D . . . *n*D) fibrous structures;
— the possibility of adjusting the matrix composition particularly in the case of a metal matrix;
— the reduction of residual porosities especially when the composite is densified by CVI or by precursor impregnation followed by a pyrolysis;
— the minimization of the mechanical or chemical degradation of the fibrous reinforcement due to high pressures and temperatures;
— the control of the fibre–matrix bonding;
— the reliability of the processing technique;
— the possibility of further transformation of the composite by heat treatments, metal working or machining;
— the suitability for industrial adaptation and moderate costs.

Despite these constrictive conditions, a great number of processing methods have been proposed and have begun to be involved in mass production, particularly through automotive industry.

2.1 Powder Metallurgy

In the case of discontinuous reinforced composites, one of the easiest ways to incorporate short fibres within metal or ceramic is by using powder metallurgy techniques. Combining successively or simultaneously pressure, temperature and sometimes reaction bonding, the sintering methods enable the reinforcement of complex and refractory matrices at moderate temperatures, according to well-known technological conditions. However, only short fibres and whiskers can be associated with those matrices which are only available in the powder state. Otherwise, the use of high pressure as well as relatively high temperatures can result in mechanical and chemical degradation of the reinforcement although a significant volume fraction of porosity may remain in the composites.

The usual procedure consists of blending a powder of alloy or ceramic with short fibres or whiskers, whose volume fraction can be rather high, followed by uniaxial or isostatic compression and sintering. The procedure can be improved by mixing powders and fibres within a liquid in order to prevent any segregation, then removing the liquid and applying pressure at high temperature. This enables the moderation of pressure and related reinforcement degradation (hot isostatic pressing, HIP). However, the use of quite high temperatures (sometimes in the liquid–solid range for metals) during sintering requires much attention to be paid to the chemical compatibility between fibre and matrix.

The techniques illustrated in Fig. 2 have been tested for reinforcing superalloys and have been successfully used for the aluminium matrix composites (AMC) reinforced by SiC whiskers (N.B. continuous fibres have also been coated with refractory alloy powders). The possibility of transforming by plastic forming the AMC processed by powder metallurgy, giving rise to preferential orientation of the initially oriented fibres (extrusion, swaging, forging, rolling) has been very attractive for mass production so that those composites have already been manufactured on an industrial scale.[2]

The use of such a process for reinforcing ceramics is not yet extensive, but an increasing number of studies deal with the incorporation of SiC whiskers within matrices such as Al_2O_3, Si_3N_4 and ZrO_2.[3–7]

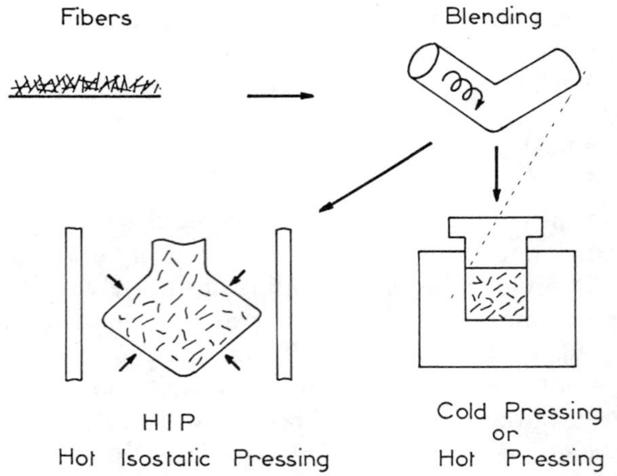

Fig. 2. Schematic of composite processing by powder metallurgy.

2.2 Fibre–Matrix Liquid Compounds

Another method allows the incorporation of short fibres or whiskers within metals or ceramics. It is based on the introduction of discontinuous reinforcement into either a liquid alloy, a slurry of ceramic powder, or a sol–gel precursor of the matrix. In fact, preparing a slurry of ceramic powder is nothing but the first step of the sintering technique summarized previously. (With regard to the sol–gel route, it is also and more often used for the impregnation of fabric layers or preforms, and therefore the features of this technique will be described in Section 2.5.) Thus, preparing fibre–matrix compounds is mainly related to processing of metal matrix composites by gravity casting, injection, squeeze casting and plastic working.

The direct incorporation and dispersion of short fibres within liquid metal is not so easy because of the poor wettability of the ceramic elements as well as their tendencies to flocculation and sedimentation. In addition, the high apparent viscosity of the liquid compounds leads to difficulties in casting when the fibre volume fraction is higher than 10%. Among the conditions which provide an improvement in the fibre–matrix compound preparation, a vigorous stirring of the liquid metal is certainly the most efficient. This effect is more pronounced as the alloy is in a semi-solid state. Thus, combining rheocasting techniques as illustrated in Fig. 3 and fibres incorporation leads to the so-called

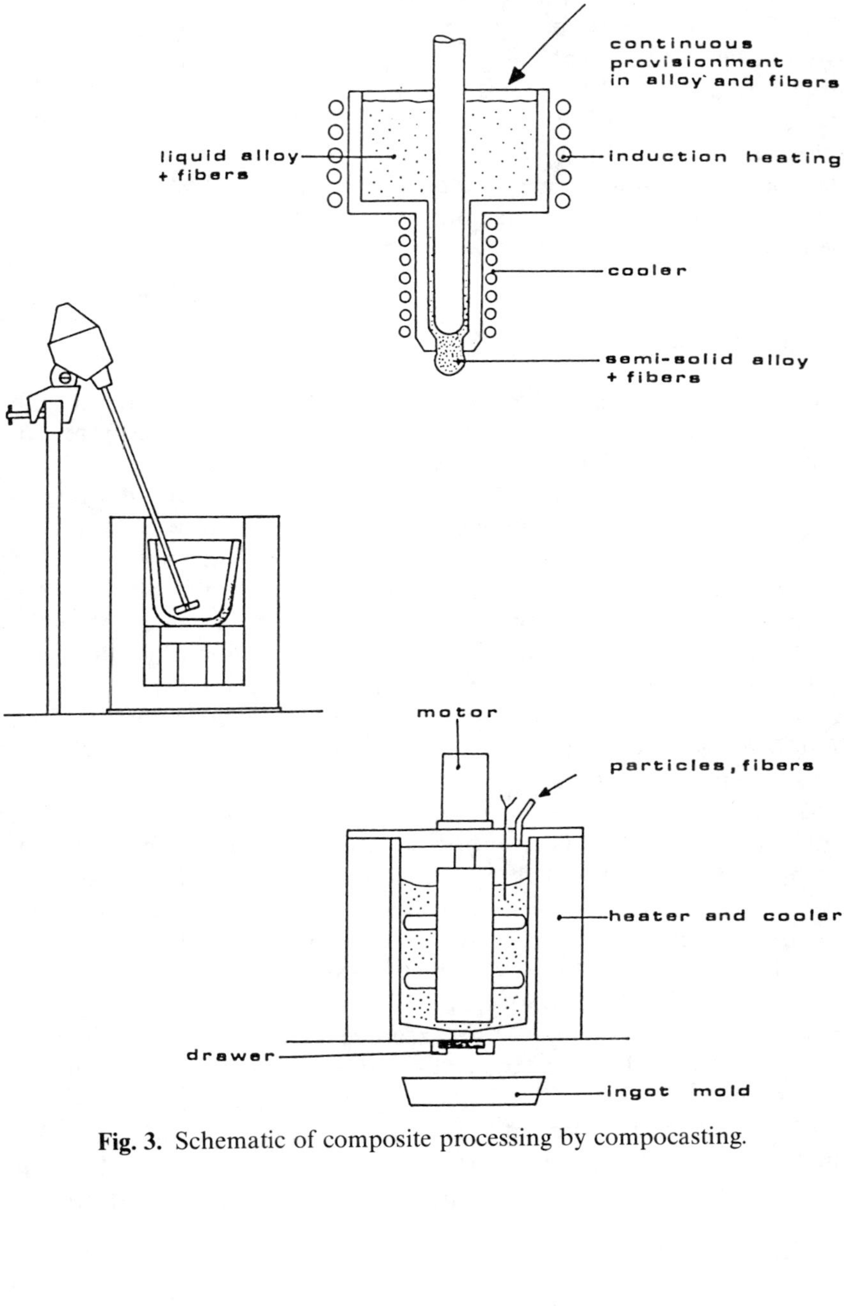

Fig. 3. Schematic of composite processing by compocasting.

compocasting which has been used until now for the reinforcement of aluminium alloys.[8,9] After degassing the liquid alloy at about 700–750 °C, the temperature is decreased to the liquid–solid range. The vigorous mixing prevents the primary phase from forming dendrites so that the solid–liquid mixture is kept fluid. Then, the reinforcement can be incorporated and dispersed without any flocculation or rejection phenomena. The thixotropic behaviour of the resulting compounds enables either immediate use according to net shape casting or further plastic forming after re-heating. Nevertheless, the reinforcement volume fraction remains limited by the castability of the fibre matrix mixture (25–30%).

2.3 Metal Coating

Metal can also be associated with the reinforcement by coating the fibres by either vapour, liquid or solid routes. In the last case, fibres are covered by a thin layer of alloy powder with the help of a liquid which is eliminated before sintering under pressure. In fact this technique relates to powder metallurgy. Liquid metal coating must be regarded as one of the impregnation techniques and as such will be described in Section 2.5. However, metals can be also projected onto fibres by plasma spraying or deposited on the fibres from electrolytic solutions. Unfortunately, the first method is suitable only for large diameter filaments and the last one, which was applied to copper and nickel deposition, leads to residual porosities in the coating. Furthermore, adjusting the matrix composition is not possible and the slow kinetics of coating give rise to rather high processing costs.[10–12]

Similar disadvantages also relate to vapour deposition; however, several methods have been proposed and could lead to metal matrix composites with a high volume fraction of fibres. The PVD (physical vapour deposition or thermal evaporation) and IVD (ionic vapour deposition) methods transfer atoms of metals from a liquid or solid metal source towards fibres.[10–12] The process is more efficient when the fibres are well separated and when a potential difference is applied between the metal and the fibres to ionize the metal vapour. The IVD process illustrated in Fig. 4 has been applied to aluminium coating of thin carbon fibre layers. Chemical vapour deposition (CVD) could be advantageously used for infiltrating a refractory metal within multi-directional pre-forms. The method has in fact been applied for depositing aluminium from a thermal decomposition (tri-isobutyl-

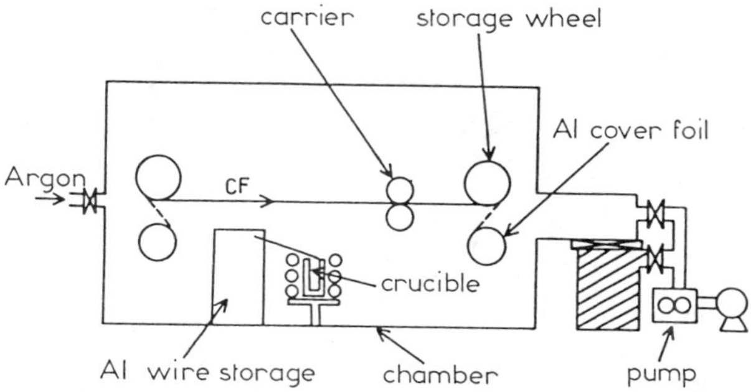

Fig. 4. Schematic of 1D carbon/aluminium composite processing by ionic vapour deposition.[12]

aluminium) or a chemical reaction (reduction of aluminium chloride by hydrogen). The major advantages of the previous methods are related to moderate chemical interaction between metal and fibre whilst, on the other hand, these techniques need additional processing steps to bind the metal–fibre elements.

2.4 Plastic Working and Diffusion Bonding

Several techniques can be used to join semi-products such as those processed by the above-described metal coating or by liquid impregnation as described in Section 2.5. All of them combine both temperature and pressure in order to induce diffusion bonding and the contribution of either plastic deformation in the case of a metal matrix or viscous deformation if the matrix is for instance a glass ceramic. In any case, a compromise has to be found to prevent mechanical degradation due to high pressure or excessive chemical interaction due to high temperature.

2.4.1 Hot Compression

The hot compression technique is not only used to bond composite semi-products but it is quite suitable for directly reinforcing refractory alloys with large diameter (140 μm) continuous CVD filaments [B, B(B_4C), B(SiC) or SiC filaments].

The reinforcement is generally arranged in unidirectional layers obtained by weaving or by drum winding as illustrated in Fig. 5(a). Then an alternate stacking of filament monolayers and metal foils (500–200 μm thick) constituting the matrix (for instance, titanium or superalloys) can

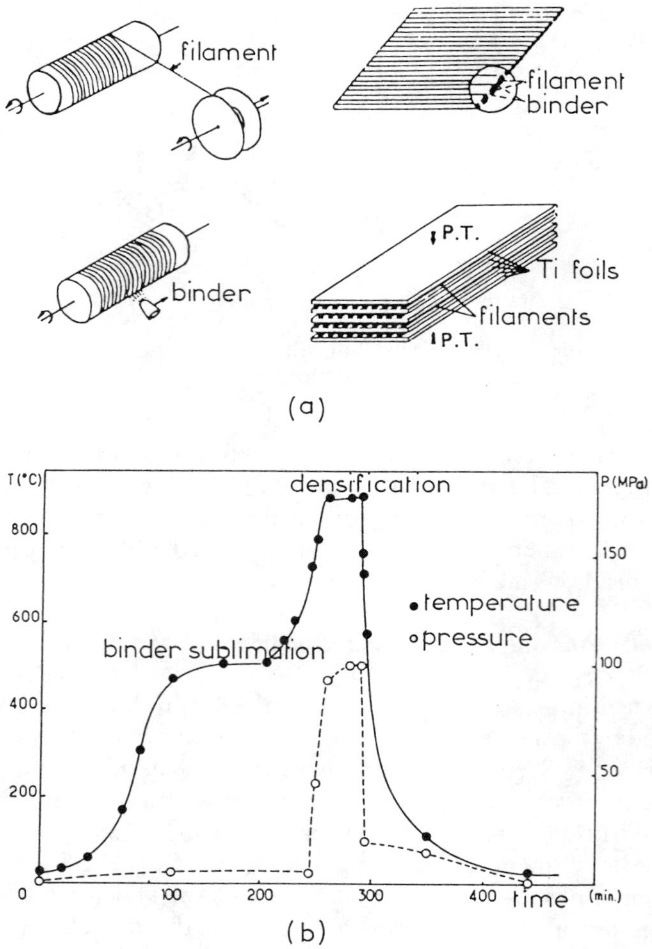

Fig. 5. Schematic of 1D SiC (CVD)/Ti 6Al 4V composite processing by hot compression.[13] (a) Arrangement of the reinforcement within the titanium matrix; (b) temperature and pressure cycle.

be placed in a flexible metallic bag. Any excessive oxidation is avoided by performing the bonding sequences schematically represented in Fig. 5(b) under vacuum or an inert atmosphere. The method has also been used for reinforcing aluminium alloys but remains unsuitable for incorporating by the solid route, a metal within yarns or fabric layers made of small diameter fibres (about 10 μm). Otherwise, the same apparatus can be used for liquid impregnation of either ceramic or metal matrix composite plates as shown in Section 2.5.

2.4.2 High-Speed Co-lamination

In order to reduce the duration of the fibre–matrix chemical interaction the bonding operation can be performed at high speed, by the continuous tape method of manufacture illustrated in Fig. 6. Ti(75 A) matrix composites with 25 vol% boron filament processed according to this method have shown that the amount of fibre–matrix interphase (TiB_2) is quite well controlled. Thus, despite a necessary increase in the temperature (980 °C over 1 s) and pressure applied to the composite during the rapid processing compared with hot compression (850 °C over 20 min), a significant improvement in the filament reinforcing is observed after relaxing the residual stresses by annealing. However, the technique is difficult to optimize in terms of preventing extended mechanical degradation of the reinforcement or the matrix delamination.[14]

2.4.3 Tube Processing

A method for manufacturing composite tubes has been proposed based on the hot pressing principle. It consists of winding metal–filament semi-products [for instance $B(B_4C)$/Al monolayer obtained by plasma spraying] on an expandable drum, placing the whole device in a cylinder and simultaneously applying pressure and temperature. Such tubes have found structural uses in aeronautic and space applications (space shuttle).[15]

2.4.4 Pultrusion

The consolidation of several impregnated fibrous yarns or coated filaments into rods can be performed by pulling the metal matrix

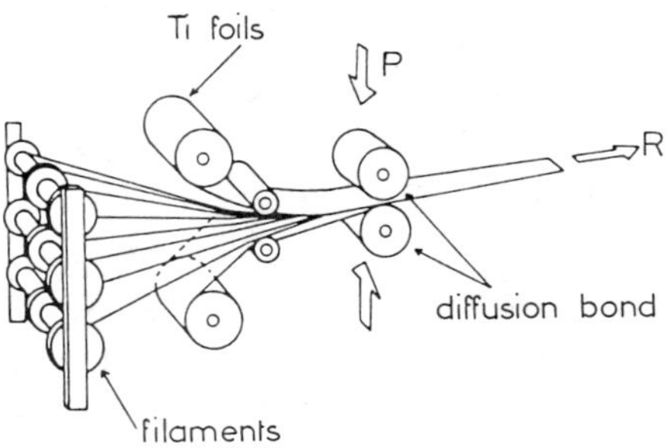

Fig. 6. Schematic of 1D boron/Ti composite tape manufacturing.[14]

composite semi-products through a die at high temperature.[16, 17]

In all the techniques briefly described above, the main parameters controlling the metal plastic deformation (or the viscous deformation of a glass ceramic matrix), the diffusion bonding and the fibre–matrix chemical and mechanical interaction, are the temperature, the pressure and the time. The quality of the resulting composites depends strongly on the adjustment of these parameters during the processing cycle.

2.5 Liquid Impregnation of Fibrous Pre-forms

The impregnation of fibrous pre-forms (one-, two- or three-dimensional arrangement of yarns) by a matrix in the liquid state concerns mainly the metal matrix composites for which the alloy melting temperature is rather low and the ceramic–ceramic composites whose ceramic matrix can be derived from a liquid precursor.

In both cases the lack of wettability of the ceramic pre-forms by the liquid metals or ceramic precursors is a major difficulty. For the metal matrix composites the wettability problem has to be related to the liquid metal–fibre reactivity at high temperature. Indeed, as illustrated in Fig. 7, the contact angle θ between a ceramic substrate and a liquid metal in a specific vapour environment represents the wettability. It may be poor at quite low temperature when the ceramic–metal chemical interaction is the smallest. However, the contact angle may change drastically over a short range of temperature if the increase in temperature activates the metal–ceramic interaction — for example, by dislocating a thin oxide layer coating on the liquid metal surface.[18–20] This means that improving the wettability may result in a chemical degradation of the fibrous reinforcement, particularly at high temperature when the reaction kinetics speed up. In addition, the corresponding formation of new phases such as the carbide Al_4C_3 for aluminium–graphite composites can embrittle the matrix and/or the fibres by facilitating microcrack initiation and propagation at the fibre surface.

Various methods have been proposed to overcome these two antagonizing difficulties simultaneously.[21]

- The addition of alloying elements (Mg, Ca or Li for aluminium matrix composites) improves the fibrous ceramic wetting but does not prevent the fibre–matrix interaction so that the duration of the liquid impregnation must be carefully controlled.[22–24]
- Coating the fibres by CVD with titanium, titanium carbide or

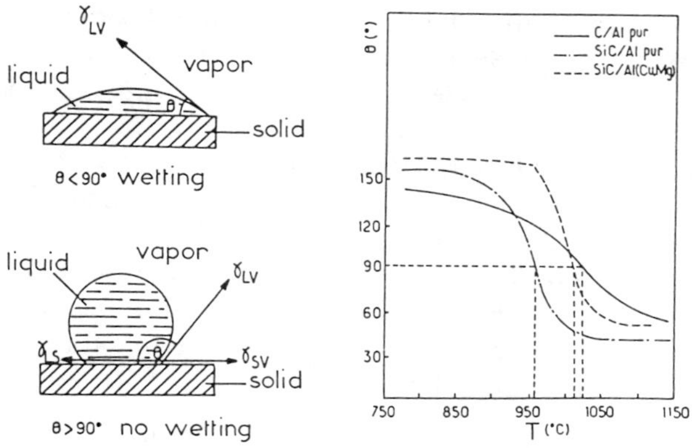

Fig. 7. Evolution of the contact angle with temperature for the systems Al/C and Al/SiC.[18-20]

boride leads to an improvement of the fibre wettability and protects the reinforcement against the liquid metal reactivity. This method has shown sufficient success to justify the first step of its industrialization (Fig. 8).[25]

— Silver or nickel coatings on the fibres facilitate and simultaneously moderate the interaction between liquid aluminium alloys and ceramic reinforcement. In fact, the method has been tested out mainly for graphite fibres and leads to a significant increase in the resulting composite density.[26-28]

— Improvement of the wettability of carbon fibres can also be

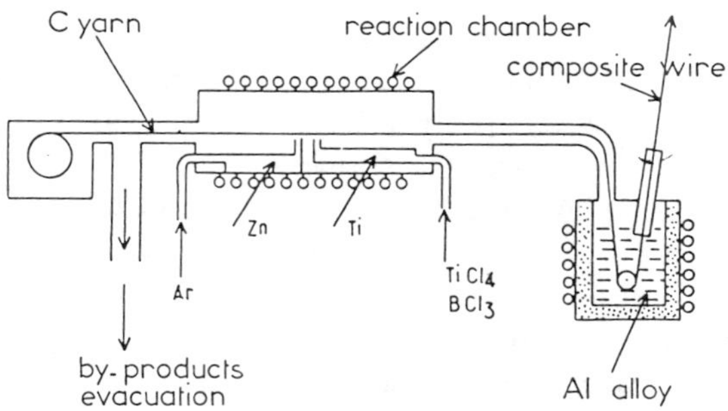

Fig. 8. Schematic of the wettability improvement of carbon fibres by Ti–B coating.[25]

obtained by treating the surface with liquid sodium or sodium–potassium alloys. Such a treatment is expected to render the fibre surface metallic by intercalation. However, this method is rather difficult to transfer to industry.[29,30]

— Easier industrial adaptation is possible with silica deposition from organo-metallic compounds but the method is not yet reliable enough to be extensively used.[31]

— Treating carbon or SiC pre-form with a fluoride (K_2ZrF_6) is also quite simple and very efficient when the reinforcement size is not too small. However, the process is not suitable for alumina fibre and needs more investigation.[32,33]

Whilst the wettability problem is more crucial for pre-form impregnation by liquid metals than by liquid ceramic precursors, the formation of porosity within ceramic matrices derived from precursor pyrolysis is often a problem which needs to be overcome with high pressures and, sometimes, several sequences of impregnation. In any case, metal impregnation is a one-step operation whilst the related procedures for processing ceramic–ceramic composites require a minimum of two sequences (the precursor impregnation and the pyrolysis).

2.5.1 Metal 'Pre-preg' and Semi-products

As already noted in the previous sections, inorganic matrix composites can be manufactured from semi-products by means of joining processes such as diffusion bonding by hot compression or pultrusion. In fact, the availability of a metal 'pre-preg' can be regarded as a factor of industrial development of MMC. Thus, several studies have already dealt with the processing of MMC foils or continuous tape and rods.

2.5.1.1 Continuous impregnation of yarns. Provided the difficulties related to the poor wettability of reinforcement can be overcome by using one of the methods briefly described above, yarns of fibres can be impregnated by a liquid metal as shown in Fig. 9. The process has been used for associating Al, Mg, Cu or Pb matrices with graphite fibres coated by either Ni or Ti–B compounds or SiO_2.

As detailed later on in this section, the poor fibre wettability can be partly compensated by the use of high pressure during impregnation. Thus passing a carbon yarn previously impregnated by a slurry of aluminium powder through a hot pultrusion die (800 °C), then removing the solution, enables an intimate association of the aluminium matrix with the fibres to be made.

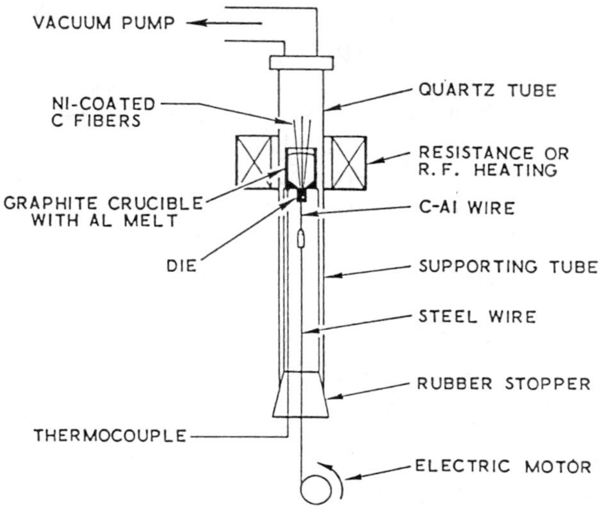

Fig. 9. Schematic of a wire-drawing apparatus.[18]

2.5.1.2 Composite foils. Apart from the use of plasma spraying for manufacturing monolayers of unidirectional CVD filaments [$B(B_4C)$, SiC] within a metal, aluminium matrix composite foils reinforced by continuous SiC fibres (Nicalon) have been processed by hot pressing an alternate stack of aluminium sheets and widely spread fibre yarns. Applying pressure at a temperature corresponding to the semi-solid state of the matrix enables the impregnation of the yarn to be achieved, while preventing any excessive fibre–matrix chemical interaction. However, the poor wetting leads to some porosity which can be reduced during the use of complementary techniques (diffusion bonding, metal forming, etc.).

2.5.2 Foundry Techniques

Net shape MMC parts or local reinforcement of castings can be performed by direct metal impregnation of fibrous pre-forms made of randomly oriented short fibres or whiskers as well as multidirectional woven (1D, 2D, 3D ... *n*D) fibres.

Impregnation by the liquid metal can be achieved either by gravity casting without any pressure, provided the wettability problem has been resolved, or by using high pressures which counterbalance the lack of wetting.

2.5.2.1 Low pressure casting. Low pressure casting is undoubtedly the most feasible process for industrial manufacturing of light alloy matrix composites. After using one of the above-mentioned wettability enhancement methods ceramic reinforcement can be placed into a permanent mould and the liquid alloy cast by gravity as illustrated in Fig. 10. In such a case, the pre-form temperature is the major parameter. Although metal feeding of the whole pre-form during the solidification is rather difficult when the reinforcement volume fraction is high, composites can be obtained with very little residual porosity.

2.5.2.2 Squeeze casting. The use of squeeze casting for MMC processing overcomes the difficulties related to the poor wettability of ceramic pre-forms by forcing the liquid metal penetration while compensating for shrinkage during solidification. Residual porosity is reduced and close tolerances can be obtained, necessary for net shape parts.[34–43]

The pressure can be applied by means of a piston acting directly on the mould containing the pre-form to be infiltrated or indirectly through runners and gates feeding several moulds as illustrated in Fig. 11.

Because this processing technique is particularly suitable for mass production the automotive industries are becoming increasingly involved in its technological development, for instance when local reinforcements are required in engine parts.

2.5.3 Glass and Glass–Ceramic Matrices

The processing techniques which have been described above for MMC can be also used for CMC provided the ceramic fluidity is sufficient at

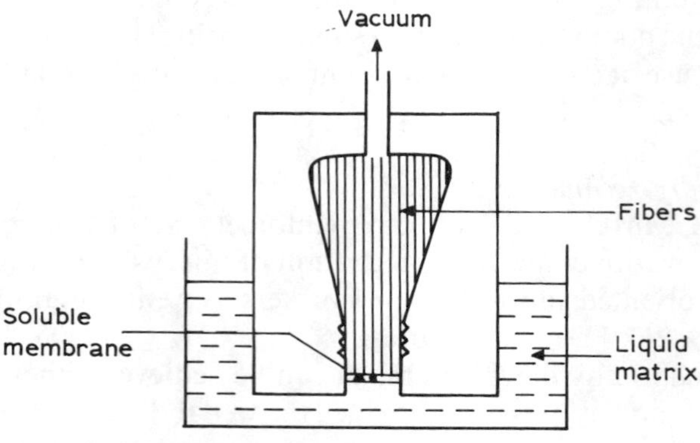

Fig. 10. Schematic of composite low pressure casting.

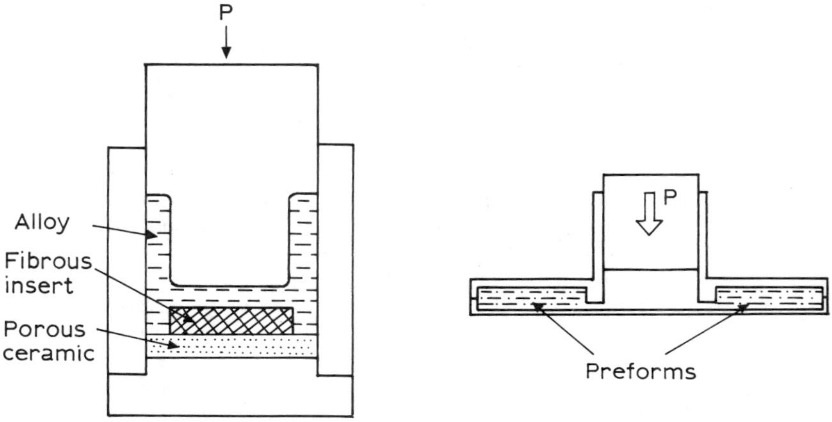

Fig. 11. Schematic representation of squeeze casting devices: (a) direct impregnation of a pre-form; (b) indirect impregnation of several pre-forms.

relatively low temperatures, allowing the impregnation of carbon, silicon carbide or alumina fibrous pre-forms. Various kinds of silica glass offer the expected low viscosity at temperatures far below 1000 °C. Furthermore, glass–ceramic matrices (for example SiO_2, Al_2O_3, Li_2O) can be processed by these methods followed by an annealing treatment of crystallization. However, for the highest temperatures generally used for glass–ceramic matrices, a slightly different procedure, illustrated in Fig. 12, is preferred and consists of:

(1) preparing a 'ceramic pre-preg' by impregnating yarns or fabric layers with a homogeneous suspension of fine glass powders within a liquid (when the particles are very small, the resulting colloidal dispersion exhibits a jelly-like consistency and hence the viscous mixture is called a gel);
(2) building a pre-form by arranging the impregnated yarns or by stacking the fabric layers;
(3) performing a temperature–pressure cycle under vacuum or inert gas in order to remove the liquid at low temperature, to enable the glass to flow (1100–1400 °C; 5–15 MPa) and then to induce crystallization at a lower temperature (900– 1000 °C).

When the pre-forms are built up before the impregnation, the entire procedure preceding the annealing treatment of crystallization can be repeated several times in order to reduce the residual porosity. However,

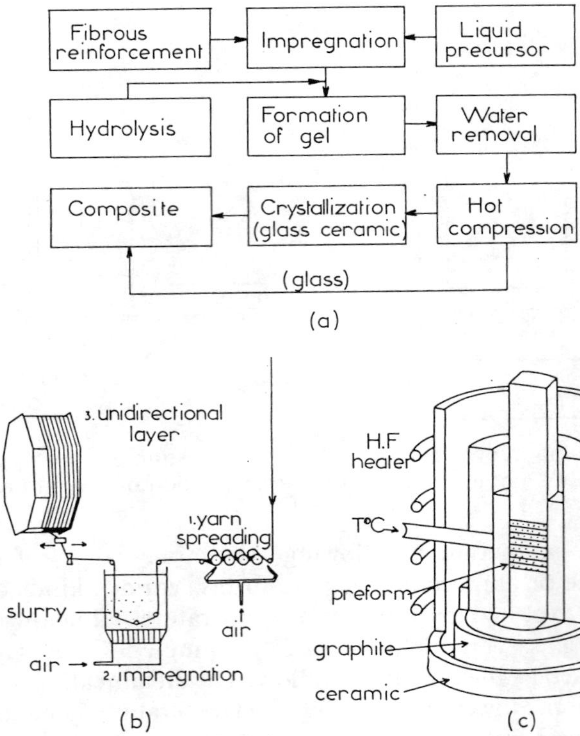

Fig. 12. Schematic of glass–matrix composite processing by the liquid route: (a) procedure; (b) impregnation;[45] (c) hot compression.[50]

the main disadvantage related to the CMC processing method previously described concerns the reinforcement degradation due to the use of high temperatures, so that an improvement of the method is expected from the decrease in the processing temperature.[44–47]

2.5.4 Firing of Organic or Organo-metallic Precursors

The impregnation techniques which have been briefly presented above can in fact be used not only for the glass or glass–ceramic matrices but also for various other kinds of ceramic matrices provided a precursor of the matrix can be found which satisfies the following main criteria:

— it exhibits a sufficient fluidity at relatively low temperatures ($<500\,°C$);

— it gives rise, after reaction and firing, to the expected ceramic matrix with a low content of impurities and residual porosities;
— it is available, inexpensive and easy to process.

2.5.4.1 Carbon–carbon composites. Due to the variety of possible applications, which are presented in later sections, carbon matrix composites reinforced by randomly oriented short carbon fibres or continuous yarns of carbon fibres arranged in multidirectional (nD) pre-forms (n = 1, 2, 3, 4...) have already led to wide industrial development.[48]

The general processing procedure can be summarized as follows.

(1) The fibrous carbon pre-form is impregnated by a liquid phenolic or furfurylic resin or a pitch.
(2) After polymerization of the resin, or pre-reticulation of the carbon mesophase, the matrix is carbonized at a high temperature (1000–1200 °C) in a non-oxidizing atmosphere. Subsequent thermal treatment at very high temperature gives rise to graphitization of the matrix.

Whatever the type of liquid precursor or the rate of pyrolysis, the insufficient densification of the pre-form requires the procedure of impregnation–carbonization to be repeated several times and in some cases under pressure. However, a nearly complete densification of the composite can be reached instead by chemical vapour infiltration, the principle of which is described in Section 2.6.

2.5.4.2 Ceramic–ceramic composites. Various types of organo-metallic precursors can lead to ceramic matrices after pyrolysis; only some of them are reported in Table 3.[49] Although most examples concern carbide and nitride as well as mixed matrices (SiC/B_4C; SiC/C; $Si_3N_4/SiC/C$), oxides can also be derived from the so-called sol–gel procedure schematically represented in Fig. 12(a) and the processing methods of which are very similar to those described in Section 2.5.3 related to glass matrices.[50–58]

In the case of silicon carbide matrices, the polycarbosilane precursors which have been prepared by different methods by Yajima and colleagues for processing SiC Nicalon fibres can also be used as precursors for impregnating fibrous preforms.[52] The firing of the precursors gives rise to gas formation, first during the polymerization at low temperatures (≈400 °C) then at higher temperatures when the

TABLE 3
Ceramic Precursors[49, 52–54, 56]

Ceramics	*Organo-metallic precursors*
SiC/C	Polycarbosilanes
SiC/C	Polysilastyrene: phenylmethylsilane–dimethylsilane
SiC	Heptamethyl-2-vinyltrisilane
SiC/B_4C	Carborane–siloxane
SiC/B_4C	Poly(borodiphenyl)siloxane
$Si_3N_4/SiC/C$	MTS/methylamine
BN	Ammonioborane

resulting organic solid is transformed into SiC matrix containing a significant amount of oxygen as well as free carbon and silicon (≈700–1300 °C). In general, the higher the temperature of firing, the more crystallized is the resulting ceramic. For the lowest temperatures of pyrolysis, the microstructure remains quasi-amorphous and exhibits rather low rigidity and strength whilst the use of higher temperatures leads to microcrystallization and an improvement in the mechanical characteristics. However, the recrystallization related to the highest firing temperatures induces a tensile strength decrease while the Young's modulus continues to increase.[52, 57]

The main difficulty concerning the methods based on the firing of a precursor is the rate of densification, so that enhancing the efficiency of the precursor is a primary concern. Finally, whatever the impregnation technique used for either MMC or CMC processing, the wettability of the ceramic fibres by liquid metals, gels or other precursors remains as the major difficulty and is most frequently overcome by using pressure.

2.6 Chemical Vapour Infiltration

Compared with the liquid impregnation techniques, the gaseous infiltration of matrices leads to a better densification of the composite due to the lower viscosity of gas. In fact, for other reasons discussed later, the chemical vapour infiltration (CVI) process is considered as particularly promising. This method has been derived from the chemical vapour deposition technique (CVD) although CVD and CVI are performed under quite different conditions. Both of them consist of the deposition of a solid by a chemical reaction occurring within a gaseous phase containing one or several chemical species. However, whilst the aim of CVD is to obtain a continuous void-free coating on a free surface, with a uniform thickness, the purpose of CVI is to fill in a porous medium so that the transfer of gaseous species through the porous solid becomes predominant and controls the processing conditions.

2.6.1 Principle of the Technique

Chemical vapour deposition or infiltration of a solid binary compound AB from a gaseous precursor can be depicted by the following general equation and by the model illustrated in Fig. 13:

$$AX_n + BX_m + \frac{1}{2}(n + m)\,H_2 \longrightarrow AB + (n + m)\,HX$$

where X can be chlorine, bromine or fluorine elements and where the refractory compounds AB as well as the corresponding gaseous presursors can be those reported in Table 4. Assuming the system is in equilibrium, a thermodynamic approach can predict the nature of the chemical species present, the thermodynamic yields, and the occurrence of by-products and deposit–substrate interaction. Figure 14 shows the conditions for which the deposit is either a single phase (SiC) or a mixture of two phases (C + SiC).

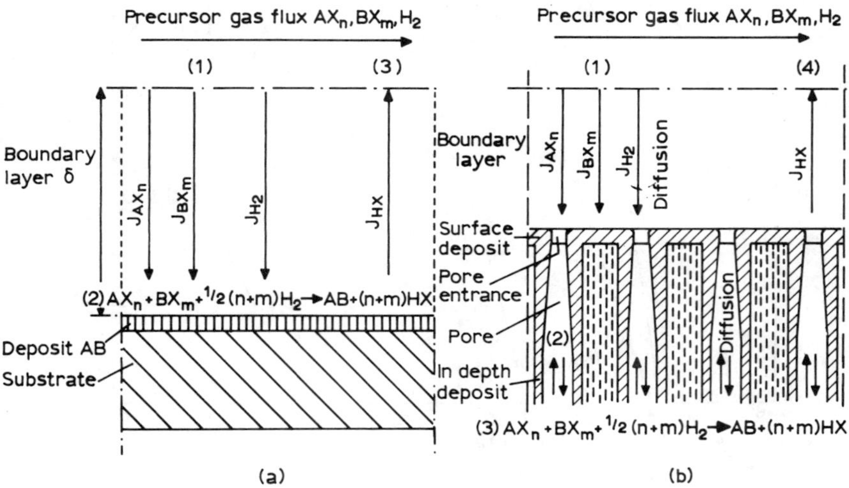

Fig. 13. (a) Chemical vapour deposition; (b) infiltration from a gaseous precursor.[59]

TABLE 4
Gaseous Precursors for CVD or CVI of Refractory Materials[59]

Ceramics	*Gaseous precursors*
C	CH_4, C_3H_8
SiC	CH_3SiCl_3–H_2
TiC	$TiCl_4$–CH_4–H_2
B_4C	BCl_3–CH_4–H_2
BN	BF_3–Cl_2 or BCl_3–NH_3
Si_3N_4	$SiCl_4$–NH_3
Al_2O_3	$AlCl_3$–CO_2–H_2

In order to prevent any homogeneous nucleation of AB within the gaseous phase leading to unsuitable deposits, the above reaction is activated in the vicinity of the porous surface by heating the pre-form alone in a cold wall reactor or both pre-form and gaseous phase in a hot wall reactor as illustrated in Fig. 15. Building up a plasma within the

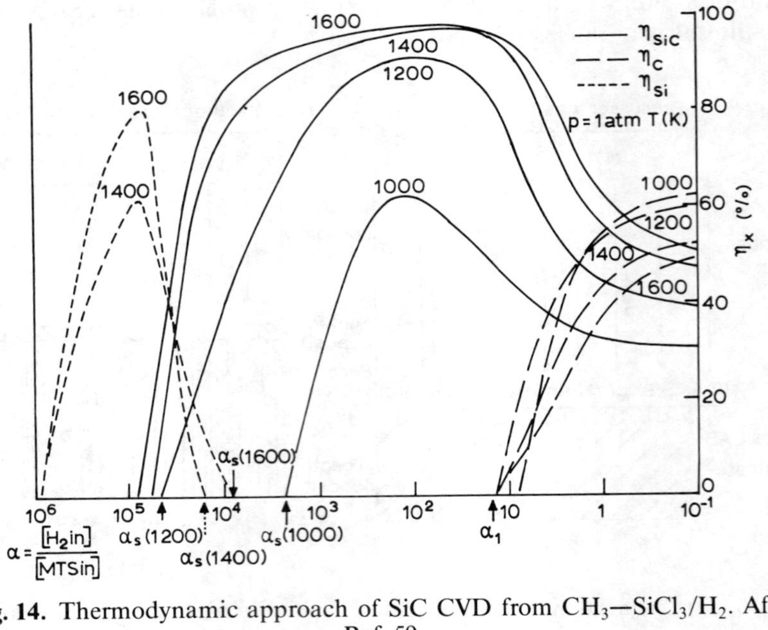

Fig. 14. Thermodynamic approach of SiC CVD from CH_3—$SiCl_3$/H_2. After Ref. 59.

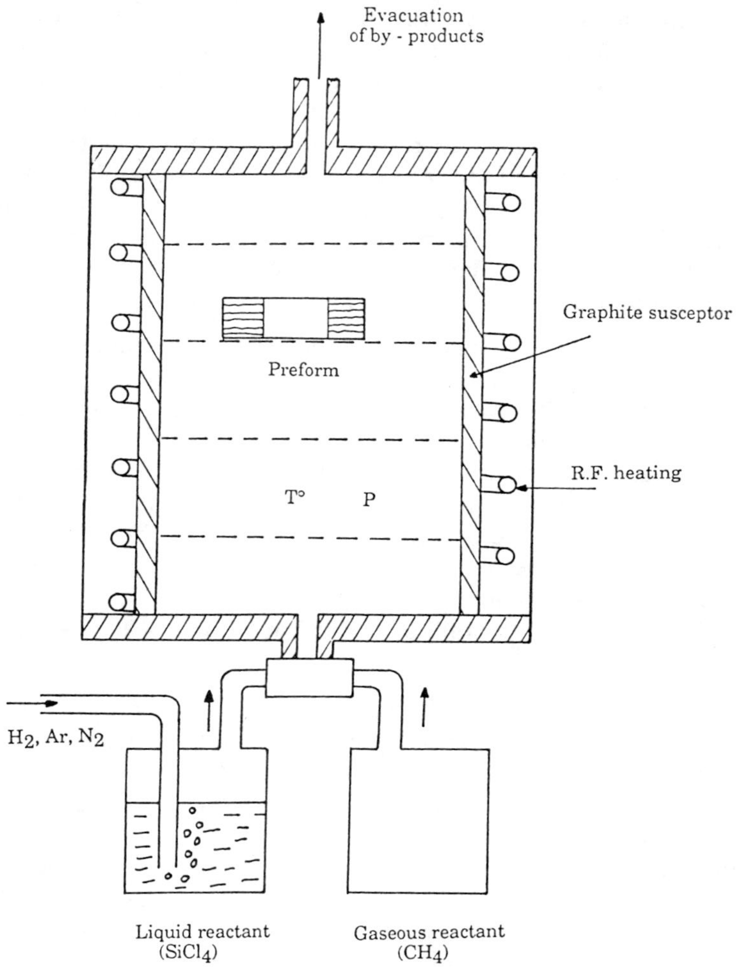

Fig. 15. Schematic of a hot wall CVI apparatus.

vapour phase by means of an electrical field is also effective in activating the reaction and is expected to enhance the CVI mechanisms.

For both CVD and CVI the deposition is controlled by either the reaction kinetics or the molecular transfer of the chemical species across the boundary layer related to the gaseous flow in the case of CVD and through the porous pre-form in the case of CVI. This means that accelerating the reaction by using high temperatures and total vapour pressure leads to a CVD operation controlled by convective gaseous flow and may prevent any suitable infiltration of thick porous pre-forms due to early sealing of the pore entrances. Therefore, for full densification of pre-forms, the reaction kinetics need to be reduced in order to enable the required transfer of the various chemical species. As a consequence, processing composites by CVI is an operation of long duration which consumes a significant amount of energy. Nevertheless, the CVI route remains very interesting particularly for CMC manufacturing due to the following main advantages.

(1) The difficulties related to the poor wettability of ceramics by liquids are avoided, as the small pores of the pre-form are more easily filled by gaseous precursors; the residual porosity is therefore smaller.
(2) A nearly perfect embedding of the fibres can be achieved.
(3) The processing temperatures can be low compared with those required by the liquid route. This reduces the fibre degradation related to either recrystallization (SiC or Al_2O_3 fibres) or chemical interaction.
(4) The fine microstructure of the infiltrated matrices gives rise to better characteristics.

The only drawbacks are the cost of the gaseous precursors, the duration of the infiltration and the difficulties in modifying at will the nature of the matrix.

2.6.2 Applications

Although CVI can be applied to MMC manufacturing as already noted for aluminium matrices in Section 2.3, the method has been mainly used for CMC processing. CVI has been used to complete the densification of carbon–carbon composites processed by successive liquid impregnations. The method has also been used to prepare fibrous pre-forms by consolidating for example a stack of fabric layers so that pre-forms can be machined into shape before CVI densification. Later on, CVI has been progressively extended to various kinds of ceramics by partly or totally replacing the carbon derived from in-situ pyrolysis of methane by carbides (SiC, TiC, B_4C), nitride (BN) or oxides (Al_2O_3).[49,59]

CVI has already shown its suitability for processing refractory matrix composites so that the industrial development of this technique is growing despite the high cost of the required chemical reactants as well as the long duration of infiltration. However, much work has still to be devoted to the interactions between the reaction kinetics and the transfer of chemical species through porous pre-forms.

2.7 In-Situ Composites

Among the methods which have been presented above, some, such as CVI and the techniques based on the impregnation and firing of an organo-metallic precursor, could be referred to as in-situ composites since the matrix is formed within the pre-form. However, the term 'in-situ' is better related to composite processing in which both components acquire their final state by reaction between the reactants or decomposition of an initially homogeneous compound.

2.7.1 Reactive Impregnation

A binary matrix composite can be manufactured by inducing a reaction between one constituent of the matrix and a part of the pre-form, provided that the first matrix constituent can be brought into contact with the pre-form (by gaseous infiltration or liquid impregnation) and that the nature of the pre-form corresponds to the second constituent of the matrix. This method is very simple but it is not easy to optimize particularly for preventing excessive degradation of fibres. However, the method has been successfully applied to manufacture C–SiC, SiC–SiC and SiC–Si composites by impregnating carbon fibre pre-forms partly densified by carbon, with liquid silicon (T_m = 1410 °C).[60]

2.7.2 Directional Solidification of Eutectic Alloys

Provided one of the two precipitating phases has a fibrous morphology, an eutectic alloy, as illustrated in Fig. 16, can be regarded as a nearly perfect in-situ composite since it exhibits the following main advantages.

(1) The microstructure is thermodynamically stable up to temperatures in the vicinity of the melting point.
(2) The fibrous phase, which is in fact aligned whiskers, offers all of the qualities required of an excellent reinforcement: crystalline perfection and small size.
(3) The fibre–matrix coupling is particularly well performed due to a coherent relationship between the whisker and matrix crystallographic network.

The processing techniques which induce the fibrous shape and

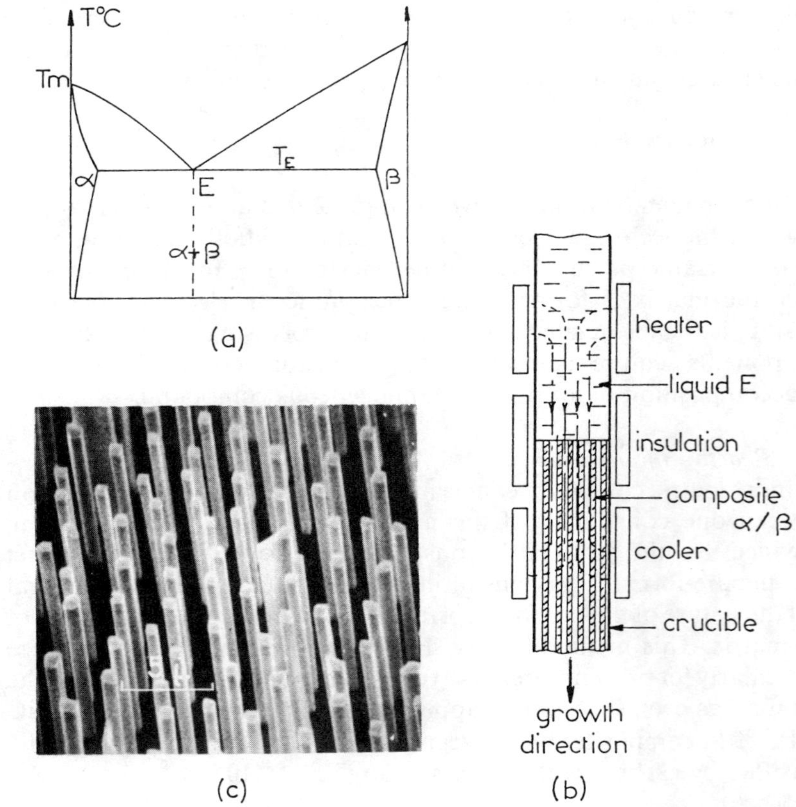

Fig. 16. Directional solidification of in-situ composites: (a) eutectic solidification; (b) directional solidification under temperature gradient; (c) SEM micrograph of Ni–Cr–TaC composite.

alignment of one of the two phases, deal with the methods of crystal growth giving rise to a directional solidification as illustrated in Fig. 16. It is noteworthy that the higher the solidification rate is, the finer the microstructure becomes and the more the 'in-situ' composite performance is improved. Unfortunately, the solidification rate is limited (in the range of 1–5 cm/h) by the need for a stable growth front which requires a higher temperature gradient as the alloy deviates from the exact eutectic composition. This is particularly the case for pseudo-binary eutectic alloys for which the matrix is complex giving rise to superalloy reinforced by refractory whiskers. Despite the attractive qualities of this type of in-situ composite, their development is impeded by the high cost of the processing technique so that such composites are only used in specialized applications.[61–63]

There is frequently no obvious distinction between the various processing methods for preparation of non-organic matrix composites. For example, silicon carbide matrix reinforced by carbon fibres can be produced by (1) preparing a slurry of resin containing SiC and Si powders, (2) impregnating a carbon fibre pre-form by this paste, (3) firing the resin into carbon and (4) inducing reaction sintering at high temperature. In such a case, the processing technique is difficult to classify among a ranking of methods. A similar difficulty occurs when SiC (whisker) reinforced Si_3N_4 composites are fabricated by using both the liquid and solid routes following the procedure illustrated in Fig. 17.[65]

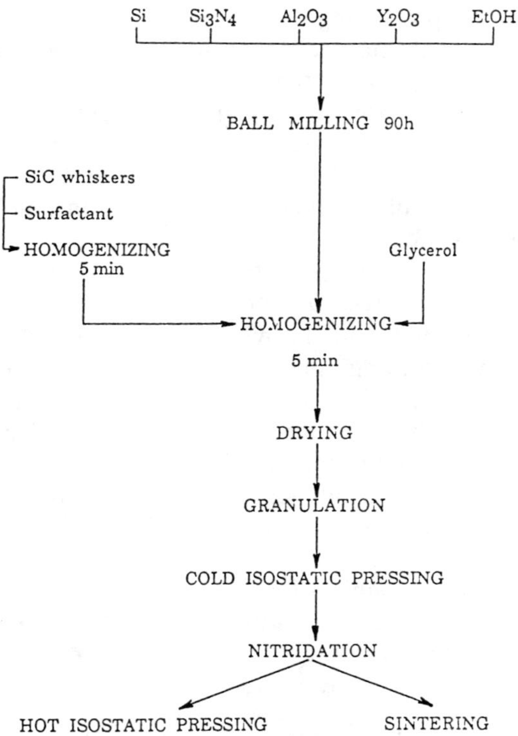

Fig. 17. Processing of SiC whisker reinforced Si_3N_4 composites. After Ref. 64.

3 PROPERTIES AND APPLICATIONS

The following sections aim to give an overview of the inorganic matrix properties through some examples illustrating their wide application field, rather than by attempting to gather data related to their continuously improving physical and mechanical characteristics.

3.1 Discontinuous Reinforced Composites

Much attention has been paid to short fiber reinforced composites which are well suited for mass production and low-cost processing. However, the efficiency of this type of reinforcement is lower than that of continuous fibres due to the following main reasons concerning both organic and inorganic matrix composites.

(1) The volume fraction of short fibres (or whiskers) is generally restricted by processing conditions.
(2) The fibre length may be too small to ensure a sufficient fibre–matrix load transfer and to take advantage of the fibre strength.
(3) The strong decrease in the reinforcing efficiency for a slight misorientation of the load with respect to the fibre direction significantly lowers the composite performance when short fibres are randomly oriented. Nevertheless, some of the processing techniques which have been presented in the previous sections enable a partial fibre orientation (extrusion, compound squeeze casting) and give rise to better rigidity and strength along the mean fibre direction. The whiskers' alignment as well as the fibre–matrix bonding can be so good in the case of a directionally solidified eutectic alloy that in-situ composites can be regarded as continuously reinforced. Furthermore, all of the aspects related to the discontinuous nature of a fibrous reinforcement can be approached following theoretical analysis proposed in the literature, independently of the nature of the matrix.[66]

3.1.1 Light Alloy Matrix Composites

Discontinuous Reinforced Aluminium Matrix Composites (DRAMC) processed by powder metallurgy, compocasting or squeeze casting exhibit the following main properties.[67]

— Whatever the nature of the reinforcement (fibres, whiskers or even particles) the composite rigidity is directly dependent on the reinforcement volume fraction. An increase of 50–100% in the modulus of elasticity can be obtained for V_f = 20–40%, as shown in Fig. 18.

— The improvement in the material strength is not as significant as for rigidity due to a strong decrease in the ductility related to the presence of the brittle phase. As a matter of fact, a ratio of about 10 is often observed between the matrix and composite failure strains.

— The unfavourable effect of ceramic reinforcements on ductility counterbalances the impeding effect of the fibres on crack growth so that toughness cannot be significantly improved.

— The hardness and wear resistance of this type of composite are better and, as the reinforcement and processing costs are low, automotive applications such as ring land area, pistons, bearings as well as cylinder liners can be foreseen.

— Applications requiring high temperature performance can also be considered as possible since a gain of about 100°C in the use temperature can be reached as illustrated in Fig. 19.

— This improvement in thermo-structural properties (high temperature rigidity and strength as well as creep resistance) related to higher flow strain is not however suitable for plastic working where

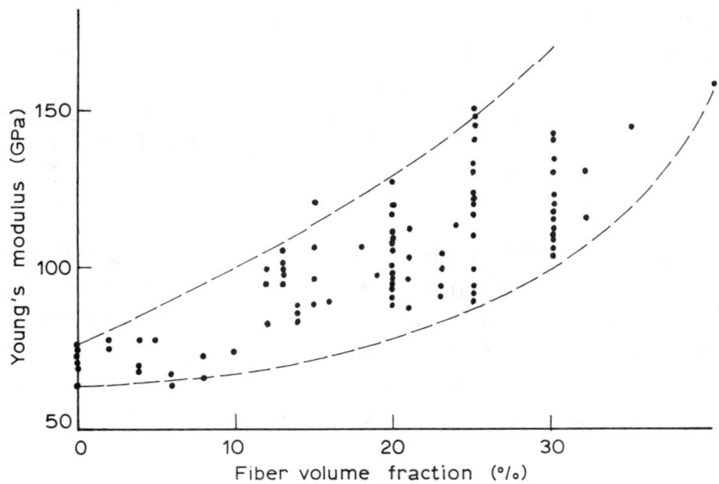

Fig. 18. Effect of reinforcement volume fraction on modulus of elasticity of discontinuous reinforced AMC.[67]

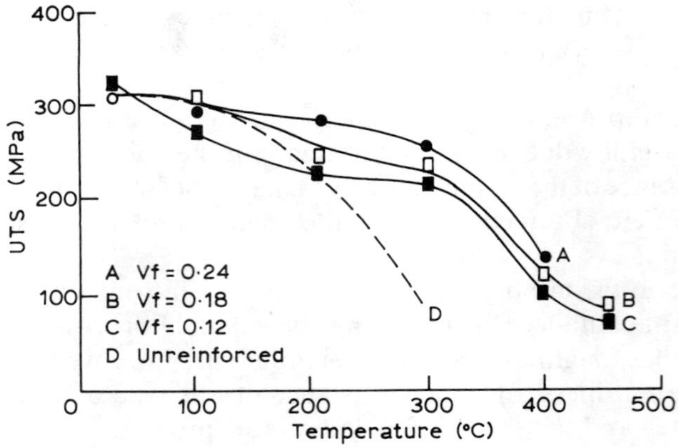

Fig. 19. Effect of temperature on the tensile strength of Al–9% Si–3% Cu based composites. After Ref. 41.

conditions must be reconsidered for AMC compared with the aluminium alloys.

— DRAMC also offer quite a noticeable improvement in the fatigue life time compared with that of the matrix. This effect is so attractive that the choice of this type of composite in the case of, for example, an automobile engine part, can be mainly justified by both fatigue resistance and low processing cost.

— Electrical and thermal conductivities or thermal expansion coefficients of aluminium alloys can be significantly reduced thanks to the ceramic component. For example, the thermal expansion coefficient of a 7075 aluminium alloy is halved with 30 vol.% of SiC whiskers and reduced to a third with 25 vol.% of Al_2O_3 fibres.

Despite the moderate performances due to the discontinuity in reinforcement, DRAMC have extended their application field by offering a possibility of mass production. For example, connecting rods have been forged from extruded bars, and antennae, space structures and tennis rackets have been made of thin walled tubing. Threaded bars and other parts can be processed by net shape casting.

3.1.2 SiC Whisker Reinforced Ceramic Composites

The mechanical performance of SiC (whisker) reinforced ceramic composites has not yet been extensively evaluated, since the processing methods are presently being investigated. However, the incorporation of SiC whiskers within alumina, silicon nitride or zirconia matrices has already shown the potential of such a reinforcement (V_f = 10–30%) for fracture strength and toughness.[68–71] For example, a toughness increase of 40% with a strength increase of 25% has been obtained by the addition of 30 vol.% of SiC whiskers in Si_3N_4, whilst the incorporation of 15 vol.% of SiC whiskers can double the toughness of alumina. The favourable effect on toughness of the whiskers due to crack deflection and bridging as well as a pull-out mechanism has been pointed out. The influence of matrix grain size and composite density on the mechanical performance remains crucial.

These types of ceramic–ceramic composites processed by combining wet techniques and sintering methods also include the reinforcement of silica base glass ceramic.

Whisker reinforced ceramic composites can be expected to be used as high temperature engine parts or screws for structural assemblies in aerospace materials.

3.2 Directional Reinforcement

It is in fact easier to classify composites according to the orientation of the reinforcement rather than by its length, since short fibres such as whiskers could be considered as a continuous reinforcement provided the whisker–matrix bonding is strong enough. Thus, this section deals with composites for which the orientation of the fibrous reinforcement is defined before processing.

3.2.1 Light Alloys Reinforced by CVD Filaments

Although cross-ply composites can be obtained by stacking unidirectionally reinforced foils in various directions and then diffusion bonding them, composites based on the reinforcement of titanium or aluminium alloy matrices by CVD boron or SiC filaments are generally used as unidirectional composites. Their high specific rigidity and strength maintained at high temperature (Fig. 20), as well as their excellent fatigue and creep resistance (Fig. 21), make these 1D metal matrix composites suitable for structural application in space or for fan blade fabrication. However, this type of composite has not been extensively developed due to high processing costs and the need to solve the compatibility problem occurring at the fibre–matrix interface when using composites at high temperatures. For example, the fibre–matrix

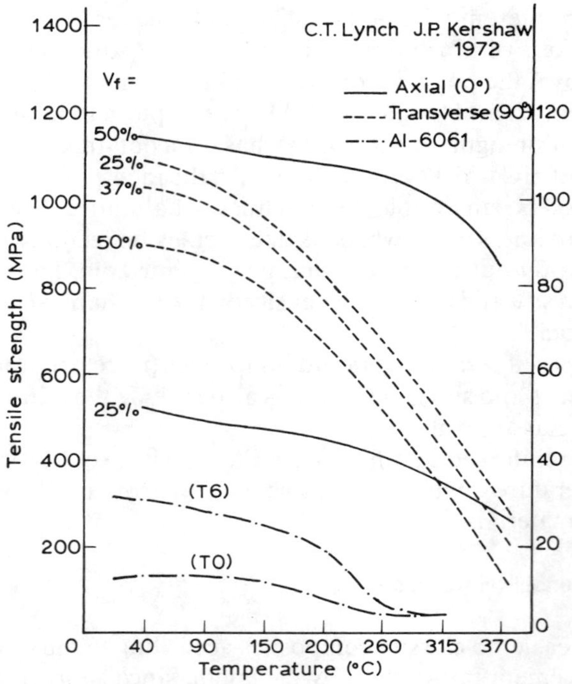

Fig. 20. Ultimate tensile strength at high temperature of 1D B/Al composites. After Ref. 72.

interaction in a 1D boron/titanium composite can be schematically represented as shown in Fig. 22. The thicker the interfacial zone, the stronger are the notch effects related to microcracks occurring in the brittle interphase, causing a significant decrease in the composite strength after thermal exposure.

Nevertheless the fibre–matrix interaction can be confined by (1) the addition to the matrix of alloying elements (Al, V in Ti) acting as inhibitors, (2) coating the filaments with a diffusion barrier (B_4C or SiC on boron filaments), (3) protecting the filaments against notch effects by a succession of thin layers impeding catastrophic crack propagation (alternate C and SiC coating on SiC filaments).

3.2.2 Metals Reinforced by C, SiC or Al_2O_3 Fibre Yarns

A decrease in performance related to poor thermo-chemical compatibility occurs when metal matrices such as Al, Mg, Pb, Cu and Zn are

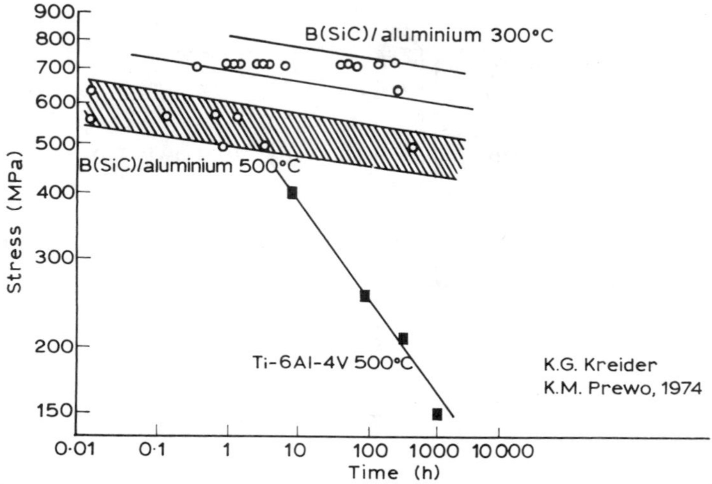

Fig. 21. Creep resistance of 1D B(SiC)/Al composites. After Ref. 73.

reinforced by carbon or SiC fibres (about 10 μm in diameter).[18] However, whilst the compatibility problem is very crucial during processing by the liquid route (particularly for Al and Mg matrices), the fibre–matrix interaction kinetics are rather slow at the possible temperatures of use. This is contrary to the case of Fe, Ni or Co based superalloys reinforced following the solid route for high temperature applications. For example, SiC fibres react with the superalloy components and carbon fibres often recrystallize at high temperatures (1000–1100 °C) in the presence of nickel.[74]

3.2.2.1 Light alloy matrix composites for structural applications. Fabricating aluminium or magnesium alloy matrix composites with a view of replacing OMC in structural applications requiring simultaneously high specific performance in the load direction, significant transverse characteristics, no material degradation by degassing as well as a conservation of mechanical performance over a wide range of temperatures, has led to composites with the following main characteristics.[20]

— As for the 1D CVD filament reinforced composites, the tensile behaviour is quasi-linear elastic up to fracture and is controlled by the fibres. A slight change in the slope of the stress–strain curves indicates matrix plastification. The evolution of the Young's

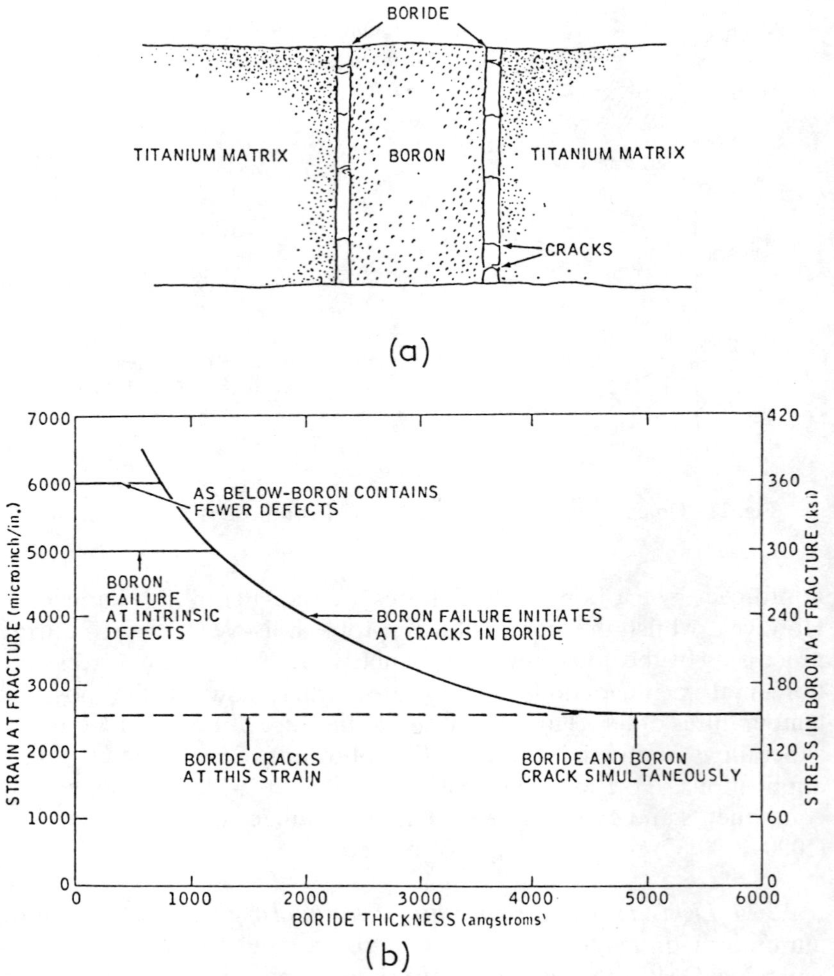

Fig. 22. Fibre–matrix chemical interaction in 1D boron/Ti composite. After Ref. 14. (a) Cracks in boride layer; (b) failure mechanisms.

modulus as well as fracture strain with fibre volume fraction approximately follows the rule of mixtures as illustrated in Fig. 23 and Table 5.

— In composites exhibiting strong fibre–matrix (FM) adhesion, the compressive strength is two to five times higher than tensile

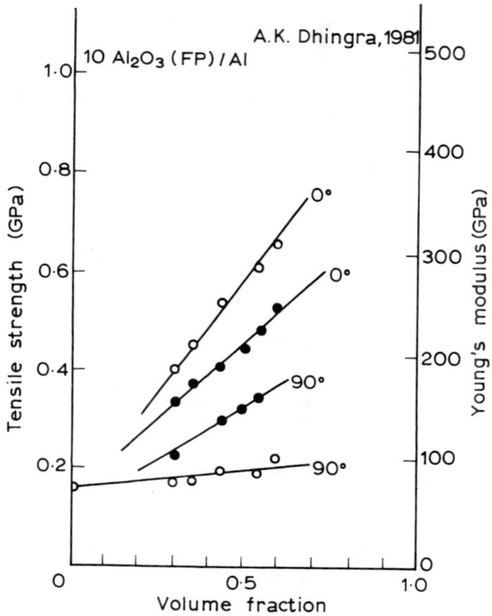

Fig. 23. Variation of the rigidity and strength in 1D Al_2O_3/Al composites versus the fibre volume fraction. After Ref. 75.

strength. In contrast, with weak FM bonding, the performance in compression is equal to that shown in tension.

— The transverse properties are mainly controlled by the matrix and the FM bonding for which an excessive interaction tends to restrict the tensile longitudinal strength by giving the composite a sensitivity to microcracks occurring in the FM interfacial zone. In the case of C/Al or C/Mg composites, achieving acceptable transverse characteristics requires such a strong FM interaction that the major part of the fibre reinforcing effect will be lost in the fibre direction.

— At high temperatures, up to about 300–400 °C, the composite properties (rupture strength and creep resistance) are controlled by the matrix, rather than by the fibre nature, particularly under transverse tension and in longitudinal compression. The improvement of characteristics with respect to the matrix is obviously very dependent on the fibre volume fraction, as illustrated in Fig. 24.

TABLE 5
Properties[a] of Unidirectional Light Alloy Matrix Composites[20]
Fibre treatment: CVD–TiB (1); H_2 cleaning + TiB–CVD (2); C deposit + TiB–CVD (3); Na treatment (4).

Composite[b]	E_L *(GPa)*	UTS_L *(MPa)*	ITS_T *(MPa)*	$\varepsilon^R{}_L$ *(%)*	UCS_L *(MPa)*	α_L *($10^{-6}K^{-1}$)*
50% Al_2O_3 (FP)/Al	210	570	190	0·30	2 900	7·2
50% Al_2O_3 (FP)/Al	207	530	207	0·26	1 760	—
50% Al_2O_3 (S)/Al	150	(860)[c]	98	0·57	1 430	7·6
50% Al_2O_3 (S)/Al 201	150	590	230	0·39	2 200	—
30–35% SiC (N)/Al 1100	90	800 [c] 850	70 15	—	—	3·5
25–38% SiC (N) Al 6061	100 110	680 780	140 60	—	—	5·8
40% C (HT 7)/Al 5056	120	883 600	—	—	800 1 080	—
30% C (T 50)/Al 201	150 160	570 690	31	—	530	—

30% C (GY 70)/Al 201	237	638	43	—	—	5·6
50% C (P 100)/Al	366	635	45	—	—	0·27
50% C (P 100)/Mg	352	635	35	—	—	0·05

Composite	*FM treatment*	*UTS_L (MPa)*	*UTS_T (MPa)*	*FM bonding*	*% RM (UTS_L)*
33% C (T 50)/Al-201	1	743	24	Weak	0·84
34% C (T 50)/Al-201	2	763	21	Moderate	0·82
37% C (T 50)/Al-201	3	920	19	Strong	0·94
21% C (T 50)/Al-201	4	407	43	Very weak	0·66

[a] E, Young's modulus; UTS, ultimate tensile strength; ε^R, rupture strain; UCS, ultimate compression strength; α, expansion coefficient. Subscripts: L, longitudinal; T, transversal.
[b] N, Nicalon; FP, Dupont de Nemours; S, Saffil—ICI.
[c] Given by the rule of mixtures (RM).

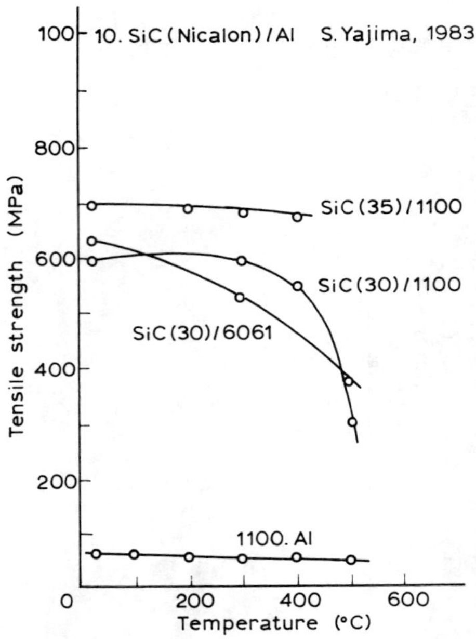

Fig. 24. High temperature strength of 1D SiC(N)/Al composites. After Ref. 52.

— As for many other types of composites, the mechanical fatigue resistance is one of the qualities most significantly improved by the fibrous reinforcement. At a stress level in the range of 50–70% of the static strength scatter band, unbroken specimens have been obtained at 10^7 cycles. The stress-cycle number curves (S–N) are less steep than those for the unreinforced material.

— Aluminium matrix composites, on the other hand, for which the difference in fibre and matrix expansion coefficients is significant are sensitive to thermal fatigue, particularly when the FM interface is relatively weak. This explains the rather good thermal fatigue resistance of Al_2O_3 (FP)/Al composites ($V_f = 50\%$) compared with carbon (T 50)/Al composites ($V_f = 70\%$).

3.2.2.2 Composites exhibiting controlled thermal expansion. Composites can offer an answer for applications requiring the control of thermal expansion. Indeed, the thermal expansion coefficient of composites can be derived from those of the components by using the rule of mixtures.

Thus, coupling two materials such as metals with carbon fibres whose longitudinal thermal expansion coefficient is negative from ambient up to about 300 °C, allows the adjustment of the composite thermal expansion by controlling the fibre volume fraction.[76] Two examples can be given to illustrate this aspect of composites:

- The reinforcement of magnesium by unidirectional high modulus carbon fibres gives rise to structures for space applications which combine both high specific rigidity and dimensional stability.
- The incorporation of ultra-high modulus (UHM) carbon fibres in copper alloys results in composites which maintain an excellent electrical and thermal conductivity while offering a suitable thermal expansion coefficient to be compatible with other materials. As illustrated in Fig. 25, this is also the case when a 2D C/Cu composite is associated with silicon, and in a glass fibre reinforced composite for power diodes.

Furthermore, the use of ceramic fibres exhibiting higher thermal expansion coefficients than those of ceramic matrices leads to build-up of a favourable compression stress within the matrix after composite processing when the fibre–matrix bonding remains sufficiently good to allow load transfer.

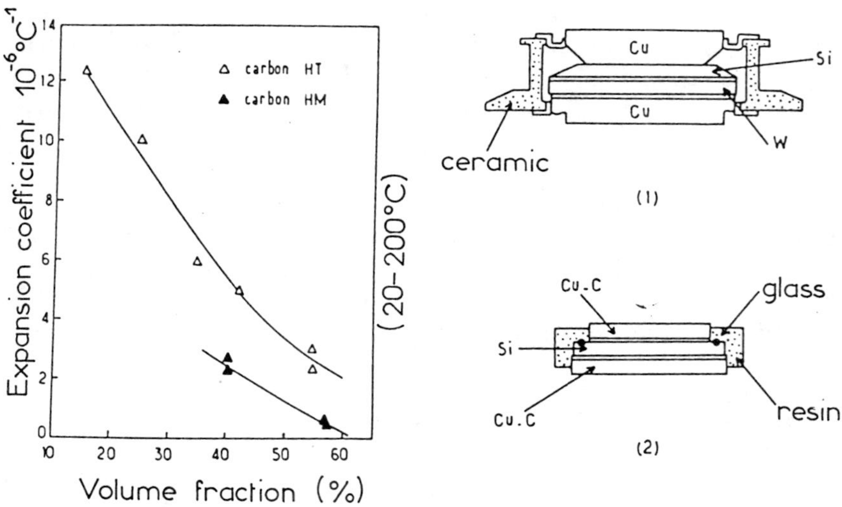

Fig. 25. Dimensional stability of C/Cu composites. After Ref. 77. (1) Diode with a W substrate. (2) Diode with a 2D-C/Cu substrate.

3.2.2.3 Reinforcement of heavy metals. In addition to the reinforcement of copper which also can be used for structural purposes (conductors with high specific resistance), various high density metals can be reinforced for the following applications.

(1) The reinforcement of iron, nickel or cobalt based alloys has been mainly expected for turbine blades, in order to improve motor efficiency by increasing the running temperatures. Due to the high thermo-mechanical performance of tungsten, MMC have been processed with tungsten filaments (100–500 μm diameter; $T_m \approx 3400\,°C$; $E = 400$ GPa; $\sigma^R = 2500$ MPa) by the solid route. However, the difficulties related to filament recrystallization and chemical interaction with matrices have complicated the optimization of those materials. As regards the in-situ composites obtained by directional solidification and reinforced by TaC or NbC whiskers, their cost has remained excessive despite their quite promising characteristics at high temperature. Carbon, SiC and Al_2O_3 fibres have also been tested as reinforcements; however, the high temperature treatments undergone by these types of composites increase the chemical compatibility problem.[78, 79]

(2) The reinforcement of lead [by carbon or oxide fibres (Al_2O_3, SiO_2)], whose mechanical resistance and rigidity are very low, has been successfully achieved by various liquid route techniques (impregnation, electrodeposition). The amazing improvement of the mechanical characteristics (rigidity and strength multiplied by 10–20 with 25 vol.% of Al_2O_3 fibres) should lead to extensive use of these Pb matrix composites in batteries.[80, 81]

3.2.3 Carbon–Carbon Materials

Initially, carbon–carbon materials were developed for manufacturing heat shields as well as nose tips and nozzle throats of rockets. For these kinds of applications the interest in carbon–carbon composites is related to the following main characteristics:

— high ablation heat;
— good thermal impact resistance;
— high specific resistance over a wide range of temperatures;
— low expansion coefficient.

In addition to the above qualities carbon–carbon composites also exhibit favourable friction behaviour, giving rise to an extending application field as brakes. Furthermore, the excellent biocompatibility of carbon allows the use of such composites as biomaterials particularly in orthopaedic surgery.[48]

3.2.3.1 Mechanical properties. The mechanical performance of carbon–carbon materials depends on the fibre nature and texture and particularly on the processing procedures (number of cycles of resin or pitch impregnation and pyrolysis, temperatures, pressure, or CVI conditions) which lead to various rates of densification (density in the range of 1·50–2·05 g/cm^3). In general, higher density results in higher composite strength and rigidity (in the range 10–100 GPa), whilst the specific heat seems to be independent of density. Some examples reported in Table 6 show the significance of the composite anisotropy and particularly the deviation between the properties along two different orthotropic directions in a 2D fibrous structure. It is noteworthy that the distinction which has to be made between the orthotropic directions of a 2D composite also concerns the thermal conductivity and expansion coefficients. However, these physical properties, which change with temperature as illustrated in Fig. 26, can be quite similar for each orthotropic direction of equilibrated 3D composites.

3.2.3.2 Ablation and oxidation resistance. The ablative resistance of a material is tested either with a plasma jet producing thermal flux of about 20×10^6 W/m^2 or on real parts of rockets. The ablative protection against thermal flux is due to radiation, heat absorption, heat consumption by chemical reaction or physical transformation as well as ejection or transpiration of gas from the material surface which gives rise to insulating gaseous layers. The ablation rate decreases with increasing material density and the protection is more effective in non-oxidizing environments. As a matter of fact the major disadvantage of carbon matrix composites is their poor oxidation resistance from 450 to 500 °C, while their mechanical characteristics tend to increase up to 2500 °C. Nevertheless, such a handicap can be partly overcome by treating the surface or the bulk matrix with a coating based on the presence of silicon (SiC CVI, for example).

In combining high specific heat, a good friction coefficient (≈0·25) and high temperature mechanical performance, carbon–carbon composites are superior to classical materials used for brakes, giving better safety with less weight.

TABLE 6
2D C/C (Sepcarb) Properties at Room Temperature[48]

		Sepcarb type			
	Direction	*3 Liquid + CVI*	*4 CVI*	*4 CVI*	*5 4D C–C (pitch)*
Density (g/cm^3)		1·45–1·55	1·40–1·73	1·7–1·9	2
Compression strength (MPa)	1	70–80	50–160	200–270	80
	3	105–115	90–135	260–300	50
Compression modulus (GPa)	1	12·1–14·9	12–21·8	47–49	65
	3	2·7–3	5–6·1		8·5
Tensile strength (MPa)	1	45–60	35–70	160–180	100
Tensile modulus (GPa)	1	10·4–12·8	12–22·7	45–72	62
Bending strength (MPa)	1		70–110	260–360	
	3			360–535	
Shear strength (MPa)	1	7–11	13–26	5–7	
Expansion up to 2000°C (%)	1	0·18–0·20	0·2–0·3		0·6–1·4
	3	0·55	0·60–0·85		0·5–4
Thermal diffusivity (10^{-6} m^2/s)	1	65–90	150		
	3	18–26	40		

Direction 1: 2D: fibre direction in the fabric layers: 4D: [111].
Direction 3: 2D: orthogonal to the fabric layers: 4D: [100].

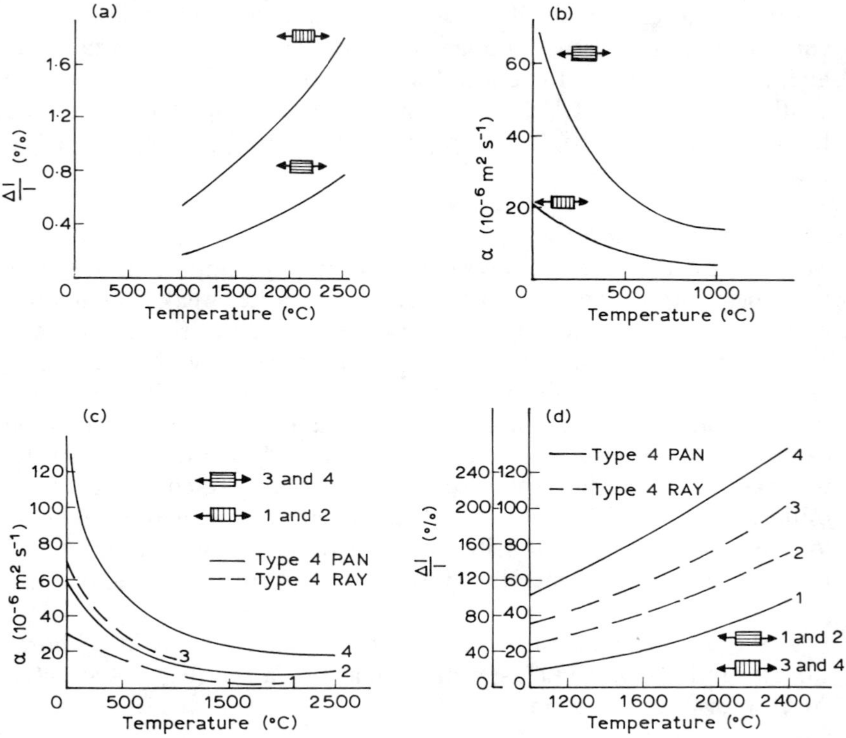

Fig. 26. Variation of 2D C/C properties vs temperature. After Ref. 48. (a) Thermal expansion (Sepcarb 3); (b) thermal diffusivity (Sepcarb 3); (c) thermal diffusivity (Sepcarb 4); (d) thermal expansion (Sepcarb 4).

3.2.3.3 Biocompatibility. In a similar manner to carbon, which has during the last two decades shown its thrombo-resistance (no fibrous outgrowth in contact with blood flow), carbon–carbon composites exhibit quite interesting biocompatible properties. The following qualities also enable various biomedical applications to be considered:

— chemical purity,
— adaptation of the rigidity to the application by adjusting the nature and volume fraction of components,
— high specific yield and failure strength,
— good fatigue resistance,
— low brittleness compared with ceramics.

Furthermore, by controlling the residual porosity particularly on the composite surface, tissues or bone may penetrate implants or prostheses giving rise to better adherence. Also, the porosity makes any surface coating by ceramic CVD easier, giving (for example) the expected friction and wear resistance properties for the upper part of a hip prosthesis.

3.2.4 Fibre Reinforced Ceramics

The main purpose of reinforcing ceramics by continuous fibres is toughness improvement. As a matter of fact, rather weak fibre–matrix bonding often enhances the impeding effect of the fibres on crack propagation so that long fibres are required to allow matrix micro-cracking to occur, while preventing any catastrophic fracture. This means that toughness improvement of a ceramic matrix can be expected only when the fibre fracture strain is larger than that of the matrix ($\varepsilon_f^R > \varepsilon_m^R$). In the opposite case ($\varepsilon_f^R < \varepsilon_m^R$), early fracture of the fibre initiates matrix failure instead of delaying crack propagation. In order to illustrate these types of brittle–brittle composites, which sometimes reach the toughness of metallic alloys, two cases are considered below.

3.2.4.1 Silica based matrix composites. The reinforcement of boro-silicate and silica glass as well as glass–ceramic (e.g. LAS: SiO_2, Al_2O_3, Li_2O) by carbon or SiC fibres has been investigated for more than a decade.[46,47,83,84] As shown in Fig. 27, the reinforcing effect is more pronounced as it concerns simultaneously the material rigidity, the failure strength, the failure strain as well as the dependence of mechanical performance on temperature. Such increases in characteristics, particularly in toughness, have been explained by the weakness of the fibre–matrix interface and the confinement of the ceramic matrix in very small interfibre spaces. Performance improves with increasing fibre volume fraction and smaller fibre diameter. Therefore the control of fibre–matrix bonding is a major aspect of this type of composite. When the fibre expansion coefficient is much larger than that of the matrix, the interface weakening is such that the transverse properties of unidirectional composites are rather poor. In contrast the use of oxide fibrous reinforcement (e.g. Al_2O_3) leads to strong FM bonding and lower toughness. Recently, the FM interface analysis by Auger electron microprobe in a SiC (Nicalon)/LAS composite has shown the existence of a thin carbon layer (about 100 Å thick) spontaneously built up at the FM interface, depending on the

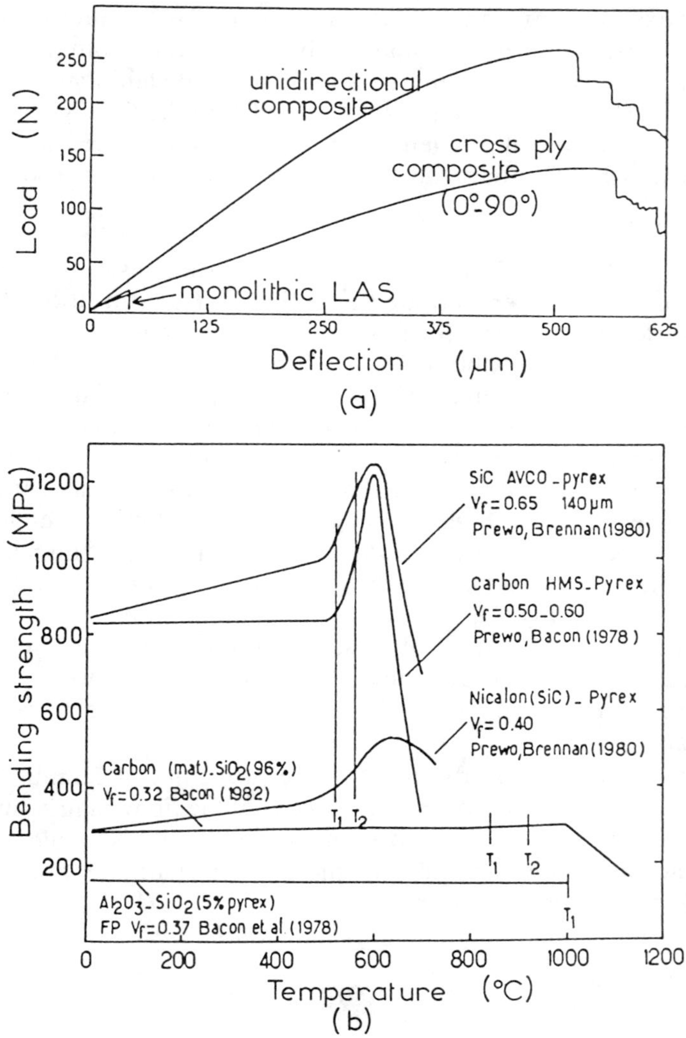

Fig. 27. Properties of unidirectional glass matrix composites. (a) Bending behaviour (after Ref. 85); (b) variation of strength versus temperature reported in Ref. 49.

processing conditions. This carbon layer has been related to the increase in toughness and the fibre pull-out mechanism.

The oxidation resistance of the matrix, and the possibility of reducing the expansion of the composite by a suitable combination of components, give the composite a great thermal stability over a wide range of temperature (up to 1000 °C). However, the fibre sensitivity to oxidation is a crucial problem which remains to be solved when these composites have to undergo long thermal exposure in an oxidizing environment.

3.2.4.2 2D SiC/SiC. The stacking of two-dimensional (2D) SiC (Nicalon) fabric layers followed by a SiC CVI densification leads to composites whose mechanical behaviour is apparently very similar to that of metal alloys, exhibiting strain hardening effects as illustrated in Fig. 28. In fact this behaviour is better related to damageable linear elastic materials, but these kinds of composites combine such high safety (similar to ductility in metals) and high temperature resistance that they are expected to be used as ceramic sheets for space shuttles or as turbine blades in high rate engines. Indeed, the very high crack growth resistance of these composites is accompanied by high thermomechanical shock and fatigue resistance with anisotropic conductivity as illustrated in Fig. 27.

It is noteworthy that all of these favourable properties are mainly dependent on the fibrous texture, on the difference between the fibre and matrix characteristics and in a large part on the fibre–matrix interface.

In summary, an improvement in the properties of ceramic matrix composites will result from acting directly at the fibre–matrix interface where, for example, the interposition of a thin layer of suitable material can increase quite significantly the efficiency of the fibre–matrix interface as a microcrack arrester.

4 CONCLUSIONS

Many processing methods are now available to manufacture inorganic matrix composites. However, only a few of them are able to satisfy the main requirements for a wide industrial development: reliability, suitability for mass production and moderate cost.

Carbon–carbon composites initially designed for rockets have already been industrialized for more than two decades, using both

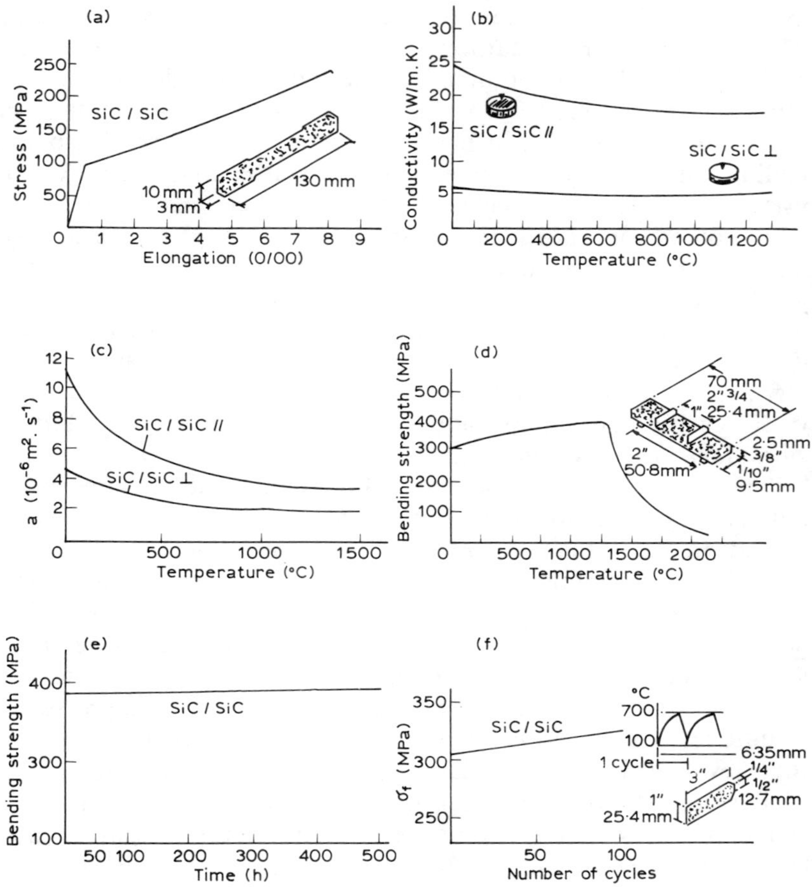

Fig. 28. Properties of 2D SiC/SiC composites. After Ref. 82. (a) Typical tensile curve; (b) thermal conductivity; (c) thermal diffusivity; (d) temperature dependence of bending strength; (e) bending strength after exposure to 1100 °C in air; (f) bending strength after thermal cycling in air.

liquid impregnation and vapour infiltration. They are now involved in various fields of applications based on their friction properties or biocompatibility. More recently, the development of ceramic–ceramic composites has not yet been so extensive at the industrial level. However, the possibility of adapting many of the processing techniques already known by ceramicists and the increasing interest expressed by aerospace and automotive industries in these composites creates much optimism. Looking to the future, much attention will need to be paid to hybrid ceramic–carbon–carbon composites already tested for space shuttles and likely to be required for nose and leading edge structures of trans-atmospheric craft.[86] They combine high strength, rigidity and stability of carbon fibres at high temperature, the energy dissipating qualities of the slightly porous carbon matrix and the oxidation resistance of coated, impregnated or infiltrated ceramics.

In another range of use temperatures, metal matrix composites are becoming increasingly important. The processing techniques based on powder metallurgy have opened up a broad domain of applications for plastic workable composites reinforced by a moderate volume fraction of short fibres. Also, the requirement of mass production has drawn the attention of automotive industries towards squeeze casting techniques which enable the impregnation of pre-forms made of short or continuous fibres by liquid light alloys.

Thus, the industrial development of inorganic matrix composites is undeniably progressing. However, much work still has to be done to take advantage of the great adaptability of these composites and the possibility of designing new composites thanks to the multiple manner of components coupling. For this purpose, a better understanding of the mechanisms which control the composite behaviour is required, particularly at the fibre–matrix interface. Equally important is the need for a precise correlation between composite properties and processing conditions. This supposes that the main characteristics of the composite can be evaluated, when unfortunately many testing procedures are not suitable for heterogeneous anisotropic materials. For instance, the choice of a method for toughness determination needs a clear description of the mechanical behaviour before any investigation can be carried out.

In fact, a new approach towards materials has to be adopted, by considering as predominant factors all of the compatibility problems due to the mismatch between the properties of the composite components: failure strain, Poisson's ratio, expansion coefficient, chemical nature as well as evolution in time.

ACKNOWLEDGEMENTS

The author has extracted information extensively from Volume 2 of *Introduction aux Matériaux Composites* edited by R. Naslain (CNRS–

IMC). Also, the author wishes to acknowledge the assistance of T. Stephenson and M. Georgy in the preparation of the manuscript.

REFERENCES

1. Naslain, R., Renforts fibreux dans les matériaux composites à matrice métallique ou céramique. In *Introduction aux Matériaux Composites*, Vol. 2, *Matrices Métalliques et Céramiques*. CNRS–Institut des Matériaux Composites, 1985, Chapter XIII, pp. 213–62.
2. Girot, F., Quenisset, J. M. & Naslain, R., Discontinuous reinforced aluminum matrix composites. *Compos. Sci. Technol.*, (1987).
3. Ueno, K. & Toibona, Y., *Yogyo-Kyokai-Shi,* **91**(11) (1983) 491–7.
4. Becher, P. F. & Wei, G. C., *J. Amer. Ceram. Soc.*, **67**(12) (1984) C 267–9.
5. Guo, J. K., Mao, Z. Q., Bao, C. D., Wang, R. H. & Yan, D. S., *J. Mater. Sci.*, **17** (1982) 3611–16.
6. Jamet, J. F., Abbe, D. & Guyot, M. H., *Proc. ICCM-5, San Diego, 1985*, ed. W. C. Harrigan *et al.* TMS–AIME, Warrendale, PA, pp. 569–685.
7. Tressler, R. E., Messing, G. L., Pantano, C. G. & Newnham, R. E., eds, Tailoring multiphase and composite ceramics. In *Materials Science Research*, Vol. 20, Plenum Press, New York and London, 1986, pp. 529–696.
8. Girot, F., Albingre, L., Quenisset, J. M. & Naslain, R., Rheocasting of aluminum matrix composites reinforced by SiC short fibers (or whiskers), *J. Metal.*, Nov. (1987) 18–21.
9. Girot, F., Quenisset, J. M., Naslain, R., Coutand, B. & Macke, T., Mechanical behavior of aluminum matrix composites reinforced by short fibers and processed by compocasting. In *Proc. ICCM VI–ECCM 2, London 1987*, ed. F. L. Mathews *et al.* Elsevier Applied Science, London, 1987, pp. 330–9.
10. Kohara, S. & Muto, N., *J. Japan Inst. Met.*, **45** (1981) 411–16.
11. Christian, J. L., In *Proc. 1st Int. Conf. Composite Materials, Boston, 1975*, ed. E. Scala *et al.* TMS–AIME, 1975, pp. 706–36.
12. Yoshida, M., Ikegami, S., Ohsaki, T. & Ohkita, T., In *Science of Advanced Materials and Process Engineering Series,* Vol. 24, Book 2, *Nat. SAMPE Symp.*, 1979, pp. 1417–26.
13. Pailler, R., Martineau, P., Lahaye, M. & Naslain, R., *Rev. Chim. Minérale,* **18** (1981) 520–43.
14. Metcalfe, A. G., In *Metallic matrix composites*, ed. K. G. Kreider. (*Composite Materials*, ed. L. J. Broutman & R. H. Krack, Vol. 4). Academic Press, New York, 1974, pp. 269–318.
15. Hearn, D., Favry, Y. & Bunsell, A. R., *Proc. JNC-1, Structures et Matériaux Composites*, Vol. 1, ed. C. Bathias *et al.* AMAC, Paris, 1978, pp. 105–10.
16. Ferte, J. P., *L'Aéronautique et l'Astronautique,* **67** (1977) 27–37.
17. Gigerenzer, A., Pepper, R. T. & Lachman, N. L., *Proc. ICCM-2, Toronto, 1978*, ed. B. Noton *et al.* TMS–AIME, 1978, pp. 175–88.
18. Kendall, E. G., In *Metallic Matrix Composites* (Ref. 14), pp. 319–97.
19. Warren, R. & Anderson, C. H., *Composites,* **15**(2) (1984) 101–11.

20. Naslain, R., Pailler, R. & Quenisset, J. M., Matériaux composites synthétiques à matrice métallique dans les matériaux composites à matrice métallique ou céramique. In *Introduction aux Matériaux Composites*, Vol. 2, *Matrices Métalliques et Céramiques*. CNRS — Institut des Matériaux Composites, 1985, Chapter XV, pp. 315–83.
21. Rocher, J. P., Girot, F., Quenisset, J. M., Pailler, R. & Naslain, R., Mémoires et études scientifiques. *Revue de Métallurgie* (Feb. 1986) 69–85.
22. Eustathopoulos, N., Joud, J. C., Desre, P. & Hicter, J. M., *J. Mater. Sci.*, **9** (1974) 1233–42.
23. Coudurier, L., Adorian, J., Pique, D. & Eustathopoulos, N., *Rev. Int. des Hautes Températures et Réfractaires*, **21** (1984) 81–93.
24. Mannig, C. R. & Gurganus, T. B., *J. Amer. Ceram. Soc.*, **52** (1969) 115–18.
25. Kendall, E. G. & Pepper, R. T., US Patent 4 082 864 (4 April 1978).
26. Lignon, R., French Patent 1 183 371 (1972).
27. Goddard, D. M. & Sexton, R. N., French Patent 2 494 260 (1981).
28. Amateau, M. F., *J. Compos. Mater.*, **10** (1976) 279–96.
29. Oberlin, M., French Patent 1 259 916 (August 1974).
30. Levitt, A. P. & Band, H. E., US Patent 4 157 409 (5 June 1979).
31. Katzman, H. A., US Patent 4376803 (March 1983).
32. Rocher, J. P., Quenisset, J. M. & Naslain, R., *J. Mater. Sci. Lett.*, **4** (1985) 1527.
33. Rocher, J. P., Quenisset, J. M., Pailler, R. & Naslain, R., Int. Patent W083/03625 (27 October 1983).
34. Abe, Y., Horikiri, S., Fujimura, K. & Ichiki, E., In *Proc. 4th Int. Conf. Composite Materials*, ed. T. Higashi *et al.* Japan Society for Composite Materials, Tokyo, 1982, pp. 1427–34.
35. Fukunaga, H. & Onde, T., *ibid.*, pp. 1435–42.
36. Fukunaga, H., Komatsu, S. & Kanoh, Y., *Bull. JSME*, **220** (1983) 1814–19.
37. Gelderloos, D. G. & Karasek, K. R., *J. Mater. Sci. Lett.*, **3** (1984) 232–8.
38. Fukunaga, H. & Goda, K., *Bull. JSME*, **27** (1984) 1245–50.
39. Dinwoodie, J., Moore, E. & Langman, C., In *Developments in the Science and Technology of Composite Materials, ECCM-1, Bordeaux, 1985*, ed. A. R. Bunsell, P. Lamicq & A. Massiah, pp. 623–7.
40. Clyne, T. W., Bader, M. G., Cappleman, G. R. & Hubert, P. A., *J. Mater. Sci.*, **20** (1985) 85–96.
41. Dinwoodie, J., Moore, E., Langman, C. & Symes, W. R., *Proc. ICCM-5, San Diego, 1985*, ed. W. C. Harrigan *et al.* TMS-AIME, Warrendale, PA, pp. 671–85.
42. Clyne, T. W. & Bader, M. G., *ibid.*, pp. 755–71.
43. Ackermann, L., Charbonnier, J., Desplanches, G. & Kolowski, H., *ibid.*, pp. 687–98.
44. Phillips, D. C., *J. Mater. Sci.*, **9** (1974) 1847–54.
45. Davidge, R. W., *Ceramurgia*, **1** (1977) 5–12.
46. Prewo, K. W. & Bacon, J. F., In *Proc. ICCM-2, Toronto, 1978*, ed. B. Noton *et al.* TMS-AIME, New York, 1978, pp. 64, 753.
47. Prewo, K. M., In *21st Annual Univ. Conf. Ceramic Soc. Tailoring Multiphase and Composite Ceramics, Pennsylvania State Univ., July 1985* (*Materials Science Research*, Vol. 20), ed. R. E. Tressler *et al.* Plenum Press, New York and London, pp. 529–48.
48. Broquere, B., Buttazzoni, B. & Choury, J. J., Les composites carbone-

carbone, leurs applications industrielles. In *Introduction aux Matériaux Composites*, Vol. 2. *Matrices Métalliques et Céramiques.* CNRS–IMC, 1985, Chapter XVII, pp. 405–38.
49. Naslain, R., Rossignol, J. Y., Quenisset, J. M. & Langlais, F., *ibid.*, Chapter XVIII, pp. 439–91.
50. Fitzer, E. & Schlichting, J., *High Temp. Sci.,* **13** (1980) 149–72.
51. Rice, R. W., *Ceramic Bull.,* **62**(8) (1983) 889–92.
52. Yajima, S., *Ceramic Bull.,* **62**(8) (1983) 893–8.
53. West, R., David, L. D., Djurovich, P. I., Yu, H. & Sinclair, R., *Ceramic Bull.,* **62**(8) (1983) 899–903.
54. Wills, R. R., Markle, R. A. & Mukherjee, S. P., *Ceramic Bull.,* **62**(8) (1983) 904–11.
55. Schilling, C. L., Wesson, J. P. & Williams, T. C., *Ceramic Bull.,* **62**(8) (1983) 912–15.
56. Walker, B. E., Rice, R. W., Becher, P. F., Bender, B. A. & Coblenz, W. S., *Ceramic Bull.,* **62**(8) (1983) 916–23.
57. Hasegawa, Y. & Okamura, K., *J. Mater. Sci.,* **18** (1983) 3633–48.
58. Jamet, J., Spann, J. R., Rice, R. W., Lewis, D. & Coblenz, W. S., In *Proc. 8th Annual Conf. Composites and Advanced Ceramic Materials, Cocoa Beach, Jan. 1984*, pp. 677–94.
59. Naslain, R. & Langlais, F., *Proc. 21st Annual Univ. Conf. Ceramic Soc., Tailoring Multiphase and Composite Ceramics, Pennsylvania State Univ., July 1985.* (Materials Science Research, Vol. 20), ed. R. E. Tressler *et al.* Plenum Press, New York and London, pp. 145–64.
60. Hillig, W. B., Mehan, R. L. & Morclock, C. R., *Ceramic Eng. Sci. Proc.,* **1** (1980) 405–12.
61. Walter, J. L., Gigliotti, M. F., Oliver, B. F. & Bibring, H., *Proc. Int. Conf., In Situ Composites III, Boston, Dec. 1978.* Ginn Custom Publish, Lexington, 1979.
62. Thompson, E. R. & Lemkey, F. D., In *Metallic Matrix Composites*, ed. K. G. Kreider. (Composite Materials, ed. L. J. Broutman & R. H. Krack, Vol. 4). Academic Press, New York, 1974, pp. 101–57.
63. Stohr, J. F. & Khan, T., In *Introduction aux Matériaux Composites*, Vol. 2. *Matrices Métalliques et Céramiques.* CNRS–Institut des Matériaux Composites, 1985, Chapter XIX, pp. 263–312.
64. Lundberg, R., Kahlman, L., Pompe, R., Carlson, R. & Warren, R., *Ceramic Bull.,* **66**(2) (1987) 330–2.
65. Rocher, J. P., Macke, T., Quenisset, J. M., Naslain, R. & Cotteret, J., *Comptes Rendus des Cinquièmes Journées Nationales sur les Composites, Paris, Sept. 1986,* ed. C. Bathias & D. Menkès. Editions Pluralis, pp. 343–60.
66. Chou, T. W. & Kelly, A., *Ann. Rev. Mater. Sci.,* **10** (1980) 229–59.
67. Girot, F., Quenisset, J. M. & Naslain, R., *Composite Sci. Technol.,* **30**(3) (1987) 155–84.
68. Hayami, R., Ueno, K., Kondou, I., Tamari, N. & Toibona, Y., *Proc. 21st Annual Univ. Conf. Ceram. Soc., Tailoring Multiphase and Composite Ceramics, Pennsylvania State, Univ., July 1985.* (Materials Science Research, Vol. 20), ed. R. E. Tressler *et al.* Plenum Press, New York and London, pp. 663–74.
69. Bhatt, R. T., *ibid.*, pp. 675–95.

70. Tiegs, T. N., & Becher, P. F., *Amer. Ceram. Soc. Bull.*, **66**(2) (1987) 339–42.
71. Buljan, S. T., Baldoni, J. G. & Hucakbee, M. L., *ibid.*, 347–52.
72. Lynch, G. T. & Kershaw, J. P., *Metal Matrix Composites*. CRC Press, Cleveland, OH, 1972.
73. Kreider, K. G. & Prewo, K. M., *Metallic Matrix Composites*, ed. K. G. Kreider. (*Composite Materials*, ed. L. J. Broutman & R. H. Krack, Vol. 4). Academic Press, New York, 1974, pp. 399–471.
74. Warren, R., Anderson, C. H. & Carlsson, M., *J. Mater. Sci.*, **13** (1978) 178–88.
75. Dhingra, A. K., *Chemtech* (1981) 600–8.
76. Wolff, E. G., Min, B. K. & Kural, M. H., *J. Mater. Sci.*, **20** (1985) 1141–9.
77. Kuniya, K., Arkawa, H., Kanai, T. & Yasuda, T., *Trans. Components Hybrids and Manufacturing Technology, CHMT-6,* **4** (1983) 467–2.
78. Warren, R., Anderson, C. H. & Larson, L. O.. *Composites* (Apr. 1979) 121–7.
79. Signorelli, R. A., *Metallic Matrix Composites*, ed. K. G. Kreider. (*Composite Materials*, ed. L. J. Broutman & R. H. Krack, Vol. 4). Academic Press, New York, 1974, pp. 229–67.
80. Dacres, C. M., Reamer, S. M., Sutula, R. A. & Angres, I. A., *J. Electrochem. Soc.,* **128**(10) (1981) 2060–4.
81. Hartmann, H. S. & Sutula, R. A., *J. Electrochem. Soc.*, **129**(8) (1982) 1749–52.
82. Lamicq, P. J., Bernhart, G. A., Dauchier, M. M. & Mace, J. G., *Amer. Ceram. Soc. Bull.,* **65**(2) (1986) 336–8.
83. Phillips, D. C., *J. Mater. Sci.,* **9** (1974) 1847–54.
84. Brennan, J. J., *21st Annual Univ. Conf. Ceram. Soc., Tailoring Multiphase and Composite Ceramics, Pennsylvania State, Univ. July 1985*. (*Materials Science Research*, Vol. 20), ed. R. E. Tressler *et al.* Plenum Press, New York and London, pp. 549–60.
85. Brennan, J. J. & Prewo, K. M., *J. Mater. Sci.,* **17** (1982) 2371–83.
86. Naslain, R., Quenisset, J. M., Rossignol, J. Y., Hannache, H., Lamicq, P., Choury, J. J., Heraud, L. & Christin, F., *Proc. ICCM–V, San Diego, 1985*, ed. W. C. Harrigan *et al.* TMS–AIME, Warrendale, PA, pp. 499–514.

Chapter 7

Advanced Composite Fabrication

Dave Woolstencroft

Composite Manufacturing Development, British Aerospace plc, Blackburn, Lancashire, UK

1 INTRODUCTION

The large growth of advanced composites can be attributed to their properties which have given performance and cost payback in their use in aerospace and other high added-value applications. Advanced composites have been accepted as structurally effective and efficient.

Growth is predicted to continue with the end uses for advanced composites expanding from aerospace to more commercial areas. As the use has increased, so the challenge facing the industry has changed and new pressure is on the manufacturing engineer to develop fabrication techniques which reduce cost, maintain or improve quality and reduce lead times.

To satisfy this demand many different automated schemes have been proposed. This chapter considers some of the successful techniques currently available in the manufacture of high performance composites and some of the new concepts which offer improvements for the future.

The selection of manufacturing methods depends critically upon the processability of the matrix material and the techniques currently utilised are aligned to thermoset systems as these dominate the aerospace market. However, much work is taking place on advanced thermoplastics and in time a whole new range of fabrication techniques will be conceived to capitalise on the much reduced make-spans possible with these materials.

2 CURRENT THERMOSET FABRICATION TECHNOLOGY

The main material formats used currently for continuous fibre pre-impregnated materials are:

(1) Unidirectional aligned pre-impregnated tape — this can be in varying widths up to 1·5 m and in nominal thickness up to 0·25 mm.
(2) Woven pre-impregnated broadgoods — this again is in widths of up to 1·5 m with varying nominal thickness, and comes in a variety of weaves (Chapter 3).

Both formats are usually pre-impregnated by a hot melt method to ensure a minimum volatile content for part production. The machinery to produce pre-impregnated materials is available off the shelf.

2.1 Unidirectional Tape Parts

All the fibres have a high degree of alignment within each lamina, and the orientations of the laminae vary according to the engineering requirements, thus ensuring an optimised level of mechanical performance of the final laminate.

Fabrication techniques vary according to the part size, curvature, thickness, ply complexity, tolerances, repeatability requirements and quality constraints.

Broadly there are four routes available for manufacture:

— Hand lay-up.
— Automated ply profiling and hand assist.
— Tape laying.
— Tape winding.

A schematic of composite part production is shown in Fig. 1.

2.1.1 Hand Lay-Up

The individual lamina (ply) is profiled and cut by hand to the required shape from a roll of material, and subsequently built ply-by-ply into a laminate.

The part is then placed on the tool and a range of ancillary materials are placed on the part to allow for volatile and excess resin extraction during cure. A pressure membrane (vacuum bag) is then placed over the part and it is ready for cure.

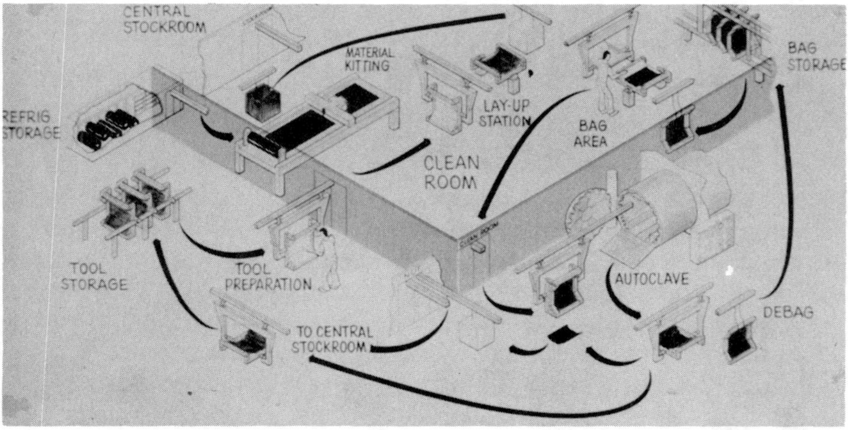

Fig. 1. Composite part flow diagram.

The cure is determined by the matrix system and involves the simultaneous application of heat and pressure. This can be achieved in either an autoclave or oven depending on the amount of pressure required (Chapter 1).

After the cure the fibres and matrix are completely bound and the part is removed from the tool and undergoes deflashing and a non-destructive quality assessment.

If it passes the quality assessment it is then ready for painting and finishing.

2.1.2 Automatic Ply Profiling and Hand Assist (Broadgoods Method)

Studies have indicated that a major proportion of the time involved in part fabrication by a hand lay-up technique is taken up by the ply profiling operation. Software nests the ply requirements together to ensure a minimum of material wastage (Fig. 2).

A device would then profile plies at high speed from a length of material to the desired configuration.

Many devices have been proposed to accomplish this task, e.g. laser cutting, water jet cutting, reciprocating knife, ultrasonic knife and pre-set steel rule dies.

The most widely used apparatus is the Gerber knife cutter. This was read across from textile technology by the Gerber Corporation and consists of a large vacuum bed on to which layers of pre-preg are placed. A vacuum membrane is finally placed over the top, the layers are

Fig. 2. Composite laminae part nesting.

vacuumed down and a reciprocating knife profiles the plies at high speed.

A device of this type is shown in Fig. 3. After profiling, the plies are stacked by hand in the required sequence and are taken to an area to have the laminae hand-laid into a laminate.

Apart from the profiling aspects the procedure is identical to the hand lay-up operation described in Section 2.1.1.

Though this reduces the time taken to fabricate a part, laminate lay-up is still done by hand.

2.1.3 Tape Laying

A tape laying machine is shown in Fig. 4. The purpose of this equipment is both to profile the plies and to lay them into a laminate simultaneously.

Much effort has been expended on machines of this nature. Initially they could only deposit tape on a flat surface slowly, but currently it is possible to purchase machines that can lay tape directly on to curved tools with up to 30° of curvature with some very healthy deposition rates for the pre-pregs.

This technique has its drawbacks in that it is limited in the level of curvature with which it can deal.

Fig. 3. Gerber knife profiling machine.

Fig. 4. Tape laying machine.

2.1.4 Tape Winding

This technique is a read across from wet filament winding. Tape is used instead of the filaments to obtain the product quality and recent studies have shown that winding is the most cost effective of all techniques for the fabrication of parts.[1]

By varying the mandrel geometry it is possible to fabricate not only curved parts but flat parts as well (Fig. 5). A machine of this type is depicted in Fig. 6.

This completes the overview of techniques currently utilised for unidirectional parts. Newer developments will be considered in Section 4.

2.2 Woven or Knitted Parts

Because of the requirement of fibre continuity within a lamina, the techniques available in the manufacture of woven parts are limited to those considered in Sections 2.1.1 and 2.1.2 as tape laying and tape winding would give fibre discontinuities in a layer.

Woven pre-impregnated material is also commonly used for part fabrication, especially where components exhibit high levels of curvature. It is easier to handle when compared with unitape, having a high level of integrity.

However, the process of weaving the tows of fibre means that the fibres are not aligned but are crimped where they cross over and under one another; this leads to a degradation in mechanical performance when compared with unitape (see Chapter 10).

Because of this crimping, knitted fabrics are currently finding wider

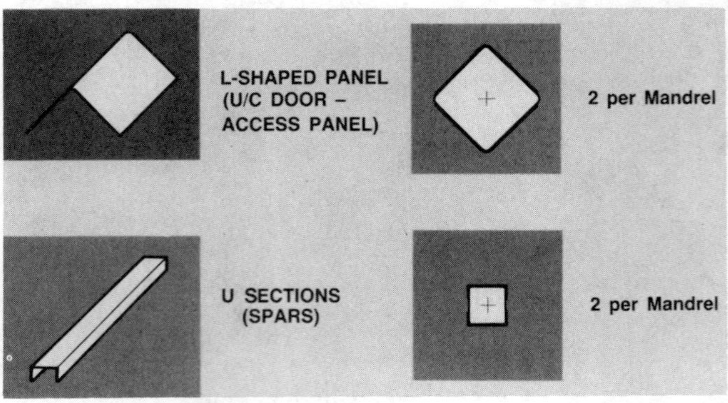

Fig. 5. Components suitable for winding.

Fig. 6. Parts being wound.

acceptance. These fabrics should not be confused with the textiles that are used in clothing — rather than being knitted themselves, the reinforcing fibres are inserted into a knitted matrix. These matrix fibres are a non-reinforcing type serving only to hold the reinforcing fibres in place.

This technology provides two significant advantages: first the reinforcing fibres are laid flat with no crimping and secondly they can be arranged in a variety of configurations from unidirectional to triaxial without a prohibitive increase in cost.

Whilst having these benefits, these knitted fabrics also retain high levels of conformability which ensures ease of fabrication.

3 TOOLING TECHNOLOGY

This is a fundamental aspect, since a laminate has been fabricated from the individual lamina it is placed upon a tool to give it the necessary profile.

The selection of the type of tooling medium is central to cost-effective quality part production. The tool finish is moulded in the part and for small batches the cost of tool manufacture can be the dominant feature in the cost breakdown.

Several options currently exist:

— Fabricated metal tooling.
— Composite tooling.
— Nickel electroform tooling.
— Machined metal tooling.
— Solid graphite tooling.
— Ceramic tooling.
— Cast resin tooling.

New materials in development offer potential use and payback as tooling media and as new materials with improved performance for structural applications are developed these in turn demand tooling systems with improved performance.

3.1 Considerations in Selecting a Tooling System

Before selecting a tooling system and designing tools for a product several parameters need to be defined and understood:

— Number of parts to be made.
— Temperature and pressure requirements of part cure.
— Timescales.
— Cost boundaries.
— Curvature complexities.
— The component dimensional tolerances.

By judicious consideration of these variables an appropriate system can be selected, and the tooling designed and manufactured. Table 1 summarises the advantages and disadvantages of various tooling materials.

3.1.1 Fabricated Metal Tooling

A range of metal profiles (headers) are manufactured and positioned together on a base; the mould surface is then rolled to shape and fastened to the headers.

Tooling of this type is widely used and can have a long life, but allowance must be made in its design for differential thermal expansion and hence the temperature at which fibre 'lock-on' occurs in the matrix system to be moulded needs to be established.

3.1.2 Composite Tooling

Several options are currently available, from wet lay-up glass reinforced tools to high performance carbon fibre reinforced tools made from tooling pre-preg.

The choice of material is governed by the factors mentioned in Section 3.1, but the logic path is similar and involves the manufacture of a master model from which either the tool or an intermediate tool is moulded.

If an intermediate tool is made from the master, the actual tool will be splashed off the intermediate. This depends on the tooling philosophy adopted, i.e. whether the tool is male or female and the information available to define the master tool surface and the surface complexity.

Master modelling media are many and their prime requirement is stability. Materials used include plastic faced plaster, filled resins, foams, wet laid composite, wood and cast ceramic, to name a few.

Again the engineering of the master is a case of materials selection against the performance requirements.

The prime advantage of composite tools is their relatively low weight which means easier handling in the manufacture area.

The life of a tool is hard to define as it is a function of the materials used and the materials to be cured on them, but up to 1000 cycles have been claimed.

3.1.3 Nickel Electroform Tooling

This type of tooling requires a plastic master. Once this has been fabricated, a thin layer of silver is deposited on the surface and the master is then placed in a bath and the nickel tool is electrolytically deposited on the silver coated surface.

When the desired thickness has been deposited it is removed from the bath, and the backing structure is attached to the tool.

3.1.4 Machined Metal Tooling

This type of tooling is manufactured by taking a large block of metal and 'numerically controlled' machining the required profile into it.

Such metal tooling can have a long life, although again consideration must be given to the differential thermal effects during the design phase and care must be taken in the tool cleaning operations as sometimes tool erosion can occur as a result of this.

3.1.5 Solid Graphite Tooling

This is similar to the description given in Section 3.1.4 but, rather than using a block of metal, a block of graphite is used as the start point.

TABLE 1
Advantages/Disadvantages of Various Tooling Materials

Tooling form	*Advantages*	*Disadvantages*
Steel/aluminium tooling	Robust tooling — relatively stable Good durability/damage tolerance	Often heavy — difficult to handle/transport. High tool mass results in slow heat-up rate. High TCE makes it impossible to accept complex contour carbon composite parts. Single-curvature carbon composite parts can be made if a composite 'slip sheet' is used between tool and part. Relatively high cost of fabrication. Tool porosity often a problem with cast and welded tooling.
Electroform nickel tooling	Complex tool shapes possible — accurate reproduction of detail Good thermal response–heat-up rate	Some detail design geometry limitations. Model needed for electroplating bath — plating process relatively slow. High TCE not compatible with carbon composite parts. Porosity can be a problem. Modifications to the finished tool can be difficult. Tool size limited by the availability of a suitable plating bath. Difficulties in attaching backing structure.

Monolithic graphite	TCE characteristics close to carbon composite. Easily machined to good finish. High operating temperature capability for polyimide/PEEK cures.	Relatively high-cost material. Porosity problems — effective tool sealing a necessity.
Composite tooling (carbon-epoxy/ glass–epoxy)	TCE fully compatible with composite part. Complex shapes can be easily formed. Rapid tool manufacturing process. Good retention of accuracy with high stability. Good strength/weight and stiffness/weight ratio performance. Relatively lightweight tooling with good thermal response — fast heat-up. Can be the lowest-cost tooling method if good tool design techniques followed.	Tool quality is process and material dependent. Poor control on processing can result in porous/badly consolidated tool. Careful selection of materials necessary to achieve long-life tooling.

The advantage of this type of tool is its stability and low thermal expansion (see Table 2 for a list of thermal expansion coefficients).

3.1.6 Ceramic Tooling

This is a newer kind of tooling medium. A pattern is made in a material which can withstand 100 °C and then the ceramic tooling material is cast, dried and post-cured. The use of tooling of this type for advanced composites is in its infancy.[2]

3.2 Tool Cleaning Techniques

Once a tool has been used to mould a part it usually requires some cleaning before a subsequent cure to remove excess resin spew, etc. This is often done manually; but it is also possible to clean the surface using mechanised means.

A system which is currently gaining favour in this regard involves blasting the tool surface with solid CO_2 pellets, which dissipate into the atmosphere after debris removal.

Cryogenic cleaning systems of this type have the ability to remove surface contamination without damaging the tool surface and offer much promise.[3]

3.3 Future Tooling Technology

Current tooling systems address the manufacture of parts from the current materials. The new trend is for materials with higher temperature capability to be developed and this will give a challenge to the tooling engineer to develop tools that can undergo more rigorous cures and forming environments to keep production cost-effective. They key to this, as always, will be materials selection and work is already underway looking at the suitability of the newer thermoplastics in this application.

4 FUTURE FABRICATION TECHNOLOGY

Having considered current manufacturing techniques it is worthwhile giving some time to the newer techniques under development which will come to the fore in the future.

Again, materials are central to this since the range, format and processing requirements dramatically affect manufacturing. It is

TABLE 2
Thermal Expansion/Specific Heat Capacity Characteristics of Tooling Materials

Material	*Density (lb/ft^3)*	*Thermal expansion coefficient ($in/in\,°F \times 10^{-6}$)*	*Thermal expansion of 16 ft tool heated to 350°F (inches)*	*Specific heat capacity (BTU/lb°F)*	*Energy needed to heat 1 ft^3 of tool to 350°F (BTU)*
Steel	490	6·7	0·36	0·10	13 380
Aluminium	170	12·5	0·66	0·22	10 200
Monolithic graphite	104	2·0	0·11	N.A.	N.A.
Electroform nickel	540	7·4	0·39	0·11	16 200
Carbon fabric–epoxy laminate	100	2·0	0·11	0·18	4 900
Glass fabric–epoxy laminate	120	7·3	0·38	0·09	2 950

worthwhile making a distinction between thermoset and thermoplastic matrix systems.

4.1 Future Thermoset Fabrication Techniques

Areas where developments are occurring are:

- Hand lay-up enhancements.
- Broadgoods enhancements.

4.1.1 Hand Lay-Up Enhancements

Whatever the changes in materials, hand fabrication will always be suitable for some products and study has been directed to enhancement that will allow hand fabrication to be integrated into the computer controlled manufacturing systems of the future.

The presentation of lay-up information and the position of individual plies are areas where improvements can be made. Recent work by Assembly Guidance Systems has provided an elegant solution:[4] a laser raster from a unit over the tool or work bench projects down on to the work area instructions relating to part fabrication and an outline of the current ply profile, thus speeding up lay-up time.

This system also offers the possibility of automating the quality assurance of part lay-up by comparing edge recognition data with lay-up reference graphics.

Whilst still in their infancy, systems of this type will enhance the efficiency of the hand fabricated part.

4.1.2 Broadgoods Enhancements

The broadgoods manufacturing route as described in Section 2.1.2 has been used extensively within the aerospace industry for manufacture of parts. Studies on integrated ply laminating centres have been undertaken with polar robots being used to lift profiled plies with a vacuum hand and stack them.

Work on centres of this type has successfully proved the concept but the use of polar robots has limitations in the weight they can manipulate and their positional tolerances.

A new concept study which promises to overcome these limitations has recently been undertaken by Composite Automation Engineering Inc. (Fig. 7).[5] This is a continuous 'flow-just-in-time' system, comprising several individual computer controlled machines that are integrated

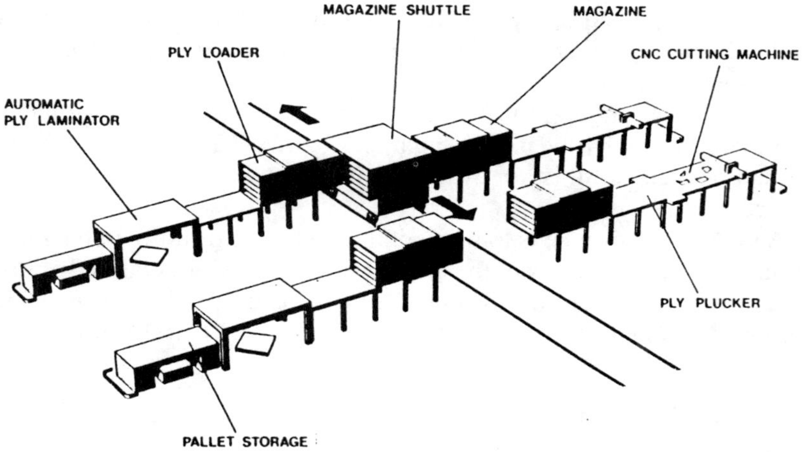

Fig. 7. Automated laminating system.

into a system with a supervisory computer directing the activities.

The material is cut one ply deep with a device which only penetrates the pre-impregnated composite material, leaving the backing paper intact. The backing paper is removed prior to the plies leaving the cutting machine.

There is no covering paper as vacuum is not used during the cutting process. Once cut, the profiled plies are conveyed to the next machine, the ply plucker, and plies are removed from the cut nest one at a time on a 'first-in, first-out' basis. As the plies are picked out of the nest they are conveyed into a magazine which has storage for at least 40 plies.

When the magazine is loaded with its correct number of plies it is automatically moved to either a temporary storage area or to another cutting machine to receive different material or directly to the next machine in the process. The purpose of the magazine is to collate the pack of plies that constitutes a finished laminate, regardless of the material mix.

The next machine is the laminator. The plies are conveyed out of the magazine individually in the correct sequence. As they enter the laminator a linear array camera inspects the ply for position. Information concerning the ply size and the appropriate location is known since the ply has been tracked through the system by the cell map. The measured dimensions are compared with the programmed data for this ply in the data base.

The ply is picked up by the laminator and transported to the lay-up table (during this operation automatic check inspection occurs), and is laid down in its correct location.

This operation is repeated for every ply until the laminate is complete, at which time it is automatically conveyed on to an automatic guided vehicle and is ready for transportation to the bagging-up centre.

A system of this type is to be available in the near future and should enhance the productivity of the broadgoods route with a claimed ability to handle a ply every 8 s. These plies can be as small as 1 in × 3 in (2·5 cm × 7·5 cm) by three inches and as large as 5 ft × 12 ft (1·5 m × 3·6 m).

4.2 Future Thermoplastics Technology

This is a category not considered under current fabrication techniques, as the use of high performance thermoplastics is still under development (Chapter 2). There are many current uses for thermoplastics but these are not primary structural applications.

However, advantages are foreseen in the use of thermoplastic matrix systems and development of appropriate fabrication technology is underway, with possible manufacturing pay-back in the future. These efforts fall into two broad categories: studies to read across thermoset techniques, and studies aimed at developing new and novel techniques.

4.2.1 Thermoset Read Across Studies

Most effort has been expended on developing tape laying for thermoplastics. The material is available in a unidirectional pre-impregnated form which is similar to the format of the thermosets. It is, however, stiff and dry in comparison and does not form readily to contours.

The concept is good in that a tape-laid part from thermoplastic material will be ready for the finishing operation once the tape laying is completed.

The basic problem is the melting and consolidation of the material as it comes off the tape laying head to be infused into the bulk of the laminate being fabricated.

Different methods of introducing the heat to induce a melted region at the point of tape laying have been investigated but all methods have the problem that, on local cool-down, distortion occurs in the local area of heat application and any laminate part fabricated will be unbalanced with a tendency to distort.

Restraint of the laminate in a partially fabricated condition is required to counter out-of-balance and thermal stresses and this further induces stress in the finally fabricated part.

Hence, at best, tape laying of thermoplastic parts can be viewed as only an intermediate operation to be followed by a post-forming cycle to fully finish the part.

4.2.2 Novel Techniques

Advanced thermoplastics are in their infancy and new and novel material formats are currently being developed. One format that shows promise is the matrix system which is itself being spun into a fibre that is subsequently intermingled with the structural reinforcing fibre to give a dry fibre tow or yarn of reinforcing fibres and matrix.

The material in this form can be subsequently processed by textile techniques to generate dry fibre forms of a variety of 2D and 3D reinforcing patterns which possess the ability to drape over curvature.

Once in this condition, it is easy to heat and consolidate the material to give a finished part with the required shape and properties cost-effectively.

Work in this area is at an early stage but by adopting techniques from different areas it will be possible to exploit cost-effectively the new and novel material forms that are becoming available.

5 GENERAL ASPECTS

We have covered aspects of current and future advanced composite manufacturing. Whilst it is always possible to devise techniques to fabricate a product, to manufacture a low-cost part to a high standard and meet programmes it is essential to select the correct material and have a suitable design.

If consideration is given to the mutual interdependence of materials, design and manufacture then it should always be possible to fabricate an advanced composite part that out-performs its competitors cost-effectively.

REFERENCES

1. Krolewski, S. & Gutowski, T., The effect of the automation of advanced composite fabrication process on part cost. *SAMPE J.* (May 1987).

2. Kronsberg, T. & Miller, L., Ceramic tooling for composite fabrication. *Adv. Compos.* (May 1988).
3. Crane, N., Cold-jet: the cryogenic cleaning system. Paper presented at ICAC88, Sept. 1988.
4. Blake, S., Laser guidance for pre-preg lay up. Paper presented at ICAC88, Sept. 1988.
5. Evans, G., The processing of pre-preg composite materials. Paper presented at ICAC88, Sept. 1988.

Chapter 8

Design of Bonded Structures

R. J. Lee[a] & J. C. McCarthy[b]

[a]Materials Development Division, [b]Engineering Division, Harwell Laboratory, UK Atomic Energy Authority, Didcot, Oxfordshire, UK

1 INTRODUCTION

Advanced composites can now be regarded as a maturing technology making significant contributions to a number of industries where weight reduction, increases in stiffness and strength or chemical resistance are prime design drivers. Ideally any structural design for composites should consist of a one-piece (co-cured) fabrication moulded in a 'one-shot' operation with, in effect, no joints. This approach, although desirable, is seldom possible for a number of technical, commercial or practical reasons. For instance:

— There may be size restrictions on the mould, press or autoclave used to form the composite.
— There may be a requirement to disassemble the component for transportation, inspection or repair.
— Requirements for structural fittings or bearings may exist.

Although the technology for joining materials using structural adhesives is by no means a new idea, and has been exploited in the aircraft industry for over 40 years,[1,2] the challenge of using adhesive bonding as a method of construction is that it offers some potential advantages when compared with other, more conventional, fastening methods such as brazing, welding, riveting or bolting. These include:

— The ability to join dissimilar and thin sheet materials.
— Improved structural efficiency leading to weight reduction and fewer component pieces.
— Superior fatigue performance compared with mechanically fastened joints.
— Smooth external appearance.
— The sealant action of the adhesives, preventing electrochemical contact between dissimilar materials and also reducing moisture ingress.
— The good damping characteristics generally possessed by bonded joints, which when combined with reinforced plastics can be effective at reducing noise and vibration.
— Avoidance of the problems associated with machining bolt holes which act as severe stress concentrators in composites.

There are however a number of disadvantages:

— Adhesives have the same limitations as any group of polymers; their properties are dependent on temperature, time and rate of loading. Careful consideration must be made during initial design to take account of these effects.
— As initial bond strengths can be low, assembled joints will need support both prior to and during cure.
— Surface preparation is crucial if joints with the required durability are to be obtained.
— Good process control and well trained personnel are essential because a badly made joint is often impossible to correct.
— Structurally bonded components are not easily dismantled for in-service repair.
— Non-destructive examination presently available is only adequate for low volume, high quality component manufacture and will not provide an indication of joint strength.

Although the basic principle of an adhesive is extremely simple, to joint surfaces together and resist separation, the science of adhesion is incompletely understood and consequently has lagged behind the applied technology. There are numerous reasons for this, one of which is that the basic understanding of the nature of the chemical and physical forces between materials is not sufficient to explain measured joint strengths. A further complication is that no simple relationships exist between bulk mechanical properties of a cured resin and their properties as an adhesive. Adhesion science is a complex multi-disciplinary subject encompassing the following areas:

— Surface chemistry and physics.
— Polymer chemistry and rheology.
— Polymer physics and fracture phenomena.
— Engineering design and stress analysis.

In this chapter it is not possible to consider these topics in detail and the reader should refer to the numerous texts,[3–23] reviews[24–33] and recent developments[34–56] for a more thorough treatment of specific aspects of adhesion and adhesives. A series of Engineering Sciences Data Unit (ESDU) items[57–62] dealing with bonded joints has been published describing a number of computer programs for calculating the elastic/plastic shear strains in bonded double-lap shear joints and also giving a comparison of analytical methods and their relevance to joint design.

2 REQUIREMENTS OF ENGINEERING ADHESIVES

Engineering adhesives are generally considered to be those that find use in an engineering process either in production or repair. The term covers a wide range including thread-locking compounds, adhesive sealants and structural adhesives. It is this last group that are of real interest here and perhaps they may be broadly defined as those adhesives which by their action alone hold the structure together. The adhesive must be able to adhere to a variety of surfaces (although currently certain adhesives are being optimised for specific adherends) and must also possess the following combinations of properties with regard to both manufacture and service.

Manufacture

— Good shelf-life.
— Good working (pot) life.
— Ease of application (manual or robotic).
— Freedom from toxic and other safety hazards (flammability).
— Reasonably quick setting times, preferably at ambient or low temperatures.
— Rapid development of strength.
— Requirement for minimal adherend surface preparation.

— Tolerance to surface contamination and deviations from optimum cure conditions.
— Yielding consistent results.
— Low cost.
— Sag-free during curing process.

Service

— Permanent adhesion to both metals and plastics.
— Resistance to applied stresses (high shear, tensile and peel strengths).
— Adequate toughness with good resistance to impact and thermal shock.
— Maintenance of strength under dynamic loading (good creep and fatigue performance).
— Resistance to environmental factors and contaminants, e.g. water, humidity, oils, greases, fuels, corrosive fluids, cleaning agents, etc.
— Maintenance of strength in the presence of contaminants over a wide temperature range (−40 °C to +180 °C for some applications).

3 SELECTION OF ENGINEERING ADHESIVES

For certain applications bonding requirements may be modest and it may be pointless using a high performance adhesive which may be stronger than the materials to be bonded. In this case many adhesives will have adequate properties and selection will be made on the basis of cost, availability and ease of application. There are, however, many applications which place severe demands on adhesives (e.g. aerospace and to a lesser extent automotive applications) where long-term performance is a prime consideration. Careful attention must then be given to surface pretreatments and the use of appropriate primers. Bonded joints must be designed to carry the expected loads, with appropriate safety factors, by maximising shear stress transfer and reducing as far as practically possible the peel stresses present at the ends of overlap joints (see Chapter 9). Selection of a suitable adhesive from the large numbers commercially available can be a difficult and time-consuming operation and may incur expensive material and component testing.

A general approach to selection of an adhesive for a particular application will consider the following:

(i) Cure or setting time.
(ii) Toxicity:
— 1974 Health and Safety at Work Act;
— Dermatitis;
— Ventilation of harmful vapours.
— Control of substances hazardous to health (COSHH legislation).
(iii) Fire hazard:
— Particularly for solvent-based adhesives.
(iv) Joint strength:
— Closely linked to joint geometry which influences adhesive strain distribution;
— High toughness and peel strengths desirable.
(v) Service conditions:
— Temperature limitations;
— Creep;
— Fatigue;
— Durability.
(vi) Cost:
— Area and thickness to be covered;
— Labour cost of application;
— Special mixing or dispensing equipment;
— Energy requirements for curing.
(vii) Special considerations:
— Electrical properties;
— Optical properties.

4 ADHESIVE MATERIALS — CLASSIFICATION

Modern structural adhesives are generally composed of mixtures of several complex materials, the components of which are added to satisfy certain fabrication properties or improve properties required in the final joint, e.g. toughness and peel strength. The base resin is generally blended with a variety of constituents, serving a number of purposes, such as:

(i) Catalysts and hardeners.
(ii) Accelerators, inhibitors and retarders.
(iii) Modifiers — fillers, extenders, plasticisers, wetting agents.
(iv) Rubber toughening agents.
(v) Diluents.

Structural adhesives can be classified in many ways, e.g.

— Physical form, e.g. film, paste, liquid and tape.
— Chemical structure, e.g. polyurethane, acrylic, epoxide, phenolic.
— Conditions of use, e.g. single-component, cold setting, solvent-based, heat curing and spray.
— Polymer type, e.g. thermosetting, thermoplastic, polymer blends.
— Materials bonded, e.g. metal-to-metal, wood, etc.

Generally, the stronger adhesives solidify by chemical reaction whereas weaker adhesives rely on some physical change or natural surface tackiness. New formulations and applications are constantly in demand and since there is not yet a universal adhesive, no individual can be fully acquainted with new developments. Fortunately, most manufacturers supply guidebooks to provide information on rudimentary joint design and surface preparation with rather limited information about in-service performance. Recent developments have been towards compiling a data bank of adhesives and selection procedures either in a desktop data bank[63] or as an expert system.[64] It is clearly not possible to cover the materials aspects in detail other than to indicate their individual uses and merits. A useful description of the composition and formulation of adhesives in common use has been produced by Wake.[16]

4.1 Chemically Reactive Adhesives

4.1.1 Phenolics

These are examples of the earliest types of structural adhesives for metal-to-metal and metal-to-wood bonding.[65] They are commonly used in a modified form (e.g. blended with vinyl or nitrile rubber) to overcome inherent brittleness. They have extremely good environmental durability and high temperature resistance. They are generally less sensitive to variations in surface preparation than the epoxides but their cure schedules require high temperatures and pressures to avoid porosity caused by water vapour, which is produced as a reaction by-product, and they thus require specialised equipment and complex procedures. Long-term high temperature exposure tests have shown little deterioration and they are commonly used in bonding brake and clutch linings.

4.1.2 Epoxides

These are available as a one- or two-component system, a resin usually cured by reaction with a polyfunctional amine or anhydride hardener.

They allow great versatility and can be tailored to provide bond strengths amongst the highest currently achievable; consequently they tend to dominate the adhesives field. They are available in liquid, paste or film forms (both supported and unsupported) and have virtually replaced the earlier phenolic-based adhesives because of their 100% reactive formulation. They are cured at moderate temperatures, do not require high pressures and do not normally emit volatiles. The cured, unmodified epoxides generally have low peel and impact strengths compared with other adhesive types and sometimes produce less durable joints than older generation phenolic adhesives. For these reasons a considerable amount of development work has been done, particularly in blending with thermoplastics and elastomers to produce toughened or high temperature variants (see Chapters 1 and 4). Typical blends include epoxy–phenolic, epoxy–vinyl, epoxy–nylon, epoxy–CTBN, or epoxy–polysulphide.

The epoxide field is large and varied and for further information a number of texts are available covering basic principles,[66] recent advances[67] and practical aspects of toughening.[68]

4.1.3 Polyurethanes

Adhesives based on polyurethane elastomers,[69] which are formed by the reaction of polyisocyanates with certain alcohols or diamines, are now well established in industrial use either as single-component solvent-based, as thermoplastic hot melts, or as two-component non-solvent-based reactive systems. They provide strong, flexible joints with good peel strengths and good resistance to shock and vibration although they are prone to moisture attack. The fast curing conditions, difficulty of handling and possible toxic hazard often require specialised mixing equipment.

4.1.4 Acrylics

These adhesives are relative newcomers but are likely to play a more prominent role in the near future because of their versatility, tolerance to surface contamination and low temperature cure. Their formulation, chemical properties and future trends have been recently reviewed[70,71] and there are a number of references on bonding to metals[72] and to reinforced plastics.[73,74] A systematic study of bonding aluminium alloys with acrylic adhesives for aircraft repair has been reported jointly by Lucas Aerospace and RAE Farnborough.[75] The acrylics can be classified into three categories:

(a) Anaerobics	These cure in the presence of metal and absence of atmospheric oxygen. Traditionally used as locking compounds.
(b) Cyanoacrylates	One-component, low viscosity systems that are rapidly cured when brought in contact with residual surface moisture. Commonly referred to as 'superglues'.
(c) Toughened acrylics	Reactive fluid comprising rubber particles dispersed in methacrylic monomer and cured by an initiator which is usually an amine–aldehyde condensate. The resulting product is a crosslinked network of poly(methyl methacrylate) chemically reacted with rubber particles. Sometimes referred to as 'second generation acrylics'.

4.1.5 High Temperature Adhesives

Although these materials are expensive and in limited supply at present, they will make a significant contribution in the near future because of their superior temperature resistance. Adhesives capable of surviving long-term exposure to 350 °C have been reported. Major drawbacks are that they are difficult to use, require complicated cure schedules and contain expensive and toxic precursor chemicals (some contain arsenic derivatives). The two systems that are currently attracting most attention are based on polybenzimidazoles (PBI) and polyimides (PI) although some recent work has been reported on rubber toughened bismaleimides (BMI).[76] A considerable amount of effort has been made by NASA Langley, USA, in developing polyimide formulations for aerospace use, e.g. LARC 13 and LARC-TPI.[77] NASA Lewis has developed a number of polyimide systems, e.g. PMR-15, and developments have led to systems which can be cured at lower temperatures.[78] Useful reviews of recent developments in high temperature adhesives are given by technologists from Rolls-Royce,[79] Boeing,[80] NASA[81] and UK MoD.[82] An evaluation of the high temperature shear properties of a polyimide adhesive has also been reported jointly by Boeing Aerospace Company and the University of Delaware.[83]

4.2 Non-reactive Adhesives

The materials in this broad class are generally less strong than previously described adhesives but nevertheless are important industrially because they are simple to use and are usually tolerant to variations in processing conditions.

4.2.1 Plastisols
These are based on heat-cured plasticised poly(vinyl chloride) dispersions which are used quite widely in the automotive industry for non-structural applications, e.g. the attachment of car bonnet stiffeners. They are usually hardened during paint stoving processes and whilst they do not have outstanding properties they are usually reliable and can be applied to metals without surface preparation.

4.2.2 Hot Melts
Hot melt adhesives are thermoplastics which are heated until liquid and then applied to the surfaces and allowed to cool. This is one of the most rapid methods of bonding as neither solvent loss nor chemical reaction is required. They are commonly used for fast, repetitive assembly of packaging materials and plastic films where they are designed to be lightly loaded. Although developed from one of the oldest methods (hot wax), current materials are polyethylene, vinyl polymers and copolymers, and polyamides.[84]

4.2.3 Tapes
Based on pressure-sensitive adhesives, they are usually adequate for bonding small components but are not able to withstand sustained loads or adverse environments. A number of papers describing tack and morphology[85] and principles of formulation[86] have been published. General aspects of pressure-sensitive adhesives have been reviewed by Aubrey.[87]

5 ADHESIVE TEST METHODS

A very large number of test methods have been developed and have been described adequately in a number of review articles.[9, 21, 25] The most widely used have been standardised and issued by the British Standards

Institute (BS series) and the American Society for Testing and Materials (ASTM series). Data provided by adhesive manufacturers, usually in the form of lap shear strengths, provide useful information on the performance of the adhesive for that particular geometry but are generally not suitable for design purposes. Adams & Wake[5] have pointed out that most, if not all, of these standard tests involve joints in which the adhesive stresses are far from uniform and their value lies essentially in comparing adhesives and surface treatments rather than validating a specific joint. They also point out that controversy still exists as to whether adhesive properties in the thin film are the same as when bulk specimens are prepared. The standard test procedures are:

(1) *Axially loaded butt joints*

BS 5350: Part C3: 1979 ASTM D897–78	Tensile properties of adhesive bonds
ASTM D2094–69 D2095–72	Specifically for bar and rod-shaped specimens
ASTM D429–73	Rubber-to-metal bonding
ASTM D1344–78	Cross-lap specimen specifically designed for glass substrates

(2) *Lap joints loaded in tension*

BS 5350: Part C5: 1976	Single- or double-lap joint test
ASTM D1002–72	Basic metal-to-metal single-lap joint test
ASTM D3528–76	Double-lap test
ASTM D3162–73 ASTM D3164–73	Specifically for polymer substrates
ASTM D2295–72	Single-lap joint test for metal-to-metal joints at elevated temperatures
ASTM D2557–72	As above, but at low temperatures
ASTM D3983–81	Thick adherend shear test

(3) *Peel tests*

BS 5350: Part C9: 1978	Floating roller test
ASTM D3167–76	As above
BS 5350: Part C10: 1976 BS 5350: Part C14: 1979	90° peel test
BS 5350: Part C11: 1979 ASTM D903–49	180° peel test
BS 5350: Part C12: 1979 ASTM D1876–72	T peel for flexible adherends

ASTM D1781-76	Climbing drum test for sandwich structures
ASTM D429-73	Rubber-to-metal bonding
ASTM D2558-69	Shoe soleing materials

(4) *Cleavage strength*

ASTM D1062-72	
ASTM D3433-75	Parallel/tapered DCB to determine fracture energy
ASTM D3807-79	Plastic adherends

(5) *Block, shear, impact resistance*

ASTM D950-78

(6) *Disc shear strength in compression*

ASTM D2182-72

(7) *Creep resistance*

BS 5350: Part C7: 1976	Various geometries permitted
ASTM D1780-72	Single-lap joint loaded in tension
ASTM D2294-69	
ASTM D2293-69	Long overlap single-lap loaded in compression.

(8) *Fatigue strength*

ASTM D3166-73	Single-lap joint loaded in tension

(9) *Environmental resistance*

ASTM D2918-71	Peel joint subjected to stress, moisture and temperature
ASTM D2919-71	As above, but uses single-lap shear test
ASTM D3762-79	As above, but uses wedge test
ASTM D1151-72	Exposure to moisture and temperature
ASTM D1183-70	Exposure to cyclic laboratory ageing conditions
ASTM D904-57	Exposure to artificial and natural light
ASTM D896-66	Exposure to chemical reagents
ASTM D3632-77	Exposure to oxygen
ASTM D1828-70	Natural weathering
ASTM D1879-70	Exposure to high energy irradiation

(10) *Pressure-sensitive tack*

ASTM D2979-71	Inverted probe test
ASTM D3121-73	Rolling ball test

(11) *Flexural strength of laminated assemblies*

ASTM D1184-69

(12) *Torque strength*

ASTM E229–70	Determining pure shear strength and modulus
ASTM D3658–78	Ultraviolet light curing glass–metal joints

Although there are many test methods for structural adhesives, only a few are useful in generating quantitative engineering design data. To design accurately a bonded structure it is first necessary to characterise a number of adhesive properties including pure shear and tensile behaviour. Unfortunately most test techniques are not suitable for this because of geometric discontinuities and large peel stresses that are generated. Two pure shear test methods which have been developed for adhesives are the Butt Torsion or Napkin Ring and the Thick Adherend Shear Tests (Figs 1 and 2).

5.1 Butt Torsion/Napkin Ring Test

The napkin ring or tubular torsion specimen was pioneered by Rutherford and his colleagues at Singer.[88] The basis of this test is a narrow, annular ring of adhesive subjected to uniform shear loads around the circumference. This specimen geometry achieves the purest of all shear loads but is costly to manufacture and, since the bond area is very narrow, it requires specialised and sensitive equipment to

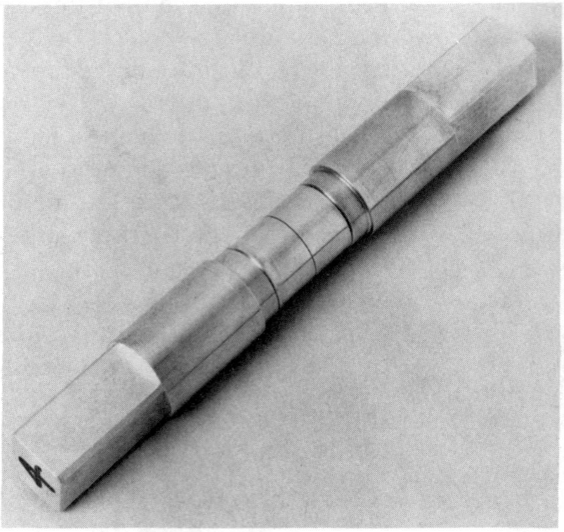

Fig. 1. Adhesive butt torsion specimen.

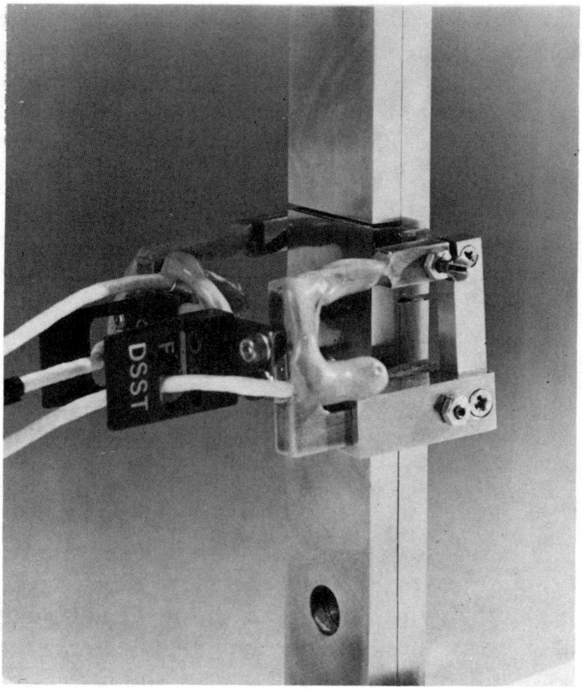

Fig. 2. Thick adherend lap shear specimen with extensometer mounted.

manufacture and test. Axial alignment in the test machine is critical and a uniform bondline thickness is essential. Sage[89] at City University has discussed the practical aspects of this test and also fatigue testing. A recent ESDU technical memorandum[90] discusses torsion testing of solid bars and napkin rings and provides a guide to obtain adhesive shear stress–strain curves for use with ESDU and other computer programs. Improved torsion testing of adhesives has also been discussed by Bell.[91] A comparison of adhesive properties measured in bulk and in thin films using torsion testing has been discussed by Ishai[92] and Fischer[93] who concluded that a good correlation existed between in-situ and bulk properties of shear yield strength and elastic modulus.

5.2 Thick Adherend Shear Test

Because of practical difficulties encountered in the manufacture and testing of the torsion test specimens, Frazier at Bell Helicopters,[94] Krieger at American Cyanamid[95–99] and Althof at the DFVLR in

FRG[100, 101] have developed independently a simpler, less pure shear test, known as the Thick Adherend Shear Test (TAST). This test was used to characterise the behaviour of a number of candidate adhesives for an EEC-sponsored programme on composite-to-metal jointing technology for vehicle weight reduction.[102] The design of specimen is such that if the adherends are made rigid, and the overlap length is short, the loads needed to fail the adhesive are incapable of distorting the adherends and the shear stress is reasonably uniform. The specimen's main weakness is also its greatest virtue in that the square cut on the adhesive at each end of the test section gives rise to localised stress and strain concentrations which are inherently more severe than those which occur in any real structural application with a natural fillet, which makes the results inherently conservative.

6 ENGINEERING DESIGN AND ADHESIVES

The process of design for an adhesive joint is similar to that for any structure. It should begin with a specification setting out the required characteristics of the joint. This should cover the adherend materials and possibly the thickness of one or both adherends; the applied operating conditions such as the magnitude and direction of applied loads together with appropriate safety factors; the stiffness of the bonded structure; temperature requirements; fatigue life and environmental considerations. After an initial estimate of the appropriate joint geometry (overlap length, thickness of adherends, if not known, and bondline thickness) which may be based on previous experience, an analysis of joint strength can be performed. The results of such an analysis may indicate that an iterative process is required to optimise the strength (and possibly stiffness) of the joint.

To date, most analyses concentrate on strength prediction as the stiffness contribution to the complete bonded structure is usually small. Prediction of joint strength is accomplished by performing a stress/strain analysis and applying a suitable failure criterion to the results. However as Adams & Wake[5] rightly point out, to predict joint strength from first principles is as difficult as 'trying to predict the fatigue life of a steel bar from a knowledge of its crystal structure and the strength of individual crystals'.

The problem is complicated because the distribution of stresses within a bonded lap shear joint loaded in tension is uneven along the bondlength with a marked stress concentration at either end. This stress concentration is more pronounced for joints between dissimilar adherends. It is in the highly stressed regions that failure usually occurs and hence it is important to calculate adhesive stresses (or strains) to ensure that the appropriate factor of safety may be applied.

The two main techniques which can be used to predict stresses within adhesive bonds are finite element analysis and closed form or continuum mechanics. For the general case of a composite-to-metal bond, analysis by either of these techniques should be capable of taking into account factors such as adhesive non-linearity, orthotropic adherends, shear deflections in the composite and thermally induced strains. The use of such techniques as part of a design procedure should also allow for changing geometrical parameters. It is in the latter respect that the two approaches differ in their capability.

A computer program based on a closed form analysis is written to solve a problem of a particular geometry (although the dimensions may vary). Hence the user need only supply details of the materials, dimensions and loadings. A computer program based on finite element analysis however, requires that the geometry is also specified along with details of a mesh used to effect the solution. Hence a greater awareness of the solution technique is required of the user. The detailed results which it provides are used mainly for analysis of bondline stresses for research purposes rather than a routine design tool. The results from both types of analysis on a given problem generally agree well. Figure 3 illustrates a comparison between finite element results and a continuum mechanics solution for a loaded lap shear joint.

6.1 Classical or Closed Form Analysis

6.1.1 History

The closed form analysis was pioneered by Volkerson[103] in the late 1930s; he developed a shear lag analysis for a lap joint where the adhesive deformed only in shear, whereas the adherends deformed only in tension. This approach was developed further in 1944 by Goland & Reissner[104] who recognised some shortcomings of the simplified analysis which arise from the fact that the opposing tensile forces are not collinear: this introduces a bending moment and causes joint rotation in the region of the overlap. More recent analyses for a variety of shear joints have been performed by Hart-Smith[105] as part of a US Air Force sponsored programme, the Primary Adhesively Bonded Structure

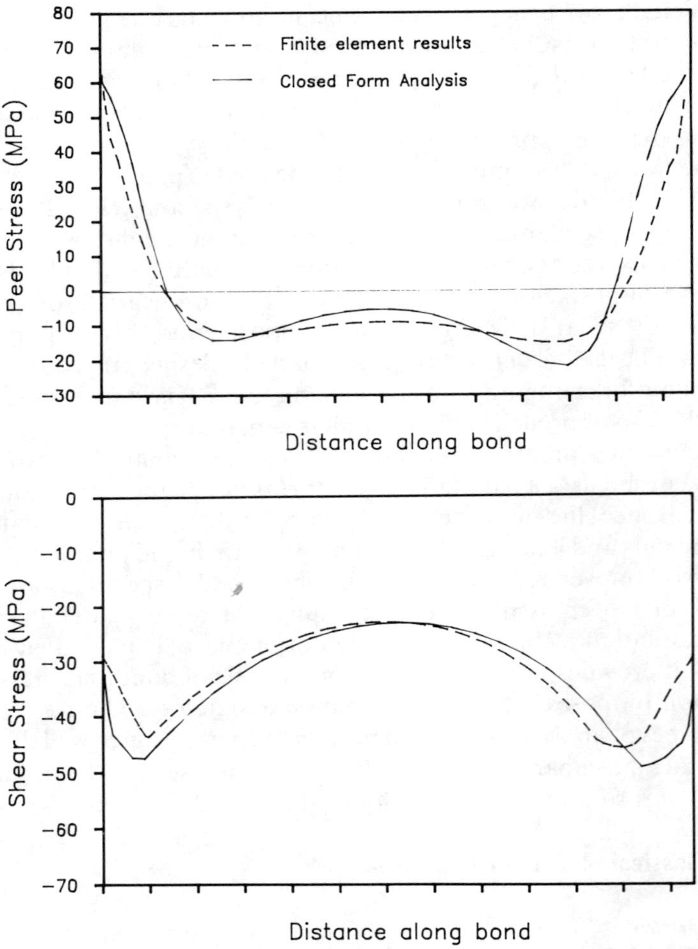

Fig. 3. Comparison of results from finite element and closed form analyses for a single lap shear joint.

Technology Program (PABST), aimed at developing the technology for bonded fuselage structures.

6.1.2 Technique

The technique essentially involves setting up a series of differential equations describing the static equilibrium and physical displacement of an element of the bondline and adjacent adherends. These equations are then combined, using known material properties of adhesive and adherends, to form a differential equation which can be solved either analytically or, for more complex problems, by computer methods. The equation which is set up depends on the assumptions and complexity of the composite-to-metal joint. For unbalanced adherends (i.e. those where the adherends are not identical) and where no account is taken of variations of stress through the bond thickness, it is usually a seventh-order differential equation.

6.1.3 Current Solutions

The elastic solution could be solved by hand calculations, but it is better suited to a computer. A number of computer programs based upon these techniques are now becoming available commercially for lap shear and tubular joints. The method, in its most recent and sophisticated form, can account for shear and peel stresses and strains along the bond length, orthotropy in the adherends, non-linear effects in the adhesive and thermally induced strains arising from dissimilar adherends, although there are many gaps in solutions available for design purposes. However some solutions make the assumption of ignoring peel stresses in the adhesive; others only allow simple tensile loads to be applied; most solutions make the assumption that the adherends are in simple bending; and many ignore variations in adhesive stresses across the bond thickness.

6.1.4 Non-linearity

Linear analysis does not predict ultimate strength accurately because of the extensive yielding in the adhesive. Non-linear analysis of lap joints has been approached by Grant[106] using an iterative finite difference technique which ignored bending and was therefore more applicable to double-lap joints. Hart-Smith[107] also developed a non-linear technique which enables a fairly rapid evaluation of stresses to be made and, despite the approximation implicit in the analysis, appears to give an accurate prediction of the onset of failure using a maximum strain criterion. He also examined thermal effects and concluded that thermally induced strains are more severe for thicker or stiffer adherends.

6.1.5 Further Refinements

Recently a number of workers including Renton & Vinson,[108] Allman[109] and Delale *et al.*[110] have produced solutions where the adherends have been modelled to account for bending, shear and normal stresses. In two of the approaches, adhesive shear stresses adjacent to the overlap end have been set to zero: this assumption is physically more realistic and its omission has been one of the main criticisms of this form of analysis. In addition, Allman made a further refinement by allowing for a linear variation in peel stress across the bondline thickness. At present no commercial program is available based on these techniques.

6.1.6 Conclusions

The greatest criticism levelled at these analyses is their limited geometry and oversimplification of the micro-geometry at overlap ends. Most only deal with single- or double-lap shear joints although some can cope with stepped-lap or profiled adherends. Their greatest virtue is their ease and speed of use. The geometrical details of joints can be rapidly entered into a computer program in terms of bondlength. adhesive and adherend thickness, material properties and applied loads and a solution can be obtained very rapidly.

6.2 Finite Element Analysis

Developed into a practical tool during the late 1960s and early 1970s, this technique is now the most widely used tool for the analysis of problems relating to structural mechanics, fluid dynamics and electromagnetics. It is not the aim in this section to enter into a detailed description of its workings and the reader is referred elsewhere for such detail.[111] Suffice it to say that virtually any commercially available finite element code can be used in the detailed analysis of adhesive bonds of any form.

6.2.1 Bond Geometry

The closed form methods previously described suffer from the limited geometries upon which they can be used. The finite element technique can be used on virtually any bond geometry and even allows a full three-dimensional analysis to be performed if required. However, this freedom is gained at the expense of a more complex data entry, as the bond geometry must be entered as a series of co-ordinates and then divided up into a number of elements.

For simple joints, this makes it much more time-consuming to use than computer programs based upon the limited geometry, closed form techniques. Deciding upon the number of elements and their shape requires experience. Attempting to consider a relatively small change in geometry, as may be done to try to reduce a stress concentration, will often result in a great deal of work in re-defining the mesh. Other computer programs called 'preprocessors' can remove some of the tedious data entry (replacing it with a drawing on a screen) and perform automatic meshing but they themselves require experience and are expensive.

Furthermore the large difference in elastic modulus between the adhesive and adherends necessitates the use of more than one element across the adhesive thickness to reduce the stress gradients across a given element, but this then requires a large number of elements to be used along the bond to keep the elements from becoming long and thin (best accuracy is usually achieved when the elements are square).

One further advantage of the finite element technique is that it provides information on all stresses and strains in the bond (shear, peel and direct stress along the bond). This allows a better prediction of plasticity and failure in both the adherend and adhesive.

6.2.2 Future Developments

A way forward to a more general approach is, perhaps, a purpose-written finite element code for adhesive bonds, in which the element formulations are optimised for a high aspect ratio and a preprocessor allows the bond geometry to be entered in a relatively simple format, as for a closed form solution.

6.3 Failure Criteria and Joint Strength

Having performed a stress/strain analysis using one of the two methods previously described, it is then necessary to apply some form of failure criterion to the results to determine if the bond has failed. For an adhesive bond there is no single criterion which can be applied. Rather each part of the bond — the adherends, the adhesive and the interface — must be examined separately, and the relevant failure criteria applied.

6.3.1 Material Failure Criteria

Considering the adherend, the correct criterion will depend upon the material. For a metal, von Mises or Tresca may be used. For a brittle ceramic it may be appropriate to use a maximum stress criterion. For a composite material maximum stress, maximum strain or Tsai–Hill criteria might be used in the plane of the fibres, but it will be essential also to consider failure through interlaminar tensile and shear stresses. For the adhesive no definitive failure criterion has yet been accepted. The problem of finding such a criterion is that an adhesive will exhibit very different failure properties when tested in bulk form outside the joint from those exhibited in a real joint. One reason for this is that stress concentrations occur near the interface and it is here that failure actually begins. Figure 4 shows a failed lap shear specimen where fracture initiates within the spew fillet at the end of the joint overlap. This effect has been described in detail by Adams and co-workers at Bristol University.[5] These concentrations however will not be shown by any but the most detailed finite element analyses.

What can be done therefore if we are using a relatively crude closed form analysis which gives only the stress/strain value at the centreline of the adhesive? Probably the most widely used criterion for single-lap shear joints is to calculate the shear strain in the adhesive (ignoring the peel strain) and to compare this with some measured shear strain to failure for the adhesive. For toughened epoxide adhesives this gives a

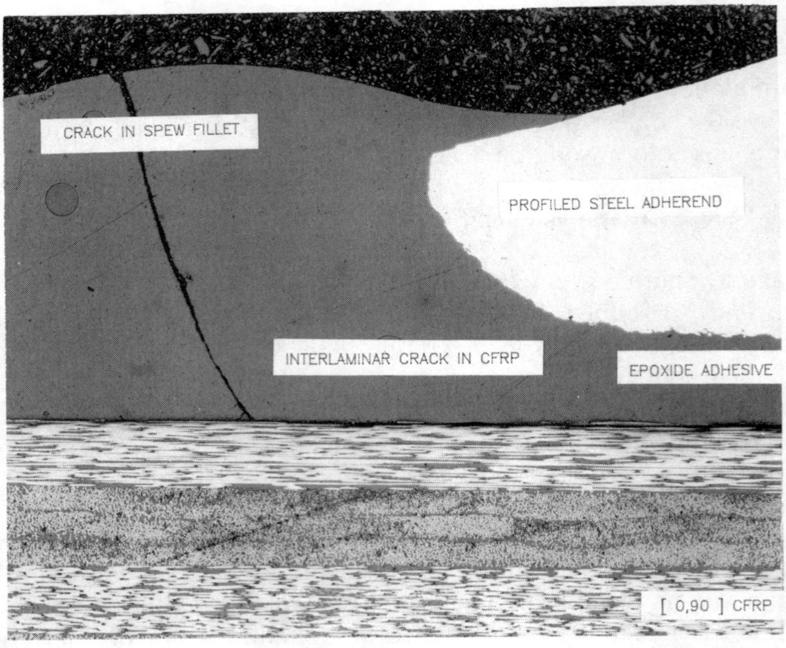

Fig. 4. Crack in adhesive fillet at end of overlap (bondline thickness 0·12 mm).

surprisingly good prediction of joint strength given the factors discussed above, but for other adhesive types such as toughened acrylics or polyurethane the agreement is often less satisfactory.

6.3.2 Interfacial Failure Criteria

The other potential area for failure is the adhesive/adherend interface. The common philosophy adopted when considering failure in this region, i.e. adhesive failure rather than cohesive failure, is that it should not occur in a properly made adhesive joint. If the adherends are correctly pretreated and the adhesive compatible with the adherends, then it is indeed not generally the failure mechanism seen. Exposure of the joint to any aggressive environment (e.g. high humidity) will often lead to corrosion of this interface and hence the joint will fail adhesively (see Chapter 1).

Even if some usable data for the adhesive strength existed, their use would be limited by the inability of the stress/strain analyses to predict values at the interface. Hence realistically the designer is left to rely upon the material scientist to ensure that adhesion is sufficient throughout the life of the joint.

6.3.3 Implementation in Commercial Codes

When designers have performed their analysis on a joint they must then apply a failure criterion to the stress and strain figures predicted. Few analysis codes will apply to them directly, although most can be made to combine the stresses or strains in a predefined way to give perhaps a von Mises criterion or an effective strain. The nature of the adhesive bond is such that different failure criteria must be applied to different parts and most codes are unable to do this. No codes will say, 'This joint has failed because the peel stress is too high here ...'.

6.3.4 Optimising the Design

Finally then the designer will decide if one of the parts of the joint will fail and, if so, look for ways of reducing the stresses or strains in this area by changing the thickness of the adhesive or the adherends, changing the bondlength, profiling the ends of the joint or changing the adhesive. The analysis will then be repeated. This iterative procedure will probably be repeated many times before a satisfactory solution is attained.

Some automation of the procedure is certainly possible by means of an 'expert system' that can select an adhesive and recommend a joint type and initial dimensions based upon information fed in and then call an analysis program automatically. Guidance can then be given on how to improve the joint's performance based upon the results and after the changes have been made by the designer, the analysis re-run. This procedure can be repeated until the designer is satisfied with his work. Such a system is currently under development at Harwell.[64]

7 APPLICATIONS OF ADHESIVES IN ENGINEERING STRUCTURES

There are currently many examples of bonded structures where the adhesive layer is the primary load-bearing element. Often the details of the joint design are commercially sensitive and in many cases the in-service loading conditions on the joint are not always defined precisely. Designers of structures containing bonded joints may decide to use a combination of practical experience and detailed analysis depending on the cost, complexity and importance of the component. Qualification of the design for safety critical components must be proved by full-scale prototype testing under conditions likely to be seen in service. Widespread use of adhesives, particularly for aerospace structures, occurs in bonded honeycomb sandwich structures where aluminium or CFRP skins are bonded to aluminium alloy or Nomex honeycomb cores. This type of construction produces stiff, lightweight panels that are used extensively in modern aircraft for a variety of purposes such as rudders, floor panels, engine cowlings, elevators and flaps. A typical Boeing 747 may contain up to 1300 m^2 of adhesive covering 65% of the aerodynamically wetted surfaces.[112]

A few examples of bonded joints are briefly described.

7.1 Williams–Honda FW11 Grand Prix Car

O'Rourke[113] describes how high structural efficiency is obtained in Formula 1 racing cars by choosing woven CFRP skinned sandwich structures for the major chassis components. The parts are autoclaved and bonded with a two-component toughened epoxide paste adhesive cured at room temperature (Fig. 5). These chassis components are designed to resist the demanding inertial forces imposed during acceleration, braking and cornering, which can produce load factors in excess of 3*g*, yet must weigh only half as much as the driver. An example

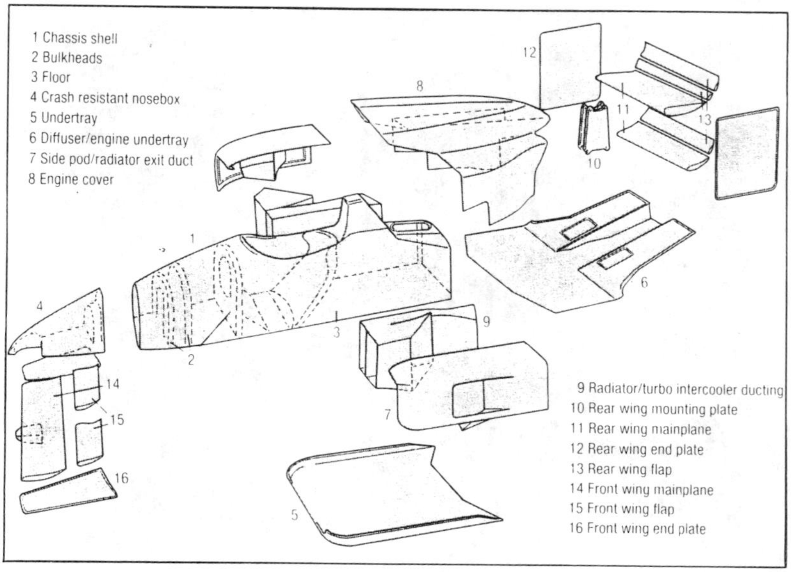

Fig. 5. Composite chassis components of a Formula 1 racing car.

of a safety-critical bonded member is the nose box which must be able to withstand a 39 kJ frontal impact without damaging the section in front of the driver's feet. Experimental results obtained from the large pendulum impact rig at Cranfield Institute of Technology, and the occasional excursion from the racing circuit, have largely allayed fears that bonded structures are not suited to impact situations.

7.2 Dowty Rotol Composite Propeller Blades for Commercial Aircraft

A method of manufacturing advanced composite propeller blades is described by McCarthy[114] in which the blade root, subjected to centrifugal and aerodynamic bending and torsion loads, is an example of a tubular composite to metal joint (Fig. 6). A circular mechanical wedge consisting of a number of annular GRP wedges inserted in the unidirectional carbon fibre spar ensures that the blade remains attached in the unlikely event of a bond failure. Extensive testing to evaluate structural integrity in terms of fatigue, harsh environmental exposure, and simulated lightning strike have proved the design concept. Certification by European and US aviation authorities allow composite

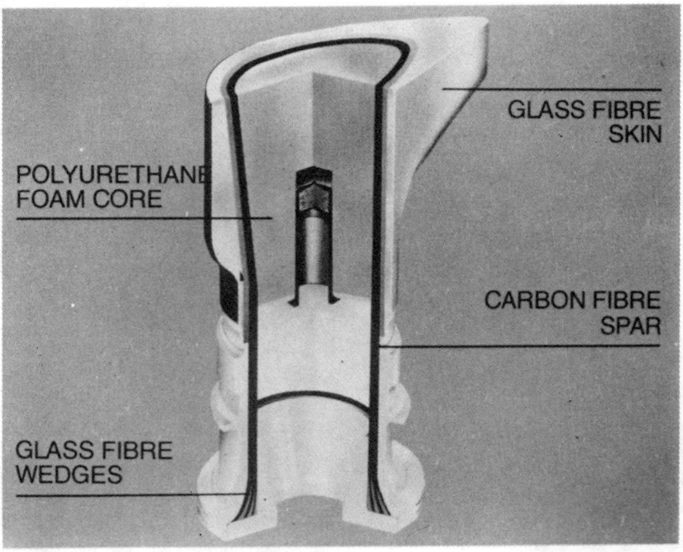

Fig. 6. Advanced composite propeller blade root.

blades, which are safety critical components and must be above suspicion, to be flown on a number of commercial commuter airliners such as the Saab–Fairchild 340 and the Piper Cheyenne 400LS.

7.3 British Aerospace Launch Vehicle Attachment (LVA) Ring

An example of a bonded and bolted lap shear joint is described by BAe (Filton),[115] where the success of the entire mission is critically dependent on the integrity of the joint (Fig. 7). The principal structural feature of the joint in the spacecraft is the central CFRP thrust cone bonded to the aluminium alloy launch vehicle attachment ring. This is subjected to complex inertial loads imposed during launch. It was felt that analysis alone was not sufficient to ensure confidence in the design, and the integrity of the attachment method was checked by static qualification to loads 40% higher than expected during service.

The baseline configuration of the LVA frame joint consisted of a simple twin-lap shear joint between CFRP-faced aluminium alloy honeycomb panels and the alloy frame assembly. Attempts to improve the strength of the joint were made by balancing the stiffness of the

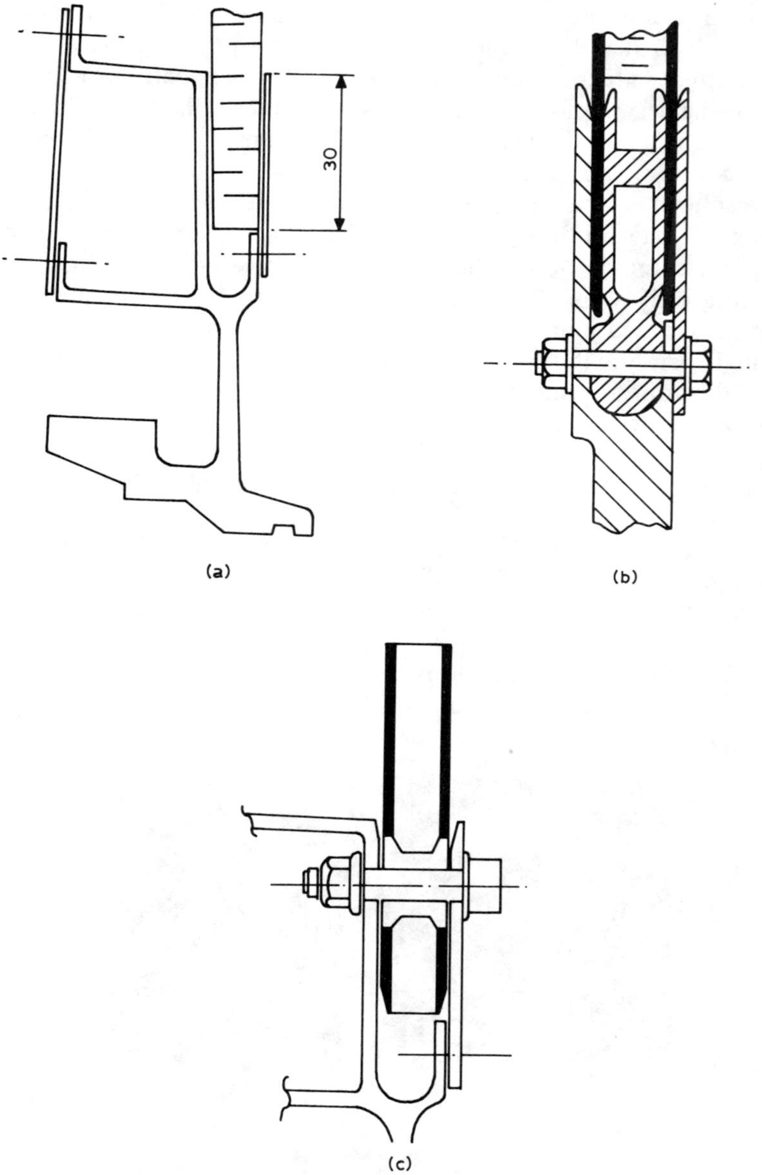

Fig. 7. Launch vehicle attachment ring. (a) base line LVA frame joint. (b) four lap shear face configuration. (c) bonded and bolted configuration.

alloy frame and the CFRP panel by matching *E/t* (stiffness/thickness) ratios which has the effect of reducing and equalising the adhesive shear stress values at the joint ends. The second feature introduced during the design was to chamfer the edges of the adherends which has the effect of locally thickening the adhesive at the joint ends and reduces peak adhesive stresses.

Despite the fact that joint selection test results gave the highest strength results for the optimised bonded design, it was finally decided to adopt a more conservative approach by using a combination of bonding and bolting. This produced panels with a 50% drop in strength over the bonded joints and incurred a mass penalty of 1·4 kg. During normal operation the majority of the shear load is borne by the adhesive but if the adhesive failed for any reason the bolts would be capable of carrying the required load.

7.4 European Space Agency/Harwell Collapsible Tube Mast

A rather unusual production process is described by Aguirre *et al.*[116] and employs a system of sequential moulding of fine-weave carbon fibre fabric for use as furlable booms or masts on satellites and other spacecraft (Fig. 8). This semi-continuous sequential moulding process is used to make two identical long channel sections, each with a circular arc cross-section and two flanges. The two channel sections are then

Fig. 8. Carbon fibre collapsible tube mast for space applications.

adhesively bonded together, using an ultra-lightweight epoxide film adhesive, along the flanges to form the collapsible tube. This tube can be flattened and reeled on to a drum for compact storage during launch and in principle a 60-m length mast of this design can be carried in a storage volume of just 24 litres. Fatigue tests during deployment and retraction between temperature extremes of −70 °C to +80 °C up to 100 times showed no strength loss in samples of the joint cut from the mast.

7.5 Rolls-Royce RB211-524 CFRP Cowl Doors

Owens[117] describes a demonstrator programme to produce from CFRP lightweight cowl doors which are directly interchangeable with aluminium alloy doors, whilst still achieving a weight saving of 25% (Fig. 9). These stiffness-critical structures are constructed from honeycomb sandwich panels with bonded CFRP top-hat stiffeners using an epoxide film adhesive incorporating a nylon fabric scrim (Redux 319A). Environmental exposure testing indicated that the adhesive and CFRP would retain satisfactory integrity under service conditions and several sets of cowl doors have now been examined after completing ~6000 flying hours. Confidence that composite components and bonded assemblies of this type could withstand normal environmental conditions was confirmed.

Fig. 9. Lightweight composite cowl door for an aero-engine.

7.6 Summary

Bonded structures are now widely used in a number of industrial applications. Their success has not been without problems, the most notable of which have been associated with poor environmental durability. If these structures are to play a major part in future designs of engineering structures (and vehicles) and become more widely accepted in a production environment, a number of practical developments are likely to occur. These developments will include robotic application, rapid curing systems, improved heat resistance, simplified surface pretreatments, rapid non-destructive examination and computerised stress analysis design codes.

REFERENCES

1. Evans, G. B., 40 years of structural adhesive bonding. *Chartered Mech. Engr* (March 1985) 23–6.
2. Albericci, P., Aerospace applications. In *Durability of Structural Adhesives*, ed. A. J. Kinloch, Elsevier Applied Science Publishers, London, 1983, Chapter 8.
3. *Engineering Design Handbook — Joining of Advanced Composites*. DARCOM Pamphlet No. 706–316, March 1979.
4. *Structural Adhesives in Engineering*. I. Mech. E. Conference Publication, 1986.
5. Adams, R. D. & Wake, W. C., *Structural Adhesive Joints in Engineering*. Elsevier Applied Science Publishers, London, 1984.
6. Shields, J., *Adhesives Handbook*. Newnes–Butterworths, London, 1984.
7. Kinloch, A. J., ed., *Structural Adhesives — Developments in Resins and Primers*. Elsevier Applied Science Publishers, London, 1986.
8. Matthews, F. L., ed., *Joining Fibre-Reinforced Plastics*. Elsevier Applied Science Publishers, London, 1987.
9. Kinloch, A. J., ed., *Durability of Structural Adhesives*. Elsevier Applied Science Publishers, London, 1983.
10. Kinloch, A. J., *Adhesion and Adhesives — Science and Technology*. Chapman and Hall, London, 1987.
11. Lees, W. A., *Adhesives in Engineering Design*. The Design Council, Springer–Verlag, Berlin, 1984.
12. De Bruyne, N. A. & Houwink, R., eds, *Adhesion and Adhesives*. Elsevier, Amsterdam, 1951.
13. Eley, D. D., ed., *Adhesion*. Oxford University Press, London, 1961.
14. Houwink, R. & Salomen, G., eds, *Adhesion and Adhesives*, Vols I and II. Elsevier, Amsterdam, 1965.
15. Bikerman, J. J., *The Science of Adhesive Joints*. Academic Press, New York, 1968.

16. Wake, W. C., *Adhesion and the Formulation of Adhesives*, 2nd edn. Applied Science Publishers, London, 1982.
17. Kaeble, D. H., *Physical Chemistry of Adhesion*. John Wiley & Sons (Interscience), New York, 1971.
18. Skeist, I., ed., *Handbook of Adhesives*. Van Nostrand, New York, 1977.
19. Shields, J., *Adhesive Bonding*. Oxford University Press, London, 1974.
20. Beadle, J. D., ed., *Design Engineering Series — Adhesives*. Morgan-Grampian, London, 1969.
21. Anderson, G. P., Bennett, S. J. & DeVries, K. L., *Analysis and Testing of Adhesive Bonds*. Academic Press, London, 1977.
22. Patrick, R. L., ed., *Treatise on Adhesion and Adhesives*, Vols 1–4. Marcel Dekker, New York, 1967–1976.
23. Cagle, C. V., *Adhesive Bonding: Techniques and Applications*. McGraw–Hill, New York, 1968.
24. Kinloch, A. J., Review — The science of adhesion: Part 1. Surface and interfacial aspects. *J. Mater. Sci.*, **15** (1980) 2141–66.
25. Kinloch, A. J., Review — The science of adhesion: Part 2. Mechanics and mechanisms of failure. *J. Mater. Sci.*, **17** (1982) 617–51.
26. Godwin, E. W. & Matthews, F. L., A review of the strength of joints in fibre reinforced plastics. Part 1: Mechanically fastened joints. *Composites* (July 1980) 155–60.
27. Matthews, F. L., Kilty, P. F. & Godwin, E. W., A review of the strength of joints in fibre reinforced plastics. Part 2: Adhesively fastened joints. *Composites* (January 1982) 29–37.
28. Matthews, F. L., Kilty, P. F. & Godwin, E. W., Load carrying joints in fibre reinforced plastics. *Plast. Rubber Proc. Applns*, **2** (1982) 19–25.
29. Bowen, D. H., McGeehin, P. & Phillips, D. C., A review of load transfer in highly stressed automotive components. *Plast. Rubber Proc. Applns*, **1** (1981) 195–200.
30. Niranjan, V., *Bonded Joints — A Review for Engineers*. Institute for Aerospace Studies, University of Toronto UTIAS Rev. No. 28, Sept. 1970.
31. Dreger, D. R., Design guidelines for joining advanced composites. *Mac. Des.*, **52**(10) (May 1980) 89–93.
32. Sturgeon, J. B., Joints in carbon fibre reinforced plastics. *Composites* (June 1971) 104–9.
33. Hart-Smith, L. J., Adhesive bonded joints for composites — phenomenological considerations. Douglas Aircraft Company, Paper 6707, March 1978.
34. Hart-Smith, L. J., Further developments in the design and analysis of adhesive bonded structural joints. Douglas Aircraft Company, Paper 6922, April 1980.
35. Hart-Smith, L. J., Differences between adhesive behaviour in test coupons and structural joints. Douglas Aircraft Company, Paper 7066, March 1981.
36. Hart-Smith, L. J., Mechanically fastened joints for advanced composites — phenomenological consideration and simple analyses. Douglas Aircraft Company, Paper 6748A, November, 1978.
37. Hart-Smith, L. J., *Design Methodology for Bonded–Bolted Composite Joints*, Vol. 1. *Analytical Derivations and Illustrative Solutions*. AFWAL TR-81-3154, February 1982.

38. Anon., Which adhesive for strength? *Engineering* (August 1982) 567–71.
39. Tennant, J., Adhesives, defeating the limitations. *CME*, **28**(8) (Sept. 1981) 67–71.
40. Beevers, A., Bridging the confidence gap in adhesive bonding technology. *Metal. Mater. Technol.* (January 1984) 26–8.
41. Lees, W. A., Use of adhesives in constructing vehicles. *Adhes. Age* (Feb. 1981) 23–31.
42. Duke, A. J., Structural adhesives to use or to neglect? *Engng Mater. Design* (June 1968) 937–47.
43. Kerr, J., Sticking with the superglues. *The Engineer* (May 1981) 34–5.
44. *Symposium: Jointing in Fibre-Reinforced Plastics*. Imperial College, IPC Science and Technology Press, 1978, UDC 621.792IMP.
45. Jointing in fibre-reinforced plastics, Imperial College. In *Composites*, **13** (July 1982).
46. Kedward, K. T., ed., *Joining of Composite Materials*. ASTM STP749 (1980).
47. *Structural Adhesives and Bonding*. Technology Conferences Associates, El Segundo, California, 1979.
48. *Bonded Joints and Preparation for Bonding*. AGARD LS-102 (1979).
49. Bodnar, M. J., ed., *Structural Adhesive Bonding*. Appl. Polym. Sci. Symposium No. 3, Interscience, New York, 1966.
50. Bodnar, M. J., ed., *Processing for Adhesives Bonded Structures*. Appl. Polym. Sci. Symposium No. 19, Interscience, New York, 1972.
51. Bodnar, M. J., ed., *Durability of Adhesive Bonded Structures*. Appl. Polym. Sci. Symposium No. 32, Interscience, New York, 1976.
52. Allen, K. W., ed., Volumes based on papers presented at annual conferences on Adhesion and Adhesives held at the City University, London — *Adhesion 1–12*. Applied Science Publishers, London, 1976–1987, UDC668.3ALL.
53. Adhesion and Adhesives: Science, Technology and Applications. Durham University, Plastics and Rubber Institute, Sept. 1980.
54. International Symposium on Adhesive Joints: Their Formulation, Characteristics and Testing. Amer. Chem. Soc., Kansas, 1982.
55. Wake, W. C., ed., *Developments in Adhesives — 1*. Applied Science Publishers, London, 1977.
56. Kinloch, A. J., ed., *Developments in Adhesives — 2*. Applied Science Publishers, London, 1981.
57. Guide to the use of Data Items in the design of bonded joints. ESDU Item 81022, July 1981.
58. Shear stresses in the adhesives in bonded joints. Single step double lap joints loaded in tension. ESDU Item 78042, Nov. 1978.
59. Inelastic shear stresses and strains in the adhesives bonding lap joints loaded in tension or shear (computer program). ESDU Item 79016, Sept. 1969.
60. Elastic stresses in the adhesive in single step double lap bonded joints. ESDU Item 80011, Aug. 1980.
61. Elastic adhesive stresses in multistep lap joints loaded in tension (computer program). ESDU Item 80039, Dec. 1980.

62. Endurance of bonded lap joints. ESDU Fat. 05.06, April 1977.
63. *Computer Aided Adhesives Selection*. Permabond Adhesives Ltd, 1982.
64. *An Adhesive Bonding Design Procedure Tutor — The Expert System 'ADEPT'*. Harwell Laboratory, Didcot, 1987.
65. De Bruyne, N. A., The strength of glued joints. *Aircraft Engng*, **16** (1944) 115.
66. Lee, H. & Neville, K., *Handbook of Epoxy Resins*. McGraw–Hill, London, 1967.
67. Garnish, E. W., Advances in epoxy adhesive technology. In Ref. 52.
68. Lees, W. A., Modified epoxides: practical aspects of toughening. *J. Adhesion*, **12** (1981) 233–40.
69. Hepburn, C., Polyurethane adhesives. In *Polyurethane Elastomers*. Applied Science Publishers, London, 1982, Chapter 8.
70. Lees, W. A., The science of acrylic adhesives. *Brit. Polym. J.*, **11** (June 1979) 64–71.
71. Martin, F. R., Acrylic adhesives. In Ref. 55, Chapter 6.
72. Zalucha, D. J., Bonding metals with acrylic structural adhesives. *SAMPE Quart.* (April 1982) 36–8.
73. Bowditch, M. R. & Stannard, K. J., Bonding GRP with acrylic structural adhesives. In *Adhesion 5*, Ref. 52, 1980.
74. Green, A. K. & Bowyer, W. H., The development of improved attachment methods for stiffening frames on large GRP panels. *Composites* (January 1981) 49–55.
75. Burrows, C., Sammes, N., Stables, G., Stone, M. H. & Tempest, C., Characteristics of toughened acrylate adhesives used with aluminium alloy adherends. In *Adhesion 8*, Ref. 52, 1983, Chapter 7.
76. Shaw, S. J., High temperature adhesives — present and future. *Adhesion 10*, Ref. 52, 1986, pp. 20–41.
77. St Clair, A. K. & St Clair, T. L., A multi-purpose thermoplastic polyimide. *SAMPE Quart.* (Oct. 1981) 20–5.
78. Delvigs, P., Lower curing temperature PMR polyimides. *Polym. Comp.*, **4** (July 1983) 150–3.
79. Fowler, J. R., High temperature adhesives for structural applications — a review. *Mater. Des.*, **3** (Dec. 1982) 602–7.
80. Hendricks, C. L. & Hill, S. G., Evaluation of high temperature structural adhesives for extended service. *SAMPE Quart.* (July 1981) 32–7.
81. St Clair, A. K. & St Clair, T. L., A review of high temperature adhesives. NASA TM83141, July 1981.
82. Shaw, S. J., High temperature polymers for adhesive and composite applications. *Mater. Sci. Technol.*, **3** (Aug. 1987) 589–99.
83. Vinson, J. R., Tschirschnitz, R. & Skoumal, D. L., Shear properties of polyimide adhesives at various temperatures. *Fibre Sci. Technol.*, **18** (1983) 137–49.
84. Carpenter, A. T. & Taylor, J. R., Melt adhesives and their applications. In *Adhesion 1*, Ref. 52, 1976, p. 289.
85. Allen, K. W., Fundamentals of pressure sensitive adhesives. In *Adhesion 3*, Ref. 52, 1978.
86. Counsell, P. J. C. & Whitehouse, R. S., Tack and morphology of pressure sensitive adhesives. In Ref. 55, 1977, Chapter 4.

87. Aubrey, D. W., Pressure sensitive adhesives — principles of formulation. In Ref. 55, 1977, Chapter 5.
88. Hughes, E. J. & Rutherford, J. L., Adhesives in a stress environment. *J. Appl. Polym. Sci., Appl. Polym. Symp.*, **32** (1977) 353–67.
89. Sage, G. N., Aspects of bonded joint design in CFRP. In *Adhesion 3*, Ref. 52, Chapter 8.
90. Sage, G. N. & Adams, R. D., Torsion tests to obtain the shear stress–strain properties of adhesives. ESDU Memorandum No. 42 (to be published).
91. McCarvill, W. T. & Bell, J. P., Torsional test method for adhesive joints. *J. Adhesion,* **6** (1974) 185–93.
92. Dolev, G. & Ishai, O., Mechanical characterisation of adhesive layer *in-situ* and as a bulk material. *J. Adhesion,* **12** (1981) 283–94.
93. Fischer, M. & Schmid, R., Shear strength of bonds and bulk properties of adhesives. In *Adhesion 3*, Ref. 52, 1979, Chapter 3.
94. Frazier, T. B., A computer assisted thick adherend test to characterise the mechanical properties of adhesives. Paper presented at SAMPE Technical Conference 1970.
95. Krieger, R. B., Stiffness characteristics of structural adhesives for stress analysis in hostile environment. American Cyanamid Report, 1975.
96. Krieger, R. B., Shear stress–strain properties of structural adhesives in hostile environments. American Cyanamid Report, 1976.
97. Krieger, R. B., Stress analysis of metal-to-metal bonds in hostile environment. Paper presented at 22nd National SAMPE Symposium, San Diego, 1977.
98. Krieger, R. B., Fatigue testing of structural adhesives. Paper presented at 24th National SAMPE Symposium, San Francisco, 1979.
99. Krieger, R. B., Shear stress strain properties of structural adhesives in hostile environments. *J. Appl. Polym. Sci., Appl. Polym. Symp.*, **32** (1977) 321–39.
100. Althof, W., Progress in structural bonding through effective bondline stress analysis. *22nd National SAMPE Meeting,* (1977) 784–95.
101. Althof, W., Klinger, G., Neumann, G. & Schlothauer, J., Environmental effects on the elastic-plastic properties of adhesives in bonded metal joints. RAE Technical Translation 1999, 1977.
102. Livey, D. T. & McCarthy, J. C., Composite/metal jointing technology for vehicle weight reduction — Summary report. Harwell Laboratory Report, AERE-R12020, Dec. 1985.
103. Volkerson, O, Die Nietkraftverteilung in zubeanspruchten Nietverbindungen mit konstanten Loschonquerschnitten. *Luftfahrtforschung,* **15** (1938) 41.
104. Goland, M. & Reissner, E., Stresses in cemented joints. *J. Appl. Mech.,* **11** (1944) A17.
105. Hart-Smith, L. J., Analysis and design of advanced composite bonded joints. NASA CR-2218, April 1974.
106. Grant, P., Analysis of adhesive stresses in bonded joints. BAC Ltd (Warton), Report SOR(P)109, March 1976.
107. Hart-Smith, L. J., Further developments in the design and analysis of

adhesive-bonded structural joints. Paper presented at ASTM Symposium on Joining of Composite Materials, Minnesota, 1980.
108. Renton, W. J. & Vinson, J. R., The efficient design of adhesive bonded joints. *J. Adhesion,* **7** (1975) 175.
109. Allman, D. J., A theory of elastic stresses in adhesive bonded lap joints. RAE TR 76024 (1976).
110. Delale, F., Erdogan, F. & Aydinoglu, M. N., Stresses in adhesively bonded joints: a closed form solution. *J. Comp. Mater.,* **15** (1981) 249.
111. Adams, R. D., Stress analysis: a finite element analysis approach. In Ref. 56, Chapter 2.
112. Kinloch, A. J., Adhesives in aerospace. In *Proc. Materials in Aerospace*, Royal Aeronautical Society, April 1986, pp. 279–89.
113. O'Rourke, B., Designing for pole position. *Adv. Comp. Engng*, Winter 1986, 6–9.
114. McCarthy, R., Manufacture of composite propeller blades for commuter aircraft. SAE 850875, Kansas, April 1985.
115. Sharkey, J. T., Nayler, G. H. F. & Reynolds, J., Design and development testing of the bonded joint between a typical launch vehicle attachment ring and CFRP thrust cone. *Composites Design for Space Applications, ESTEC, Noordwijk, Oct. 1985*; ESA-SP-243, Feb. 1986, 275–84.
116. Aguirre, M., Bowen, D. H., Davidson, R., Lee, R. J. & Thorpe, T., The development of a continuous manufacturing method for a deployable satellite mast in CFRP. In *BPF Congress, Sept. 1986*, pp. 107–10.
117. Owens, G. A., The material development, component manufacture and post-service evaluation of RB211–524 cowl doors utilising carbon fibre composite materials. In *Proc. 3rd Int. Conf. Composite Structures, Paisley, 1984*, pp. 83–99.

Chapter 9

Delamination

P. Davies*

Division Polymères et Composites, Université de Technologie de Compiègne, 60206 Compiègne, France

1 INTRODUCTION

Composite structures are subjected to a wide range of loading conditions in service. These include both static and dynamic loads and the life of a component will be determined by a failure criterion defining an 'acceptable' level of stiffness or strength. There are several mechanisms which contribute to property degradation and these have received considerable attention in recent years as they differ markedly from the failure processes observed for metal components. Thus processes such as fibre debonding and pull-out and matrix cracking have been shown to play an important role. Another failure mode, unique to laminated structures, is the interlaminar separation known as delamination. In many cases the distinct interply layer is a plane of weakness and delamination resistance becomes the critical property limiting use of composite materials. Loading conditions which promote decohesion of plies can cause a loss of stiffness and strength which may constitute failure in high performance applications. Unfortunately such loading conditions include impact, fatigue and even static tensile loading for certain lay-ups so very careful consideration of delamination resistance is required in designing components.

At present a global theory allowing the precise prediction of failure by

*Present address: Polymer Laboratory, Swiss Federal Institute of Technology — Lausanne, 32, chemin de Bellerive, CH-1007 Lausanne, Switzerland.

delamination growth in real structures does not exist. A semi-empirical analysis based on an average stress criterion is one possible approach, which will be described later. However the concept of damage tolerance, defined as the ability of a structure to resist failure due to the presence of damage for a specified period of time,[1] is also gaining popularity. This involves the use of fracture mechanics tests to enable predictions of delamination growth to be made, based on a known original defect size. Such predictions require an extensive data base covering a range of environmental conditions but initial predictions based on test specimen results appear to be justified by full-scale testing. An alternative approach has been to adopt tougher matrix materials, but whilst this provides a safety factor at present, if use of these materials is to be optimised predictive techniques must be developed.

In a single chapter there is clearly insufficient space for an exhaustive review of published literature. This is a very active area of study at present, as witnessed by a recent ASTM publication devoted entirely to *Delamination and Debonding*.[2] This review will therefore be presented as follows: in the next section the origins of the stresses responsible for delamination are described briefly, showing the limitations of classical laminate theory. The application of fracture mechanics concepts is then described, followed by a detailed review of the test methods used to characterise delamination resistance. This aspect of the subject is emphasised as practical information on these tests and is not widely available outside certain specialised laboratories. Advantages and disadvantages of the various methods are discussed and the influence of test parameters is also considered. Finally, some recent advances in materials are assessed. An extensive list of references is included to facilitate further study.

2 INTERLAMINAR STRESSES

The stress analysis of orthotropic materials has been covered in detail elsewhere[3,4] and this brief introduction is based on these references. Classical lamination theory assumes a plane stress state involving σ_x, σ_y and τ_{xy} but the existence of interlaminar stresses has been recognised for many years; early studies showed that interlaminar shear stresses existed and that these stresses reached a maximum near a free edge.[5–7] Subsequent work revealed that interlaminar normal stresses could also be generated and affect composite behaviour.[8,9]

The presence of interlaminar shear stresses may be easily shown by considering the case of a simple angle ply laminate using the approach of Pipes & Pagano[7]; at the free edge of the laminate shown in Fig. 1 the value of τ_{xy} must equal zero. From laminate theory, when the coupon is subjected to a tensile load, σ_x, a shear strain exists, γ_{12} in principal material co-ordinates. At a ply free edge τ_{xy} is zero, so another stress must exist to ensure moment equilibrium; this must be the interlaminar shear stress τ_{xz}. This stress is thus important at the edges of certain lay-ups. The classical laminate theory holds at the centre of the coupon, as shown in Fig. 2. Salamon has reviewed the methods used to calculate interlaminar stresses.[10]

Herakovich has given a qualitative explanation of the origins of these interlaminar stresses,[11] in terms of the engineering properties of individual plies and in particular the shear coupling coefficient or coefficient of mutual influence $\eta_{xy,x}$ and the Poisson's ratio, ν_{xy}.

$$\eta_{xy,x} = \frac{\gamma_{xy}}{\varepsilon_x}$$

and

$$\nu_{xy} = \frac{-\varepsilon_y}{\varepsilon_x}$$

(ε is the direct strain).

The fundamental reason for the presence of interlaminar stresses in laminated composites is the existence of a mis-match in engineering properties between layers. The magnitude of these stresses is related to the magnitude of the mis-match, the elastic and shear moduli and the stacking sequence: τ_{xz} stresses appear in $\pm\theta$ laminates, which exhibit shear coupling but in which there is no mis-match between the Poisson's ratio values for the different layers. In [0°/90°] laminates however, where there is a Poisson's ratio mis-match but no shear coupling, the non-zero interlaminar stresses are τ_{yz} and σ_z. The interlaminar normal stress σ_z must exist to balance the moment due to σ_y. In more complicated lay-ups all three interlaminar stresses can exist and it has been shown that the stacking sequence will determine whether these stresses will be detrimental to the laminate performance; for example, σ_z may be tensile or compressive thus aiding or hindering delamination.[8]

The case of the quasi-isotropic laminate under tensile loading has been considered by several authors.[12–14] The approach used by

a)

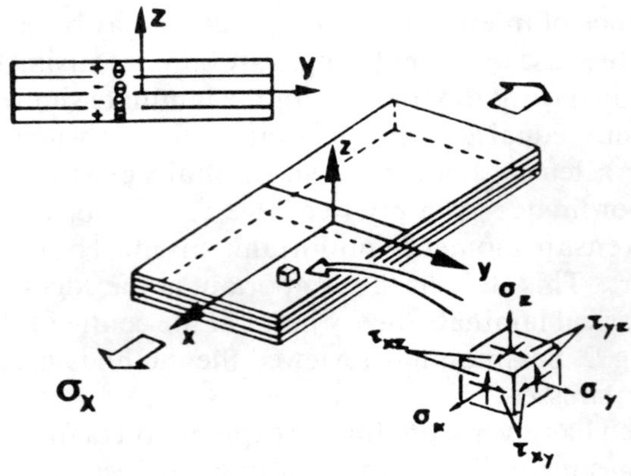

b)

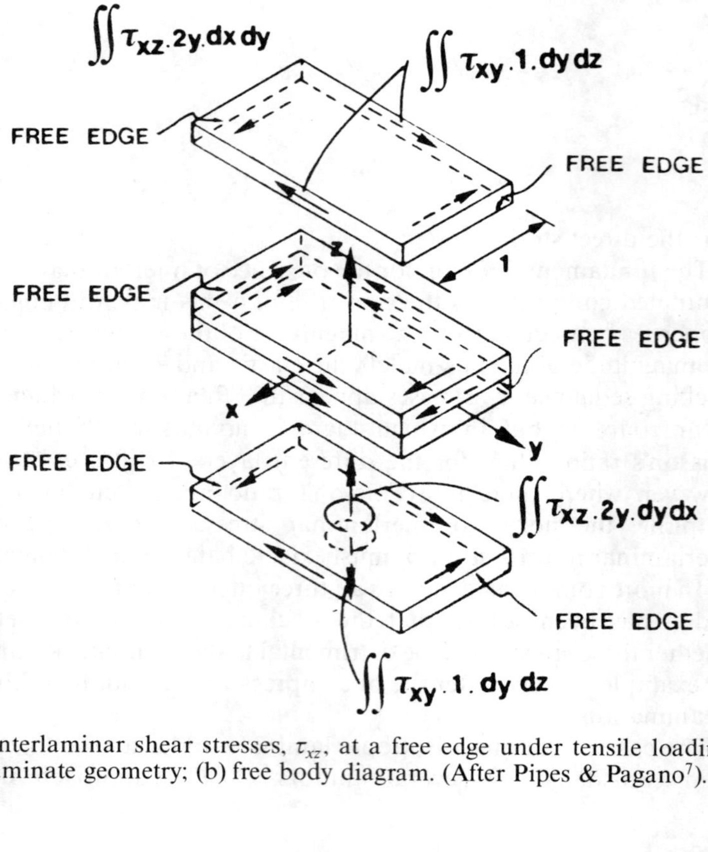

Fig. 1. Interlaminar shear stresses, τ_{xz}, at a free edge under tensile loading. (a) Laminate geometry; (b) free body diagram. (After Pipes & Pagano[7]).

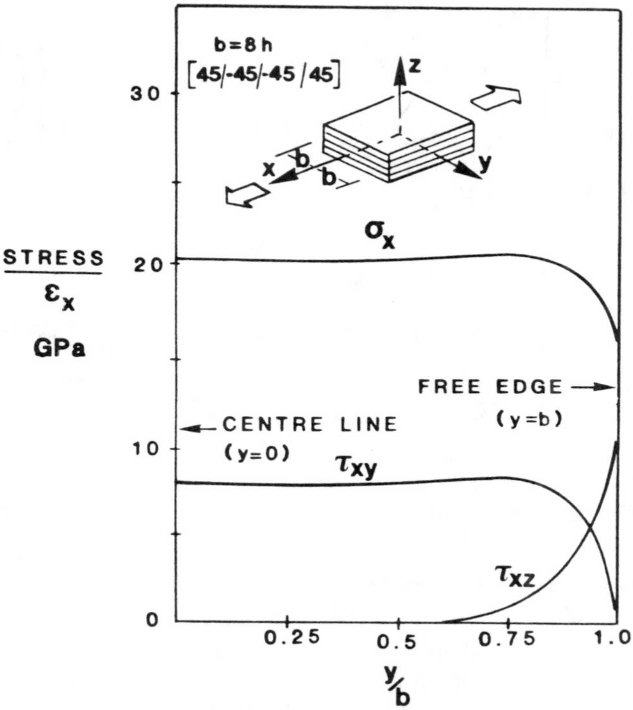

Fig. 2. Stresses at a free edge, finite difference solution. (After Pipes & Pagano.[7]) According to classical lamination theory σ_x/ε_x = 20·4 GPa and τ_{xy} = 7·9 GPa.

Herakovich, Fig. 3, can assist in the optimisation of a stacking sequence to resist delamination. Thus for the quasi-isotropic laminate containing 0°, 90°, +45° and −45° plies, Fig. 3 indicates that the +45° and −45° plies should be separated to minimise $\eta_{xy,x}$ and hence reduce τ_{xz}. Unfortunately other factors such as thermal and hygroscopic property mis-match between layers may also affect the interlaminar stresses set up.

Most of the early studies on interlaminar stresses looked at tensile test coupons. However Salamon examined the case of pure bending and found that, although the interlaminar normal and shear distributions are similar in form to those for uniform extension, for narrow sections the 'free edge problem' affected the whole cross-section.[15]

The implications of these stresses for design are considerable as, in

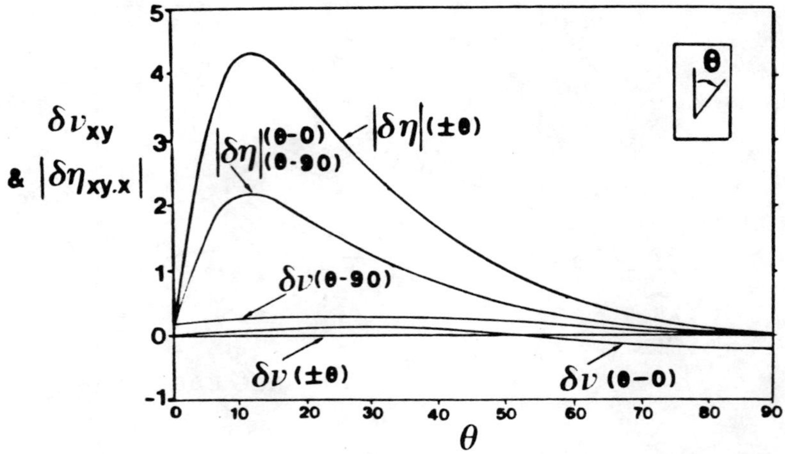

Fig. 3. The mis-match of Poisson's ratio and the shear coupling coefficient for T300/5208 graphite/epoxy as a function of angle. (After Herakovich.[11])

addition to the free edge problem, out-of-plane loads may be set up by a wide range of design details as shown in Fig. 4.

As was noted previously, two approaches to designing to take account of delamination are currently receiving considerable interest. The first is a strength of materials approach developed particularly by Kim & Soni.[16] In this method the interlaminar stress state over a critical

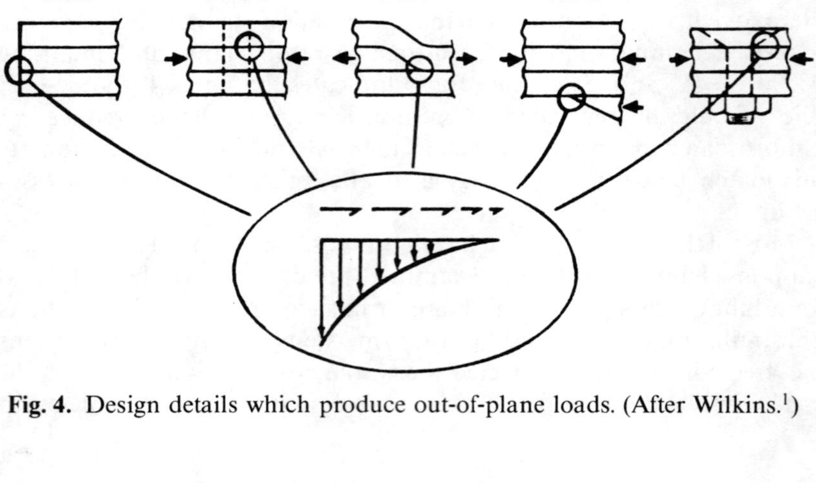

Fig. 4. Design details which produce out-of-plane loads. (After Wilkins.[1])

distance from a free edge is calculated. This stress state is then compared with a failure criterion which may be simply the transverse tensile strength or a more complex three-dimensional criterion.[17] Once the critical distance over which the stresses are to be averaged has been established, this type of analysis may be integrated into a programme for dimensioning of structures.

The second approach, which will be treated in more detail in Section 3, is based on a fracture mechanics methodology.

3 A FRACTURE MECHANICS APPROACH TO DELAMINATION

A number of excellent introductions to fracture mechanics are available[18-20] and the reader is referred to these works for a detailed background to the subject. However, these authors concentrate on the study of isotropic materials, whereas delamination is a rare example of the application of fracture mechanics to an orthotropic material. It is worthwhile therefore to consider the constraints which this imposes.

The fracture mechanics approach includes two distinct starting points from which the potential of a defect may be considered; thus either a stress (K) or an energy (G) criterion may be invoked to predict crack extension, though these approaches have been shown to yield equivalent predictions. A defect may be stressed in three different modes (Fig. 5), these modes acting independently or together.

In studying composite materials, the stress approach could be used and the applied crack tip stress intensity factors K compared with a

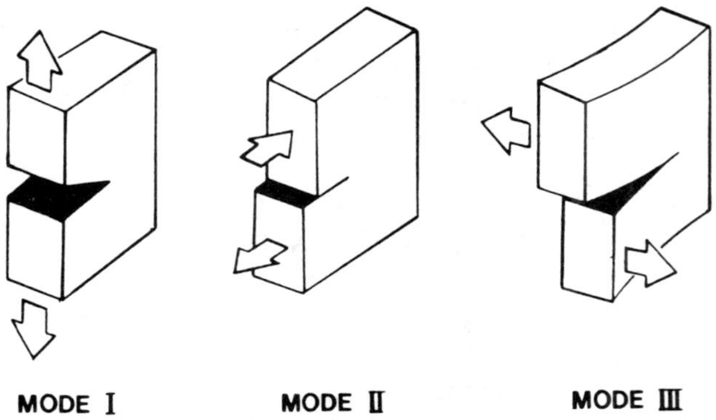

Fig. 5. The three modes of loading of a defect: opening, sliding and tearing.

critical stress intensity factor K_C. However, in composite materials the crack tip stress geometry function relating K to the stress σ and defect dimension a is not the standard isotropic factor and most delamination tests have been developed using the energy approach.

The energy criterion, first proposed by Griffiths for brittle (glassy) materials,[21] relates the energy necessary for unit area of crack propagation to the energy supplied to the stressed body. Thus for crack extension to begin to occur,

$$\frac{\mathrm{d}W}{\mathrm{d}A} - \frac{\mathrm{d}U}{\mathrm{d}A} > \frac{\mathrm{d}U'}{\mathrm{d}A} + \gamma \quad \text{or} \quad G > G_c$$

The left-hand side of this equation represents the difference between the applied work (W) and the global elastic strain energy (U), while the right-hand side includes the irreversible (dissipated) strain energy (U') and the energy required to form new crack surfaces (γ). Thus for materials showing some ductility, crack propagation will occur when the crack-driving energy release rate exceeds a critical value G_c. This critical strain energy release rate is a meaningful parameter for composites if:

(a) the crack propagates in a collinear fashion and
(b) the damage zone is small relative to the size of the specimen.

These criteria are fully developed in a number of recent papers,[22–28] but one case for which propagation is normally collinear is delamination. If the specimen or structure is sufficiently thick to allow the damage zone to be considered as confined, we may therefore adopt a fracture mechanics approach. The material may then be characterised by the Mode I, Mode II and Mode III critical strain energy release rates G_{IC}, G_{IIC} and G_{IIIC}. As laminates have been shown to generate both normal and shear interlaminar stresses during loading, an approach which can characterise the resistance of a material to opening (Mode I) and shear (Modes II and III) thus looks promising.

4 DELAMINATION TESTING

Many tests have been proposed purporting to characterise composite delamination resistance. A number of these are reviewed here, but it should be noted that at the time of writing the only standardised interlaminar test is the Short Beam Shear test, which is of limited value as will be shown below. Unless specified otherwise, all the tests are performed on unidirectional material. Mode I tests are considered first, followed by Mode II and then mixed mode.

4.1 Mode I Tests

The most widely used delamination test is that performed on the Double Cantilever Beam (DCB) specimen which characterises the interlaminar fracture toughness of a composite (Fig. 6). The origins of this test, as for many of the composite fracture mechanics tests, may be found in early work on adhesives.[29, 30] Other notched specimens have also been used for Mode I testing, such as Centre Notch (CN) and Double Torsion (DT) geometries[31–33] but unless very thick laminates are available these will yield intra- rather than inter-laminar fracture toughnesses. The interlaminar test has therefore been preferred in the past, as it represents the lowest energy case and minimises fibre bridging which can otherwise artificially raise toughness values. However the interpretation of the DCB test requires some care and will be considered in detail.

4.1.1 The DCB Test

4.1.1.1 Specimen details. The starter crack is introduced during moulding, to ensure controlled delamination. It is usually a Teflon or Vak-Pak film, although aluminium may be required for high melting point matrices such as PEEK. Specimen width is generally at least 20 mm, to minimise edge effects. Length is 200 mm or more to allow

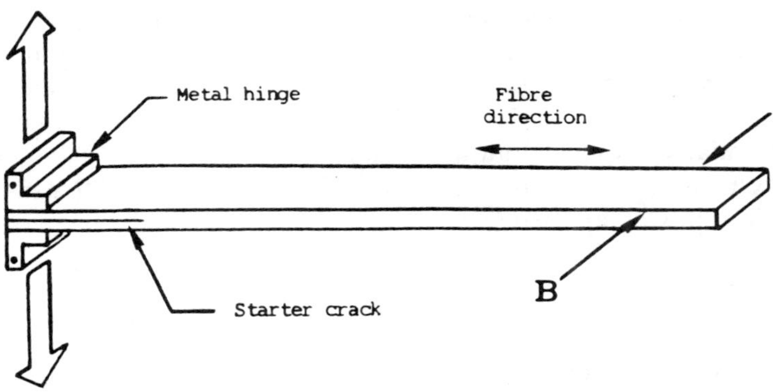

Fig. 6. Typical Double Cantilever Beam (DCB) specimen geometry.

several readings to be made on the same specimen. Thickness is 3 mm or more, as thin specimens undergo large displacements which complicate interpretation.[34,35] An alternative to the use of thick laminates is the bonding of stiff adherends to thin specimens. The behaviour of the sandwich, adherend + adhesive + laminate, is then interpreted using a well-documented procedure.[36]

Load is introduced through metal tabs or hinges which are either bonded to the composite or fixed with small screws. Tests are performed under displacement control, typically at 1 or 2 mm/min. The relative insensitivity of epoxies to test rate has up to now obviated the need for concern over test rates; recent results on composites with thermoplastic matrices suggest that the crack tip displacement rate is an additional parameter to be considered for these materials.[37]

4.1.1.2 Data reduction. Data reduction procedures are of two types, direct energy methods such as the Areas method, and compliance methods. These methods have recently been reviewed.[38] The Areas method[39] is the most widely used; a specimen is initially loaded to propagate the delamination a few millimetres away from the starter crack. It is then unloaded after marking the new crack length. A series of subsequent loading and unloading cycles between known crack lengths results in a load–displacement plot of the type shown in Fig. 7. From each load–unload cycle a value of G_{IC} may be obtained directly, using the expression:

$$G_{IC} = \frac{\Delta A}{B \Delta a}$$

The simplicity of this approach has encouraged its popularity. However, the representation of delamination behaviour by a single number can be extremely misleading and a number of alternative methods based on compliance calibrations have been proposed.

Essentially these techniques require a relationship to be found between compliance and crack length, such that the Irwin–Kies equation may be applied:

$$G_{IC} = \frac{P^2}{2B} \frac{dC}{da}$$

where P is the load required for crack initiation, B is the specimen width, C the compliance and a the crack length. The simplest compliance

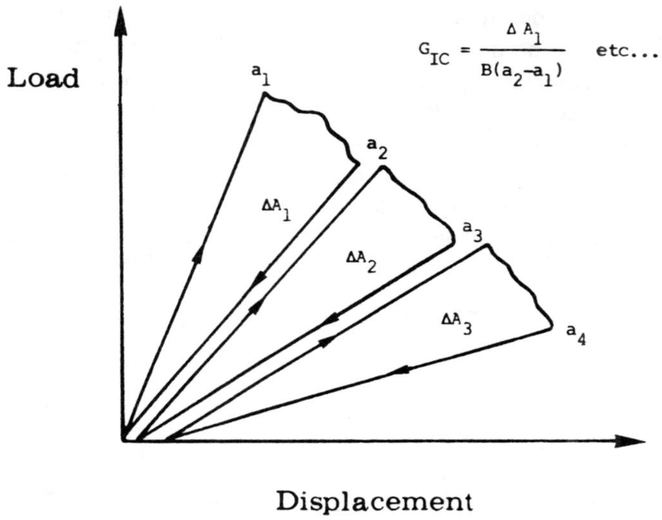

Fig. 7. The Areas method for G_{IC} determination.[39]

relation is the strength of materials expression derived from beam theory:

$$C = \frac{\delta}{P} = \frac{K a^3}{E_f I}$$

where E_f is the flexural modulus, I the beam moment of inertia, and δ is the opening displacement. This assumes that the specimen halves are rigidly clamped and that no rotation is possible at the crack tip. Shear deformation is also ignored, i.e. the crack length is supposed to be sufficiently long to promote pure bending. The simplest modification to this expression is to include a shear correction but a more useful practical approach was proposed by Berry,[40] who suggested an expression of the form

$$C = \frac{\delta}{P} = \frac{a^n}{h}$$

The parameters n and h can be determined experimentally by testing a number of specimens with different initial crack lengths.[41] The value of n obtained may be considerably lower than the 'ideal' bending value of 3.

Other compliance methods have also been proposed by Wilkins *et al.*,[42] Ashizawa[43] and others. Once the compliance calibration has been obtained, a more complete representation of delamination can be presented in the form of *R*-curves,[41] which trace the development of delamination resistance throughout a test. This resistance is plotted as a function of a', the effective crack length taking account of the influence of crack tip damage on compliance. An example is shown in Fig. 8 in which the fracture toughness increases dramatically from a relatively

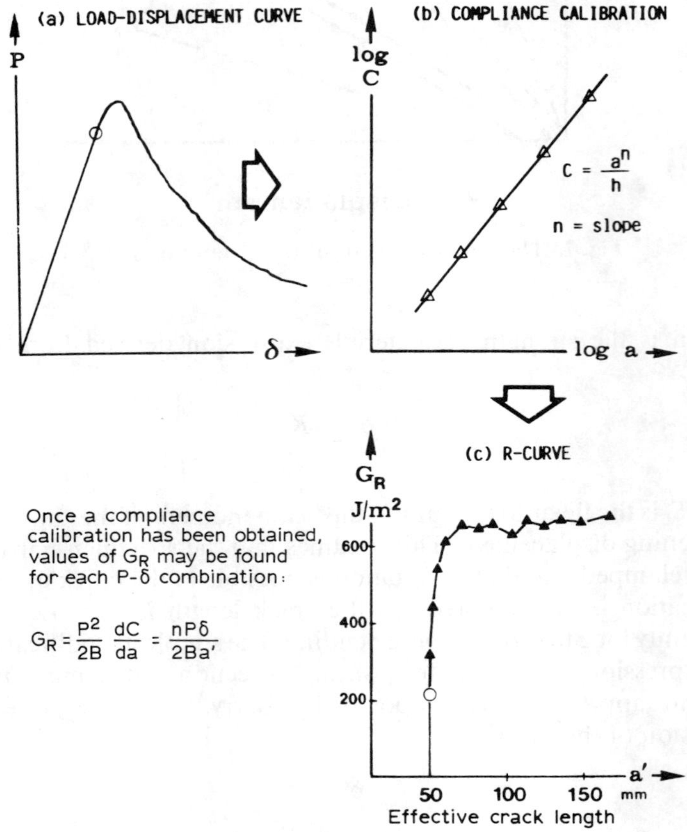

Fig. 8. Construction of *R*-curves. The compliance calibration may be theoretical or experimental, using a single specimen or several. The example shown is for IM6/6376, a carbon fibre reinforced, toughened epoxy. A number of other examples are presented in Ref. 41.

low initiation value. The rise is due in this case to fibre bridging and would be missed if a simple Areas method interpretation had been followed.

Fibre bridges are observed between the specimen halves when the delamination changes plane or if the interlaminar zone is poorly defined. Several authors have studied this phenomenon,[44–46] which is more frequent in tougher materials as their larger damage zones make out-of-plane cracks more likely. If adjacent plies in the DCB specimen are assembled at a small angle, 5° for example, the fibre bridging contribution may be reduced,[45, 47] but in tough composites it will often still be present.

Typical G_{IC} values, as determined by the Areas method, range from 50–200 J/m^2 for carbon fibre reinforced unmodified epoxy systems up to 1000 J/m^2 for toughened epoxy systems, and may reach 3000 J/m^2 for some reinforced thermoplastic composites.

4.2 Mode II Tests

The majority of the methods which have been proposed to characterise the resistance of composites to delamination under shear loading are based on bending tests, using specimens with starter cracks to promote shear failure. Figure 9 shows three such tests. Their interpretation, however, is not always straightforward.

4.2.1 The ENF Test

We first examine the End Notch Flexure test, Fig. 9(a), which has received considerable attention recently; this test is most attractive, as it may be performed on thin specimens. A minimum thickness is required to avoid flexural failure and to ensure the applicability of small displacement beam theory as indicated by Carlsson *et al.*[48] However, a thickness of 3 mm is generally adequate and so a single panel 3 mm or more thick, with a starter crack, enables values of both G_{IC} and G_{IIC} to be obtained, by DCB and ENF tests respectively.

The ENF test was proposed by Barrett & Foschi to study split wooden beams.[49] Russell & Street applied it to carbon/epoxy composites using a data reduction method based on beam theory[50] whilst Giare has studied thick glass/epoxy specimens using a compliance calibration approach.[51] Typically a specimen of length L = 125 mm and width B = 20 mm, containing a starter crack, a = 25 mm, is loaded to failure at a constant

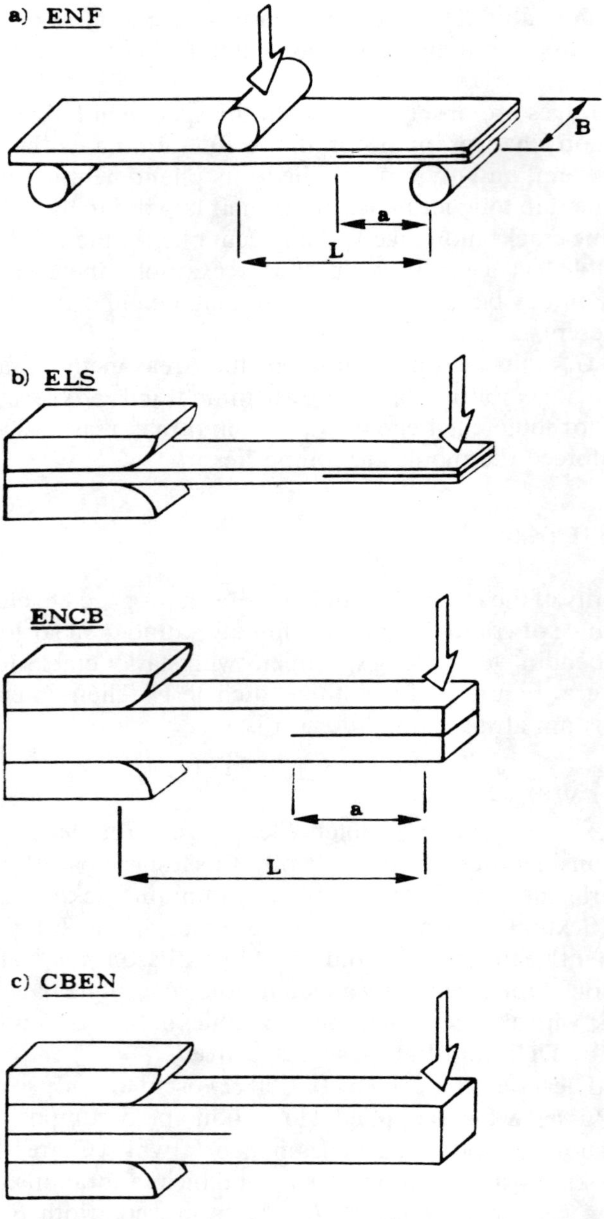

Fig. 9. Schematic diagrams of Mode II bending tests: (a) End Notched Flexure;[50] (b) End Loaded Split (thin[68]) or End Notched Cantilever Beam (thick[56]); (c) Cantilever Beam Enclosed Notch.[57]

cross-head rate. In the beam theory procedure the load P and compliance C at failure are used to obtain G_{IIC} from the expression:

$$G_{IIC} = \frac{9\,a^2\,P^2 C}{2B(2L^3 + 3a^3)}$$

This expression may be modified to account for shear deformation[48] but discrepancies have been found when results have been compared with those obtained by finite element analysis.[52]

Additional problems are caused by the influence of the starter crack; for a carbon-toughened epoxy system we have obtained a value of G_{IIC} of 1250 J/m^2 for propagation from the starter film, compared with 640 J/m^2 for propagation from a (sharp) Mode I precrack.

Typical values of G_{IIC} obtained from ENF tests on precracked carbon reinforced composites are of the order of 500–800 J/m^2 for epoxies and 1500–2500 J/m^2 for some thermoplastics. This kind of generalisation indicates orders of magnitude but tends to mask a major problem in the application of the ENF test to tough composites, best illustrated by the load–displacement curves for three different materials shown in Fig. 10. Characterisation of the relatively brittle carbon epoxy system, Fig. 10(a), is unambiguous as the maximum load P_{MAX} corresponds to the onset of unstable propagation. The toughened epoxy, Fig. 10(b), shows a slight non-linearity after P_{NL}, before the maximum load so that two values of G_{IIC} may be quoted. The increase in compliance corresponds to a crack propagation of 1–2 mm on average for this material, if all the non-linearity is assumed to be due to propagation. This alters markedly the value of G_{IIC}, as shown. Finally, Fig. 10(c) shows a more extreme case, in which not only significant non-linearity is observed before the maximum load, but subsequent propagation is also stable. By load–unload cycles and in-situ observations in the scanning electron microscope we have established that for this material the non-linearity corresponds to extensive microcracking before macroscopic propagation. The Mode II behaviour may therefore be represented by an R-curve on which G_{II} is plotted against the effective crack length calculated from the compliance, Fig. 10(d). Values increase to reach a plateau during stable propagation. It is thus clear that great care is required in the presentation of data from the ENF test to characterise the Mode II behaviour of tough composites.

4.2.2 Other Mode II Tests

The other Mode II tests shown in Fig. 9 have received less attention, but these specimens are better suited to fatigue studies than the ENF

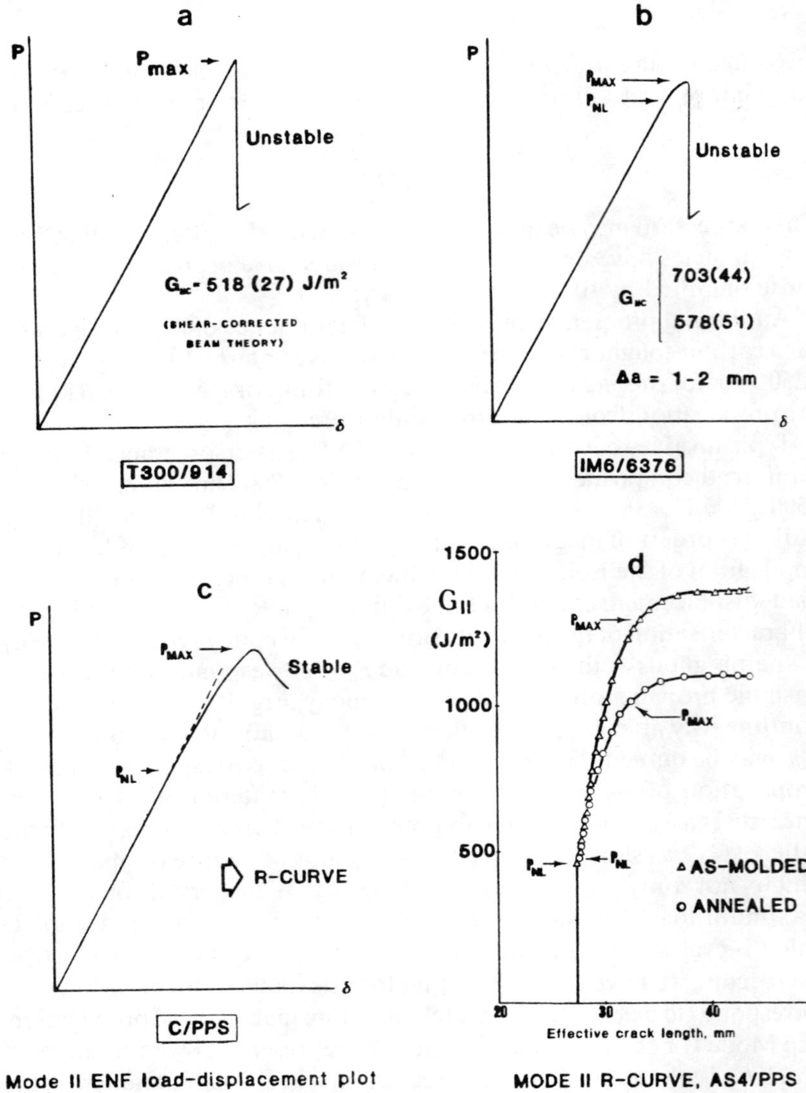

Fig. 10. Interpretation of ENF Mode II test results for different composites: (a) carbon/epoxy T300/914, linear to instability; (b) carbon/toughened epoxy, IM6/6376, slight non-linearity before instability corresponding to damage growth, two possible G_{IIC} values; (c) carbon/thermoplastic, C/PPS; extended damage zone develops before macroscopic stable propagation, for both as-moulded (low crystallinity) and annealed (around 30% crystalline[79]) specimens. An *R*-curve (d) may therefore be constructed, based on calculated crack lengths.

specimen. Bradley and co-workers have used an ELS (End Loaded Split) specimen, of the type shown in Fig. 9(b) to examine the influence of matrix toughness on shear delamination.[53, 54] Russell & Street[55] and Bathias & Laksimi[56] have examined fatigue crack propagation in specimens with open defects of this type. The latter used a thick version of this specimen, sometimes referred to as an ENCB (End Notched Cantilever Beam) specimen.

The Cantilever Beam Enclosed Notch (CBEN) specimen was introduced by Benzeggagh,[57] and is shown in Fig. 9(c). In a recent paper Prel *et al.* have presented static and fatigue results for different materials using both this specimen and the two other Mode II bending specimens described above.[58]

A fourth Mode II test, analysed by Lakshminarayana[59] and used by Donaldson[60] and others[31] is a modified Rail Shear test. A centrally notched panel is loaded and an intralaminar value of K_{IIC} may be obtained. Whilst this has the advantage of being adaptable to existing test equipment, the data reduction is based on the form factor derived for isotropic materials.

Certain authors have observed that increasing the matrix toughness tends to increase G_{IC} rather more than G_{IIC}.[31, 55] Although there is only a limited amount of reliable data available at present for tough composites, given the problems described above, it does appear likely that Mode II testing will become increasingly important if use of these new matrices is to be optimised.

4.3 Mixed Mode Tests

In real structures, defects are subjected to mixed mode loading, so a fracture mechanics approach must be able to describe the interaction between these modes. Thus while material characterisation tends to be performed using pure Mode I and II tests, there has also been considerable interest in mixed mode tests. The best known of these is the Edge Delamination Tension (EDT) test developed by O'Brien *et al.*[61]

4.3.1 The Edge Delamination Tension Test

As was shown in Section 2, tensile loading of certain lay-ups promotes very high interlaminar stresses at specimen edges. O'Brien used this phenomenon to design an 11-ply specimen with the stacking sequence $[\pm 30°/\pm 30°/90°/\overline{90}°]_s$ which delaminates under mixed mode (I and II) loading at the [−30°/90°] interfaces. In the EDT test this specimen, typically 125 mm long by 12·5 mm, is loaded and the strain at the onset of edge delamination, ε_C, is measured at the deviation from linearity on the stress–strain curve. This strain value is then substituted into an equation for the mixed mode critical strain energy release rate G_C:

$$G_C = \varepsilon_C^2 \frac{t}{2} (E_{LAM} - E^*)$$

where t = laminate thickness, E_{LAM} = longitudinal modulus of original laminate, and E^* = modulus of delaminated laminate. Subsequently different stacking sequences have been tested, giving different percentages of Mode I and Mode II.[62] The proportions of G_I and G_{II} were determined by finite element analysis. Results show that Mode I component, G_I, controls the onset of delamination under static loading but that in fatigue G_C controls behaviour. In addition it has been noted that the incorporation of a tougher matrix resin improved delamination resistance under static loading to a greater extent than in fatigue.

Johannesson and co-workers have used an EDT test on angle-ply specimens to examine the behaviour of different materials under Mode III dominated loading.[63, 64] Increasing G_{III} component resulted in higher G_C values.

The EDT test is useful as it investigates the behaviour of real composites, rather than unidirectional specimens, though for some materials the tendency for interaction with tensile interlaminar stresses resulting from 90° ply cracking appears to produce considerable scatter. The criterion of non-linearity may also not always be easy to apply, as toughened composites may show non-linearity before the onset of delamination, and recent results on carbon/PEEK suggest that this material fails by tensile failure before delaminating[65]; tests on thicker specimens should overcome this problem.

4.3.2 Cracked Lap Shear Test

A second popular mixed mode test involves the Cracked Lap Shear specimen. Developed to test adhesives, it was applied to composites by Wilkins *et al.*[42] and is shown in Fig. 11(a). By varying the thicknesses and stacking sequences of the top and bottom parts of the specimen, G_I/G_{II} can be varied between 0·2 and 0·5. A starter delamination is provided by a Teflon film. For brittle composites thin specimens may be used but for tougher composites the thickness must be increased.[66] This test has proved useful for studying fatigue crack propagation under mixed mode

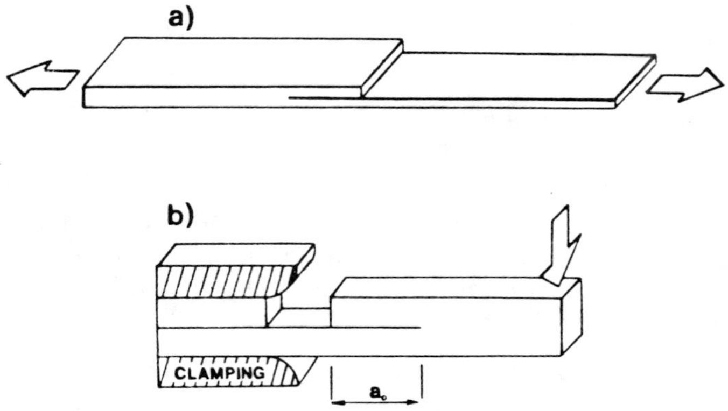

Fig. 11. Mixed mode specimens of the Cracked Lap Shear (CLS) type: (a) tensile test coupon[42] and (b) flexure specimen.[58]

loading.[42, 67] A thick version of this specimen, with a midplane defect, has also been used in bending tests[58] (Fig. 11(b)).

4.3.3 Other Mixed Mode Tests

A third test of interest was proposed by Vanderkley[68] (Fig. 12); this is a test in which mixed mode loading is applied to a DCB specimen by different loads on the two arms, an attractive idea as it allows the same specimens to be used for Mode I, Mode II and mixed mode tests. A similar idea was used recently by Hashemi *et al.*,[71] who introduced a variable ratio, mixed mode specimen.

Finally a test set-up known as the Arcan fixture[69] has been applied to composites to obtain mixed mode interlaminar fracture.[70] This involves a specimen with a starter crack, loaded in a fixture which allows the

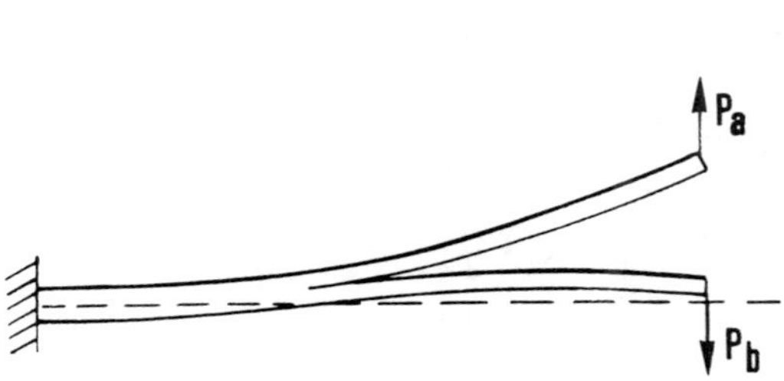

Fig. 12. Mixed mode cantilever beam specimen proposed by Vanderkley.[68]

direction of the load to be varied from pure Mode I through to pure Mode II (Fig. 13).

4.3.4 A Mixed Mode Failure Criterion

In order to apply the data from the simple tests described above to practical design problems it is necessary to be able to predict the fracture behaviour of different materials under any mixed mode loading combination. Despite this fundamental requirement relatively few studies have been published on this subject. A criterion which is often assumed is that:

$$G_C = G_I + G_{II} + G_{III} = \text{constant}$$

For many materials however, G_{IIC} is considerably larger than G_{IC}, so this constant G_C criterion cannot apply. In these cases an empirical criterion of the type

$$\left[\frac{G_I}{G_{IC}}\right]^{m/2} + \left[\frac{G_{II}}{G_{IIC}}\right]^{n/2} = 1$$

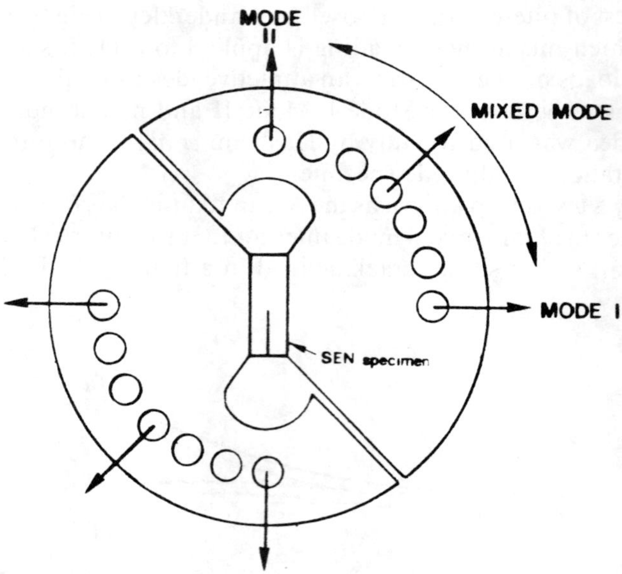

Fig. 13. Arcan fixture used on composites by Jurf & Pipes.[70]

has been shown to describe the behaviour of a number of materials quite closely,[22,70,72] although different values of m and n have been obtained.

Alternatively, a geometrical criterion based on the appearance of fracture surfaces has been proposed to model combinations of both Modes I and II and of I and III loading.[63,64,73] This has enabled differences in fracture energies to be explained on a physical basis, in terms of hackle formation and fibre debonding for example, although this approach is complicated by resin plasticity in more ductile matrices. In fact recent work suggests that in tough composites such as carbon/PEEK a constant total fracture energy may be a reasonably good description of mixed mode behaviour.[31,58]

4.4 The Influence of Test Parameters

Clearly the tests described above would be of little use if the values obtained were dependent on specimen geometry. Unfortunately, geometry independence is not easy to demonstrate. For example it is difficult to investigate the influence of specimen thickness and to be sure that other parameters such as curing rate or residual stresses have not been changed. The influence of other parameters such as loading rate is easier to demonstrate but results can be contradictory. Thus Daniel *et al.* showed that G_{IC} for an unmodified epoxy composite can either increase or decrease with loading rate, depending on the range of speeds considered.[74] Fortunately, over the normal range of testing rates the changes are very small. However, high (impact) loading rates can cause both G_{IC} and G_{IIC} to drop dramatically.[75]

Little work has been published to date on the effect of temperature on delamination resistance. Some tests on epoxy systems suggest that at temperatures up to 100°C these are relatively insensitive to temperature.[72,76] On the other hand thermoplastic composites show increasing G_{IC} values as temperature is increased over a similar range, G_{IIC} remaining largely unchanged.[77–79]

4.5 Other Tests

Finally a few words should be included on the Short Beam or Interlaminar Shear Strength test, to place it in context; while not a fracture mechanics test, its widespread use has made it a reference. However, as the standard (ASTM 2344–76) points out, this test is a quality control tool. The value of τ_{xz} obtained may reflect delamination resistance and interface properties but the complex stress distribution in the specimen can result in very large errors in its calculation.[80]

5 RECENT DEVELOPMENTS AND CONCLUSIONS

The introduction of a range of thermoplastic matrices in recent years has provoked much activity among epoxy resin suppliers. A tough thermosetting matrix allowing continued use of conventional autoclave technology is an attractive proposition and ideas have not been lacking. One solution is Interleaf Technology, the incorporation of thin layers of ductile resin between plies, to provide delamination barriers[81, 82] (Chapter 4). A similar approach is the insertion of layers of tough adhesive during lay-up, in regions likely to undergo delamination such as around holes.

Ultimately, however, the solution must be the elimination of the plane of weakness and the fabrication of three-dimensional composites. Through misalignment or fibre nesting in unidirectional composites the interply layer may sometimes be reduced, with resulting fibre bridging and increased delamination resistance; the natural extension of this idea is the deliberate stitching of plies with fibres running in the thickness direction.[83] Development of machines to achieve this type of lattice reinforcement is already underway and highly sophisticated fibre networks are being proposed.

In this review a summary has been presented of the origins of the stresses causing delamination and of the tests available to characterise the resistance of composite materials to this type of failure. The wide range of tests described underlines the conclusion that in order to establish a usable data base the standardisation of test specimens and procedures is essential. Considerable efforts are now being directed towards this goal.

ACKNOWLEDGEMENTS

The assistance of colleagues in the Division Polymères et Composites, and particularly M. Benzeggagh, M. Prel and M. de Charentenay, in the preparation of this chapter is gratefully acknowledged.

REFERENCES

1. Wilkins, D. J., Damage tolerance of composites: changing emphasis in design, analysis, and testing. Paper presented at 6th Conf. on Fibrous Composites in Structural Design, New Orleans, 1983.
2. Johnson, W. S., ed., *Delamination and Debonding of Materials*. ASTM STP876, 1985.
3. Jones, R. M., *Mechanics of Composite Materials*. McGraw–Hill, New York, 1975.
4. Tsai, S. W., Halpin, J. C. & Pagano, N. J., eds, *Composite Materials Workshop*. Technomic, 1968.
5. Hayashi, T., *Trans-Japan Soc. Aero. Space Sci.*, **10**(17) (1967) 43.
6. Puppo, A. H. & Evensen, H. A., Interlaminar shear in laminated composites under generalized plane stress. *J. Compos. Mater.*, **4** (1970) 204–20.
7. Pipes, R. B. & Pagano, N. J., Interlaminar stresses in composite laminates under uniform axial extension. *J. Compos. Mater.*, **4** (1970) 538.
8. Pagano, N. J. & Pipes, R. B., The influence of stacking sequence on laminate strength. *J. Compos. Mater.*, **5** (1971) 50.
9. Pagano, N. J. & Pipes, R. B., Some observations on the interlaminar strength of composite laminates. *Int. J. Mech. Sci.*, **15** (1973) 679–88.
10. Salamon, N. J., An assessment of the interlaminar stress problem in laminated composites. *J. Compos. Mater., Suppl.* **14** (1980) 177.
11. Herakovich, C. T., On the relationship between engineering properties and delamination of composite materials. *J. Compos. Mater.*, **15** (1981) 336.
12. Reifsnider, K. L., Henneke, E. G. & Stinchcomb, W. W., Delamination in quasi-isotropic graphite–epoxy laminates. In ASTM STP617, 1977, pp. 93–105.
13. Stalnaker, D. O. & Stinchcomb, W. W., Load history edge damage studies in two quasi-isotropic graphite–epoxy laminates. In ASTM STP674, 1979, pp. 620–41.
14. Crossman, F. W., Analysis of free edge induced failure of composite laminates. In *Fracture of Composite Materials*, Sih & Tamuzs. Sijthoff and Noordhoff, 1979.
15. Salamon, N. J., Interlaminar stresses in a layered composite laminate in bending. *Fibre Sci. Technol.* **11** (1978) 305.
16. Kim, R. Y. & Soni, S. R., Experimental and analytical studies on the onset of delamination in laminated composites. *J. Compos. Mater.*, **18** (1984) 70.
17. Lagace, P. A., Delamination in composites: is toughness the key? In *Proc. 31st SAMPE Symp., April 1986*, p. 738.
18. Broek, D., *Elementary Engineering Fracture Mechanics*, Martinus Nijhoff, The Netherlands, 1982.
19. Knott, J. F., *Fundamentals of Fracture Mechanics*. Butterworths, London, 1973.
20. Williams, J. G., *Fracture Mechanics of Polymers*. Ellis–Horwood, Chichester, 1984.
21. Griffith, A. A., The phenomena of rupture and flow in solids. *Phil. Trans. Royal Soc., London, A* **221** (1921) 163.

22. Wu, E. M., Application of fracture mechanics to anisotropic plates. *J. Appl. Mech.*, (Dec. 1967) 967.
23. Phillips, D. C., The fracture mechanics of carbon fibre laminates. *J. Compos. Mater.*, **8** (1974) 130.
24. Dharan, C. K. H., Fracture mechanics of composite materials. *J. Eng. Mater. Technol.*, **100** (July 1978) 233.
25. Kanninen, M. F., Rybicki, E. F. & Brinson, H. F., A critical look at current applications of fracture mechanics to the failure of composite materials. *Composites* (Jan. 1977) 17.
26. Parhizgar, S., Zachary, L. W. & Sun, C. T., Application of the principles of linear fracture mechanics to the composite materials. *Int. J. Fracture,* **20** (1982) 3.
27. Benzeggagh, M. L., Application de la mécanique de la rupture aux matériaux composites, exemple de la rupture par délaminage d'un stratifié. Thesis, Université de Compiègne, June 1980.
28. Caprino, G., Halpin, J. C. & Nicholais, L., Fracture mechanics in composite materials. *Composites* (Oct. 1979) 223.
29. Mostovoy, S., Crosley, P. B. & Ripling, E. J., *J. Mater.*, **2**, (1967) 661.
30. Ripling, E. J., Mostovoy, S. & Patrick, R. L., Measuring fracture toughness of adhesive joints. *Mater. Res. Stand.* (March 1964) 129.
31. Berglund, L. & Johannesson, T., Mixed mode fracture of carbon fibre/PEEK composites. In *Proc. ECCM-1, Bordeaux, Sept. 1985*, p. 259.
32. Whitney, J. M. & Browning, C. E., Materials characterization for matrix dominated failure modes. In ASTM STP836, 1984, p. 104.
33. Lee, S. M., A comparison of fracture toughness of matrix controlled failure modes. *J. Compos. Mater.* **20** (1986) 185.
34. Devitt, D. F., Schapery, R. A. & Bradley, W. L., Method for determining the Mode I delamination fracture toughness of elastic and viscoelastic composite materials. *J. Compos. Mater.*, **14** (1980) 270.
35. Whitcomb, J. D., A simple calculation of strain-energy release rate for a non-linear double cantilever beam. *J. Compos. Technol. Res.* (1985) 64.
36. *Standard Tests for Toughened Resin Composites*, NASA Reference Publication 1092, May 1982.
37. Davies, P., Benzeggagh, M. L. & de Charentenay, F. X., Delamination of continuous carbon fibre reinforced thermoplastic composites. In *Proc. ISCMS (International Symposium on Composite Materials and Structures), Beijing, China, June 1986*. Technomic, Lancaster PA, USA, p. 458.
38. Keary, P. E., Ilcewicz, L. B., Shaar, C. & Trostle, J., Mode I interlaminar fracture toughness of composites using slender double cantilever beam specimens. *J. Compos. Mater.*, **19** (1985) 155.
39. Whitney, J. M., Browning, C. E. & Hoogsteden, W., A double cantilever beam test for characterizing Mode I delamination of composite materials. *J. Reinf. Plastics Compos.*, **1** (1982) 297.
40. Berry, J. P., Determination of fracture surface energies by the cleavage technique. *J. Appl. Phys.*, **34**(1) (1963) 62.
41. De Charentenay, F. X., Harry, J. M., Prel, Y. & Benzeggagh, M., Characterizing the effect of delamination defect by Mode I delamination test. In ASTM STP836, 1984, p. 84.

42. Wilkins, D. J:, Eisenmann, J. R., Camin, R. A., Margolis, W. S. & Benson, R. A., Characterizing delamination growth in graphite–epoxy. In ASTM STP775, 1982, p. 168.
43. Ashizawa, M., Improving damage tolerance of laminated composites through the use of new tough resins. Paper presented at 6th Conf. Fibrous Composites in Structural Design, New Orleans, Jan. 1983.
44. Phillips, D. C. & Wells, G. M., The stability of transverse cracks in fibre composites. *J. Mater. Sci. Lett.*, **1** (1982) 321.
45. Johnson, W. S. & Mangalgiri, P. D., Investigation of Fiber Bridging in DCB specimens. NASA Technical Memo 87716, 1986.
46. Russell, A. J., Micromechanisms of interlaminar fracture and fatigue. *Polym. Compos.*, **8**(5) (Oct. 1987) 342.
47. Curtis, P. T., ed., CRAG Test Methods for the Measurement of the Engineering Properties of Fibre Reinforced Composites. RAE Tech. Report 85099, 1985.
48. Carlsson, L. A., Gillespie, J. W. & Pipes, R. B., On the analysis and design of the end notched flexure (ENF) specimen for Mode II testing. *J. Compos. Mater.*, **20** (Nov. 1986) 594.
49. Barrett, J. D. & Foschi, R. O., Mode II stress-intensity factors for cracked wood beams. *Eng. Fract. Mech.*, **9** (1977) 371.
50. Russell, A. J. & Street, K. N., Factors affecting the interlaminar fracture energy of graphite/epoxy laminates. In *Proc. ICCM-IV, Tokyo, 1982*, p. 279.
51. Giare, G. S., Fracture toughness of unidirectional fibre reinforced composites in Mode II. *Eng. Fract. Mech.*, **20**(1) (1984) 11.
52. Gillespie, J. W., Carlsson, L. A. & Pipes, R. B., Finite element analysis of the end notched flexure specimen for measuring Mode II fracture toughness. *Compos. Sci. Technol.*, **27** (1986) 177.
53. Bradley, W. L. & Jordan, W. M., The relationship between resin ductility and composite delamination fracture toughness. In *Proc. ISCMS, Beijing, China, June 1986*, p. 445.
54. Corleto, C. R., Mode II delamination fracture toughness of unidirectional graphite/epoxy composites. MSc Thesis, Texas A&M University, Aug. 1986.
55. Russell, A. J. & Street, K. N., The effect of matrix toughness on delamination: static and fatigue fracture under Mode II shear loading of graphite fiber composites. Paper presented at NASA/ASTM Symp. on Toughened Composites, Houston, 13–15 March 1985.
56. Bathias, C. & Laksimi, A. In *Delamination and Debonding of Materials*, ed. W. S. Johnson, ASTM STP876, 1985, p. 217.
57. Benzeggagh, M. L., Prel, Y. & de Charentenay, F. X., Instrumentation of Mode I and Mode II tests for crack tip strain profile study under static and cyclic loading. In *Proc. ECCM-1, Bordeaux, 1985*, p. 291.
58. Prel, Y., Davies, P., Benzeggagh, M. & de Charentenay, F. X., Mode I and Mode II delamination of thermosetting and thermoplastic composites. Paper presented at 2nd ASTM Symposium on Fatigue and Fracture, Cincinnati, April 1987. ASTM STP1012, p. 251.
59. Lakshminarayana, H. V., A symmetric rail shear test for Mode II fracture toughness (G_{IIC}) of composite materials. *J. Compos. Mater.*, **18** (1984) 227.

60. Donaldson, S. L., Fracture toughness testing of graphite/epoxy and graphite/PEEK composites. *Composites,* **16**(2) (1985) 103.
61. O'Brien, T. K., Johnston, N. J., Morris, D. H. & Simonds, R. A., A simple test for the interlaminar fracture toughness of composites. *SAMPE J.,* **18**(4) (1982) 8.
62. O'Brien, T. K., Mixed-mode strain-energy-release rate effects on edge delamination of composites. In ASTM STP836, 1984, p. 125.
63. Johannesson, T. & Blikstad, M., Mixed mode fracture in graphite/epoxy laminates. In *Proc. ISCMS, Beijing, China, June 1986*, p. 495.
64. Johannesson, T. & Blikstad, M., Fractography and fracture criteria of the delamination process. In ASTM STP876, 1985, p. 411.
65. Leach, D. C., Curtis, D. C. & Tamblin, D. R., In Delamination behaviour of aromatic polymer composite APC-2. Paper presented at ASTM Symp. Toughened Composites, Houston, 13–15 March 1985. ASTM STP937.
66. Mangalgiri, P. D. & Johnson, W. S., Preliminary design of cracked-lap shear specimen thickness for determination of interlaminar fracture toughness. *J. Compos. Tech. Res.,* **8**(2) (1986) 58.
67. Ramkumar, R. L. & Whitcomb, J. D., Characterization of Mode I and mixed-mode delamination growth in T300/5208 graphite/epoxy. In ASTM STP876, 1985, p. 315.
68. Vanderkley, P. S., Mode I–Mode II delamination fracture toughness of a unidirectional graphite/epoxy composite. MSc Thesis, Texas A&M University, Dec. 1981.
69. Arcan, M., Hashin, Z. & Voloshin, A., A method to produce uniform plane-stress states with applications to fibre-reinforced materials. *Exper. Mech.* (April 1978) 141.
70. Jurf, R. A. & Pipes, R. B., Interlaminar fracture of composite materials. *J. Compos. Mater.,* **16** (Sept. 1982) 386.
71. Hashemi, S., Kinloch, A. J. & Williams, J. G., Interlaminar fracture of composite materials. In *Proc. ICCM VI/ECCM2*, July 1987, London, Vol. 3. Elsevier Applied Science Publishers, London, pp. 3.254–3.264.
72. Russell, A. J. & Street, K. N., Moisture and temperature effects on the mixed-mode delamination fracture of unidirectional graphite/epoxy. In ASTM STP876, 1985, p. 349.
73. Hahn, H. T., A mixed-mode fracture criterion for composite materials. *Compos. Tech. Rev.,* **5**(1) (1983) 26.
74. Daniel, I. M., Shareef, I. & Aliyu, A. A., Rate effects on delamination fracture toughness of a toughened graphite/epoxy. Paper presented at ASTM Symp. Toughened Composites, Houston, 13–15 March 1985.
75. Smiley, A. J., Rate sensitivity of interlaminar fracture toughness in composite materials. MSc Thesis, University of Delaware, Dec. 1985. ASTM STP937.
76. Hunston, D. L. & Bascom, W. D., Effects of lay-up, temperature, and loading rate in double cantilever beam tests of interlaminar crack growth. *Compos. Tech. Rev.* (Winter 1983) 118.
77. Davies, P., Benzeggagh, M. L. & de Charentenay, F. X., Delamination behaviour of continuous carbon fibre reinforced thermoplastic composites. In *Proc. JNC5, 5th French Composites Conference*, Paris, Sept. 1986 (in French) Pluralis, p. 17.

78. Davies, P. & de Charentenay, F. X., The effect of temperature on the interlaminar fracture behaviour of tough composites. In *Proc. ICCM-6, London, July 1987*. Elsevier Applied Science Publishers, London, p. 3.284.
79. Davies, P., Benzeggagh, M. L. & de Charentenay, F. X., Delamination behavior of carbon fiber reinforced PPS. *SAMPE Quart.*, **19**(1) (1987) 19–24.
80. Berg, C. A., Tirosh, J. & Israeli, M., Analysis of short beam bending of fiber reinforced composites. In ASTM STP497, 1972, p. 206.
81. Masters, J. E., Courter, J. L. & Evans, R. E., Impact fracture and failure suppression using interleafed composites. In *Proc. 31st SAMPE Symp., April 1986*, p. 844.
82. Krieger, R. B., Advances in toughness based on interleaf technology. In *6th European SAMPE Conf., Scheveningen, May 1985*, p. 189.
83. Guess, T. R. & Reedy, E. D., Comparison of interlocked fabric and laminated fabric Kevlar 49/epoxy composites. *J. Compos. Technol. Res.*, **7**(4) (1985) 136.

Chapter 10

The Fatigue of Organic Matrix Composite Materials

P. T. Curtis

Materials and Structures Department, Royal Aerospace Establishment, Farnborough, Hampshire, UK

1 INTRODUCTION

When materials are used in structural applications they have to satisfy a range of design criteria including static strength and stiffness. In addition most materials will suffer some long-term weakening during service and in many cases this is the result of the application of cyclic loading or fatigue. In composite materials this is equally true.

Composite materials are usually regarded as possessing good fatigue behaviour[1,2] which, combined with their high specific strength and stiffness, good corrosion resistance and potentially low unit fabrication cost, has made them candidates for many structural applications, particularly where weight is at a premium. Indeed, their excellent tensile fatigue resistance has led to their widespread use in helicopter rotor blades, which are recognised as having a very demanding fatigue requirement. However, many of the newer composite materials in tensile loading, and all in compressive loading, show poorer fatigue performance. The need to understand the effect of fatigue loading on these materials is thus of great importance. In particular there is a great need to develop life-predictive capability at least as good if not better than that available to the metals designer.

In this chapter the fatigue performance of composite materials will be presented and discussed, covering the case of basic unidirectional composites and progressing through multidirectional laminates, stress concentrator effects, environmental effects and life prediction, and finishing with a section on structural design.

2 TEST METHODS

Before discussing the fatigue behaviour of composite materials, it is perhaps first worth considering the means of obtaining the fatigue data on composite materials. The main requirements for a materials fatigue test coupon are that it should fail in a manner similar to material of the comparable structural component. Ideally this should be combined with ease of use and cheapness in preparation. The composites literature contains many papers reporting work comparing different coupon configurations for fatigue testing, in efforts to meet these requirements.[3,4]

2.1 Coupon Profiling

Coupon profiles have been varied, in attempts to ensure failure away from the stress concentrations at the gripped portions of the specimens, ranging from coupons incorporating waists and cut-outs to simple parallel-sided coupons. Waisting usually ensures static failures away from the grips, but not nècessarily in fatigue. Indeed, the plain parallel-sided specimen frequently yields the longest fatigue lives (Fig. 1) and the best all-round behaviour, although failures do occasionally occur at the grips. With care in the preparation of the coupons and the application of end pads, however, the incidence of these can be minimised.

If waisting must be used, then care should be taken to avoid disturbing the lay-up. Waisting is therefore usually restricted to the width dimension of the coupons. Waisting in the thickness will disturb the lay-up of the laminate except for material with all the fibres in the same direction. Waisting unidirectional material, with all the fibres in the test direction, frequently leads to shear failures at the waists (Fig. 2). These grow back in fatigue loading to the grips and trigger failure within the grips at short lifetimes.

2.2 Compression Testing

Compression testing or tension–compression loading is more complex than tensile loading, with the added problem of stabilising the coupon

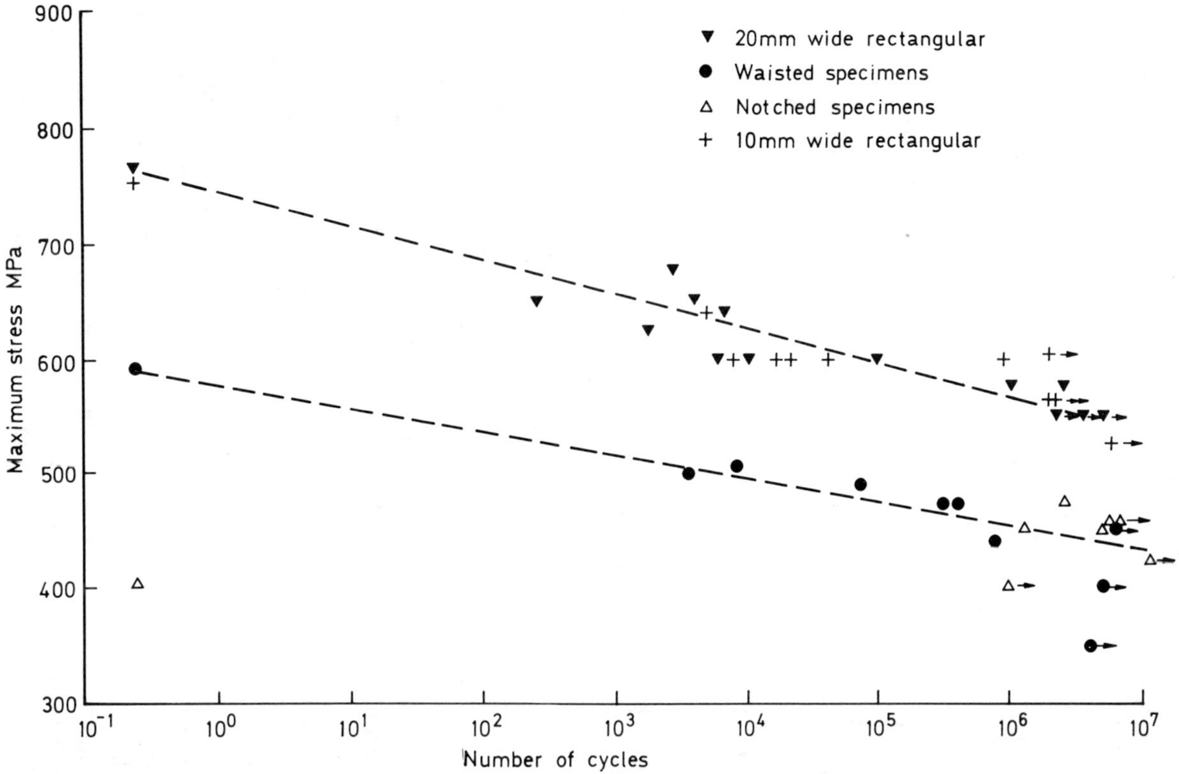

Fig. 1. *S–N* curves for $[0°, \pm 45°, 0°]_s$ laminates in zero-tension fatigue loading: high strength carbon fibre in DX210 epoxy resin.[3]

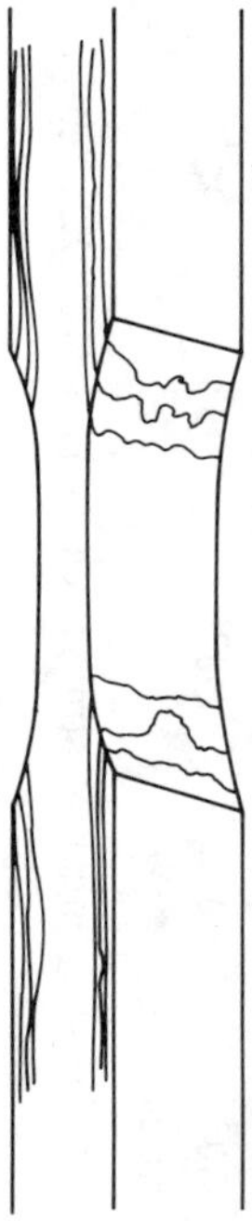

Fig. 2. 0° shear cracks at the waists of a unidirectional fatigue specimen.

during the compressive loading. This requires that either short stable coupons be used or an anti-buckling guide utilised to support the coupon during the compressive fatigue. Short stable coupons, which may be parallel-sided or waisted, suffer from the disadvantage that the stress distribution in the short free length may be affected by the restraint at the grips. Reducing the specimen width to allow for this renders the edge stresses more critical. Typically such specimens will be about 10 mm wide with a 10 mm free length and a minimum thickness of 1·5 mm.[5]

Long coupons are to be preferred but, when a compressive excursion is to be included in the fatigue cycle, it is necessary to provide supports to prevent buckling. No standard anti-buckling guide is recommended, every laboratory generally having developed their own devices. The main factor to consider when designing guides is that the free unsupported area of the specimen should be a maximum, consistent with the requirement of preventing buckling,[5] so as not to restrict any anticipated failure process. In addition, friction between the supports and the specimen must be minimised, for example by the use of PTFE tape on the contact surfaces. A typical device is shown in Fig. 3.

2.3 Flexural Testing

Many laboratories use flexural fatigue testing as an alternative to axial fatigue, since it is easier to perform; it requires no supporting guides and, generally, lower capacity testing machines. However, the data and information generated by this test are limited. This is firstly because

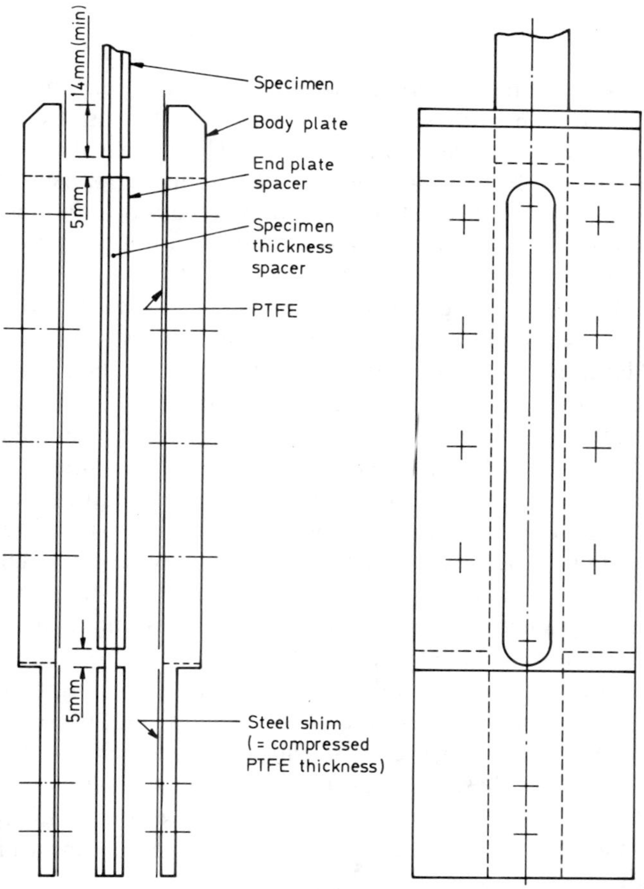

Fig. 3. Typical anti-buckling guide used to support coupons during compressive loading.[4,5]

many structural designs are based on box-like construction and therefore carry predominantly axial loads, and secondly because the results obtained from flexural fatigue tests are difficult to analyse because of the complex stress distribution through the specimen thickness. Consequently axial fatigue testing is preferable unless the end design carries significant flexural loads. Flexural test methods used for static loading are generally suitable for fatigue, but care should be taken to minimise friction at the loading rollers, and it is sometimes necessary to introduce backing rollers on the reverse of the coupons.

2.4 Shear Testing

Fatigue testing in shear loading is less common, but perhaps should be considered more than it is currently. The most widely used shear test in use is probably the interlaminar shear test.[5] This can easily be modified for fatigue use by the introduction of backing rollers opposite the main rollers, particularly if the deflection is to be reversed in fatigue. Again, care has also to be taken to avoid fretting failures at the rollers. Nevertheless, this approach has been used widely.

Alternative shear fatigue test methods that are used are also based on modifications of methods used for static testing. Firstly, the tensile test on $\pm 45°$ laminates, which induces shear along the fibres, has been used extensively as a static test for shear strength and could also be used for the generation of shear fatigue data.[5] Secondly, the rail shear test, equally widely used for static shear strength measurements, has also been used in fatigue. The rail shear specimen requires some modification for it to be suitable for fatigue testing. Recent work with this specimen has shown that the fatigue lives obtainable are very dependent upon the surface quality of the exposed edge of the coupon.[6] Polishing the edge results in a significant increase in life (Fig. 4), presumably as potential cracks are removed. However, if small slots are introduced into the coupon ends, parallel to the rails, even greater lives are obtained (Fig. 4). This may be because the failure zone is shifted from the coupon edge, where constraint and edge effects lead to complex stress fields, to a region where a simple shear stress field exists. It is also worth noting in passing that the effects of shear fatigue loading on composite materials are quite significant; in the case of the glass fibre/epoxy shown in Fig. 4 the sustainable stress corresponding to long lifetimes has been reduced to nearly 50% of the static strength.

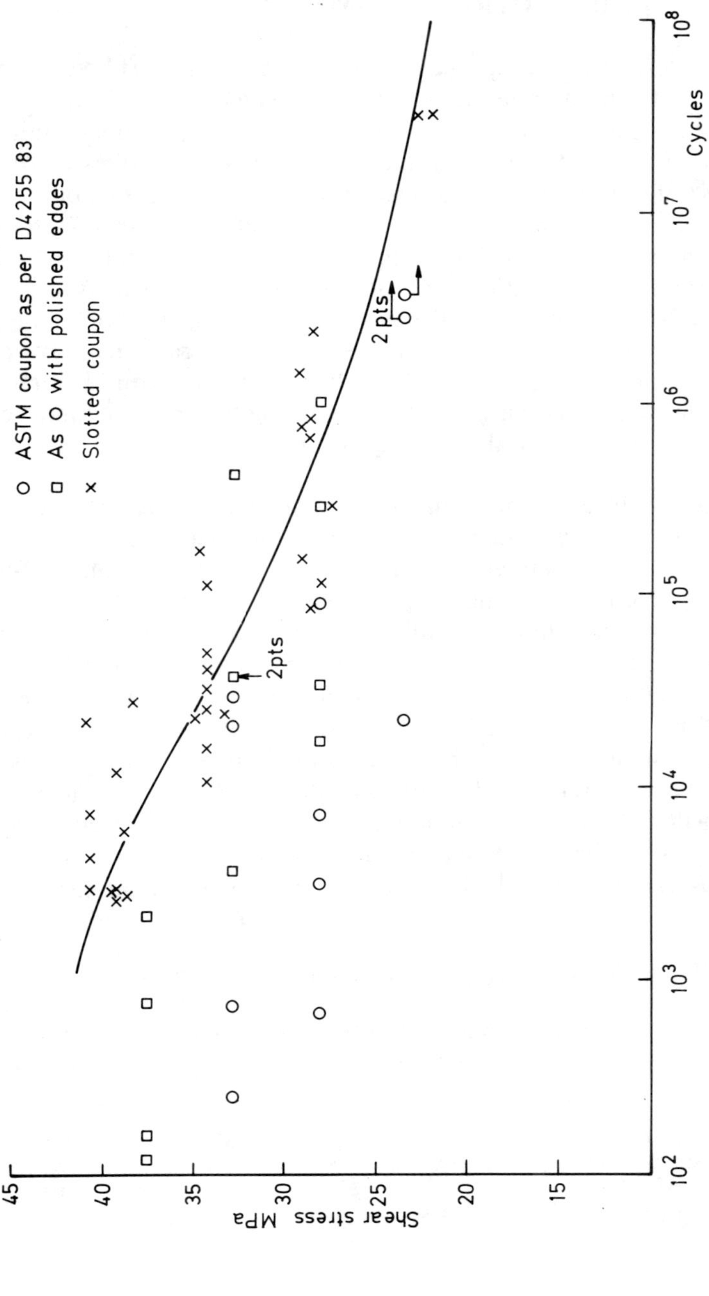

Fig. 4. Shear fatigue *S–N* curves for unidirectional E-glass fibre/epoxy rail shear coupons.[6]

3 UNIDIRECTIONAL COMPOSITE MATERIALS

The simplest form of composite material is one with all the fibres in the test direction. In these materials static strengths as high as 3 GPa can be achieved. Examples of typical unidirectional composite strengths are given in Table 1. Although application of such composites is limited, because of the poorer mechanical properties in directions other than along the fibres, use has been made of composites with large fractions of fibres in one direction, for example in helicopter rotor blades and vehicle leaf springs. Typical plots of peak tensile stress versus log(cycles to failure), the traditional *S–N* presentation of data, are shown in Fig. 5. Data are displayed for three unidirectional materials, namely epoxy resin reinforced with carbon fibre, glass fibre and aramid fibre. The ratio of fatigue stress at high life to static strength is smallest for the glass fibre reinforced material and largest for the carbon fibre reinforced material.

Since for unidirectional materials under tensile loading the fibres carry virtually all the load, the tensile fatigue behaviour of unidirectional composite materials might be expected to depend solely on the fibres, and since the fibres are not usually sensitive to fatigue loading, good fatigue behaviour should result. However, experimental evidence has shown that the slopes of the curves are determined principally by the strain in the matrix.[7,8] Consequently plots of mean strain rather than stress versus log(cycles to failure) are frequently more meaningful for composite materials.[8] All non-metallic fibres, however, have a statistical distribution of strength, determined by flaws; thus a few of the weakest fibres will fail during fatigue loading, as has been observed by many workers.[9] This gives rise to locally high stresses in the matrix and at the fibre/matrix interface, which lead to the development of fatigue damage

TABLE 1
Typical Unidirectional Composite Material Properties

	Standard CF thermoset	*High-strain CF/tough thermoset*	*IM CF/tough thermoset*
Modulus (GPa)	137	136	174
Failure stress (GPa)	1·76	2·26	2·69
Failure strain (%)	1·18	1·66	1·55
Density (g/cm^3)	1·59	1·61	1·56

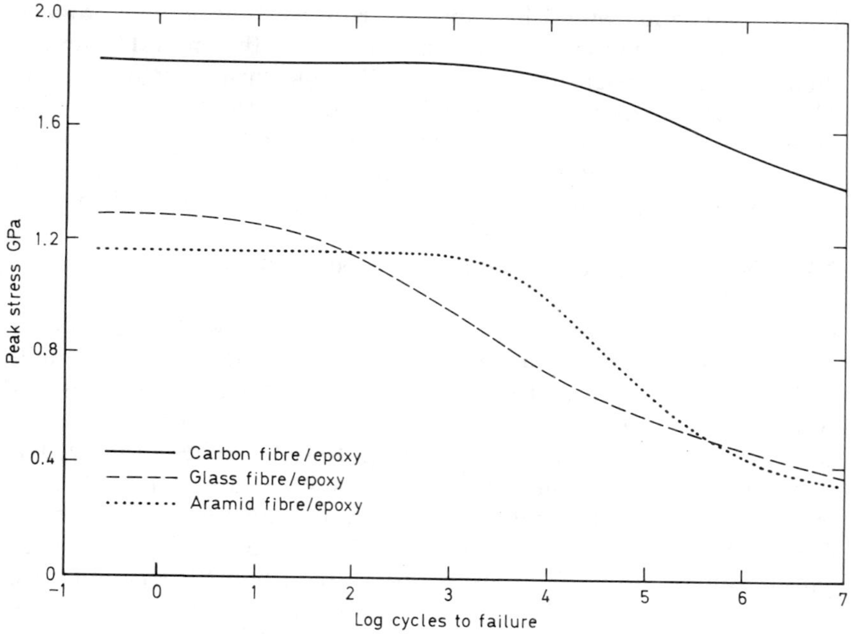

Fig. 5. *S–N* tensile fatigue data for unidirectional composite materials.

with increasing numbers of cycles. Damage may also develop at local micro-defects, such as misaligned fibres, resin-rich regions or voids.[4, 10] Fatigue-induced cracks frequently develop between the fibres, isolating them from adjacent material and rendering them ineffective load carriers. Resin damage also causes fibres to become locally overloaded and further static fibre failures to occur. Close to failure, the matrix may show extensive longitudinal splitting parallel to fibres caused by resin and interfacial damage, leading to a brush-like failure characteristic of most unidirectional materials.[4] The rate of this degradation process in the matrix and at the interface is a function of the bulk strain in the resin as well as the nature of the matrix. The use of very stiff fibres, such as carbon fibres (typically with stiffnesses of 220–700 GPa and failure strains of 0·6–1·8%), results in relatively low strains and thus shallow *S–N* curves, whereas less stiff fibres, such as glass (typically with stiffnesses of 70–80 GPa and failure strains of 2·5–3·5%), lead to greater matrix strains and steeper *S–N* curves. The aramid fibre has a stiffness between that of glass and carbon fibres (typically 120 GPa); thus intermediate fatigue behaviour would be expected. However, the fatigue damage mechanism is complicated in this material since the aramid fibres are themselves fatigue-sensitive and can defibrillate during fatigue loading. This causes the *S–N* curve to adopt the shape depicted in Fig. 5, in which the curve becomes much steeper at intermediate to long lifetimes.[11] Some composites, such as glass fibre reinforced materials, show a significant rate sensitivity and care must therefore be taken when testing them.[11, 12]

3.1 Effect of Recent Fibre and Resin Developments

In recent years manufacturers have sought to improve the mechanical properties of composite materials based on standard high strength carbon fibres in resin matrices. Much of the driving force for this has come from aerospace companies who have published sets of performance goals sought from new materials (e.g. Ref. 13). These included a requirement for a carbon fibre with a breaking strain increased to 2% from the 1·2% typical at that time. Vigorous development by the carbon fibre manufacturers ensued, inspired perhaps by potential commercial gains through the more widespread use of carbon fibre composites, particularly their potential adoption in primary civil aircraft structure. Now carbon fibres with a breaking strain of 1·8% are widely available, all the major manufacturers offering products in this class, and fibres with a 2% breaking strain have been produced (see Chapter 3).

Using higher performance carbon fibres in the same standard epoxy resin matrices generally results in little improvement in either static or fatigue behaviour, as shown in Fig. 6, which depicts plots of peak strain versus log(cycles to failure) for several composites all based on the same resin matrix but with differing carbon fibre reinforcements.[7] Only small changes in the fatigue behaviour are apparent, usually just a slight shift along the ordinate in keeping with changes in the static strength differences measured. This is undoubtedly a consequence of the fatigue behaviour being dependent upon the matrix and interfacial characteristics, rather than the fibre strength.

Concurrently with fibre developments, resin manufacturers have sought to improve the performance of the composite matrix materials. The principal goal has been increased toughness without detriment to the high temperature properties, particularly after moisture absorption. This has led to improvements in static strength, but usually poorer fatigue behaviour with steeper *S–N* curves, as shown in Fig. 7.[7] Thus fatigue behaviour in the next generation of composite materials may be

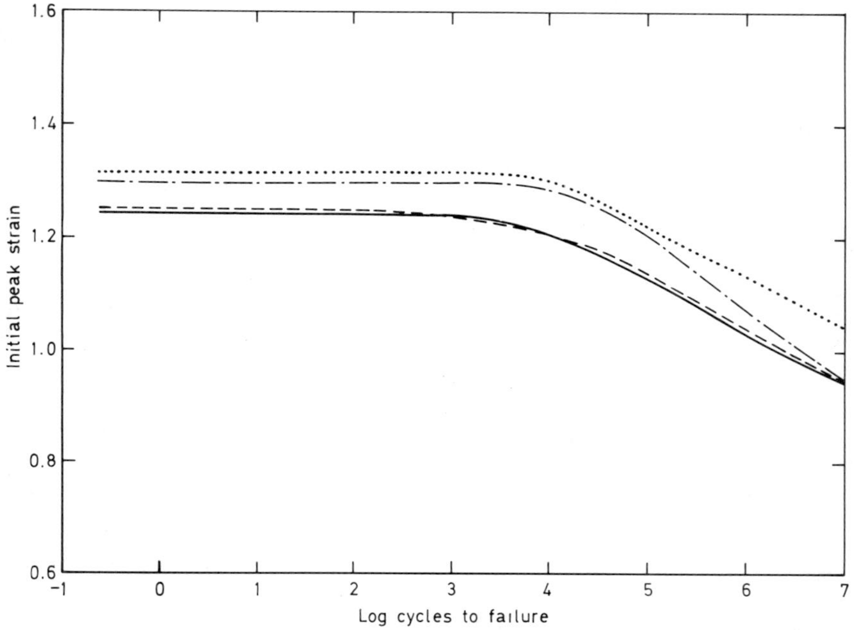

Fig. 6. Fatigue strain–life data for unidirectional carbon fibre composites with different fibre types in the same basic epoxy resin (———, standard CFRP; —·—, — — —, ········, high performance carbon fibres in a standard resin).[7]

somewhat more important than has been the case in the past. The poorer fatigue behaviour of these new materials appears to be due mainly to poorer fatigue performance of the tougher matrices used, but sometimes different failure processes may be triggered, such as excessive fibre/matrix debonding or splitting.

4 MULTIDIRECTIONAL LAMINATED COMPOSITE MATERIALS

Composite materials are usually used in laminated form, layers being arranged so that the fibres are oriented in the principal load directions. Thus typically a large percentage of the layers, perhaps 50%, will have the fibres positioned in the 0° direction to support principal tensile and compressive loads, a smaller number at ±45° to carry torsional loads and perhaps an even smaller number at 90° to impart transverse rigidity

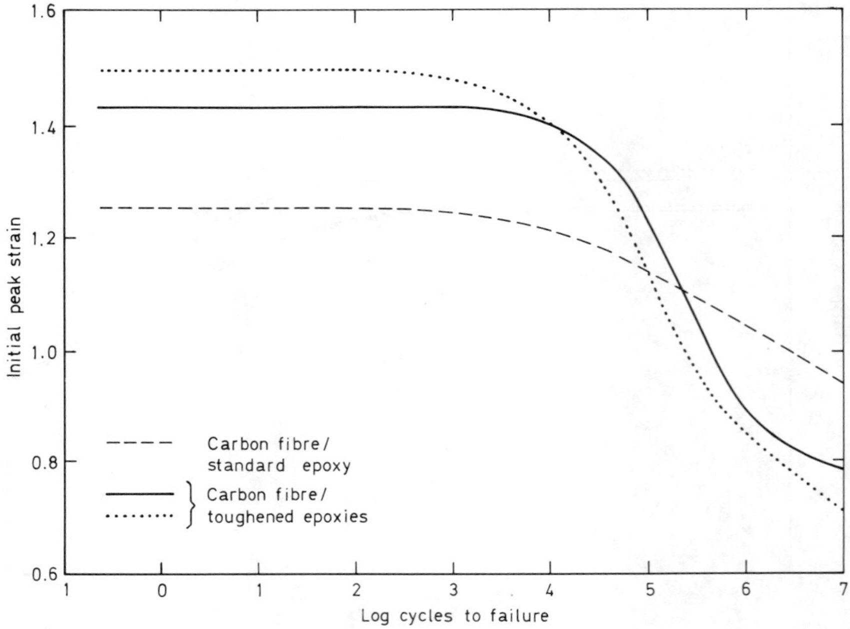

Fig. 7. Fatigue strain–life data for unidirectional carbon fibre composites with the same high strain fibre in different epoxy matrices.[7]

to the panel and limit lateral shrinkage on axial loading. On increasing the percentage of non-axial fibres in a laminate, the static tensile strength and stiffness are reduced since fewer fibres are available to support the mean applied loads. The slope of the tensile *S–N* curve increases in relation to the static strength (Fig. 8), as the layers with off-axis fibres, whose mechanical properties are resin dependent, are more easily damaged in fatigue.

Transverse layers, with fibres at 90° to the test direction, develop transverse cracks either upon the first tensile load cycle or with increasing numbers of cycles,[14, 15] but since they support little axial load this has little effect on the axial strength or stiffness of the material. Angled ply layers, with fibres typically at ±45°, can also develop intraply damage, which will cause small reductions in strength and stiffness. The stress concentration at the ends of intraply cracks can lead to the initiation of delamination between layers, usually resulting in the decoupling of the 0° principal load-bearing layers, which may lead to a general loss of integrity with potential for environmental attack and certainly to a reduction in compressive strength.[9, 14, 16, 17] Alternatively, the cracks may occasionally propagate into adjacent primary load-bearing layers and seriously weaken the material.[7] Typical intraply and interlayer damage is shown in Fig. 9. Ultimate tensile fatigue failure of composite laminates is still determined by the unidirectional layers; thus the tensile *S–N* curves for multidirectional laminated composite materials are still relatively shallow, although steeper than for fully unidirectional material. The poorer fatigue performance observed in unidirectional materials through the use of tough composite matrices is less of a problem in multidirectional laminates, as shown in Fig. 10.

In some applications final fatigue failure may not be the design criterion and the onset of intraply damage in the angled plies may determine the setting of design stresses. The study of fatigue damage development than takes on even greater importance and it becomes necessary to generate *S–N* curves based on the thresholds for the initiation of particular forms of damage, as shown schematically in Fig. 11.

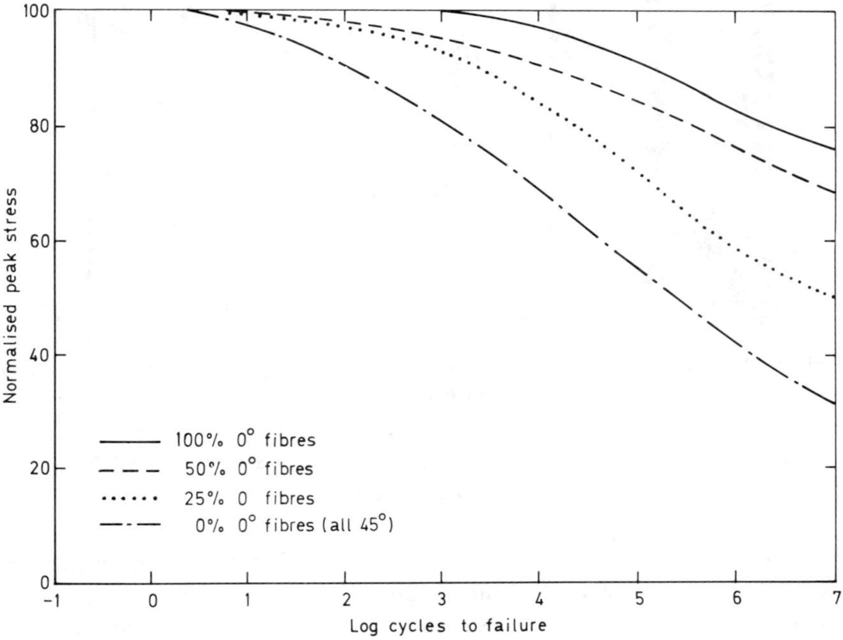

Fig. 8. *S–N* Curves for carbon fibre/epoxy laminates with varying percentages of 0° fibres, normalised with respect to static strength.

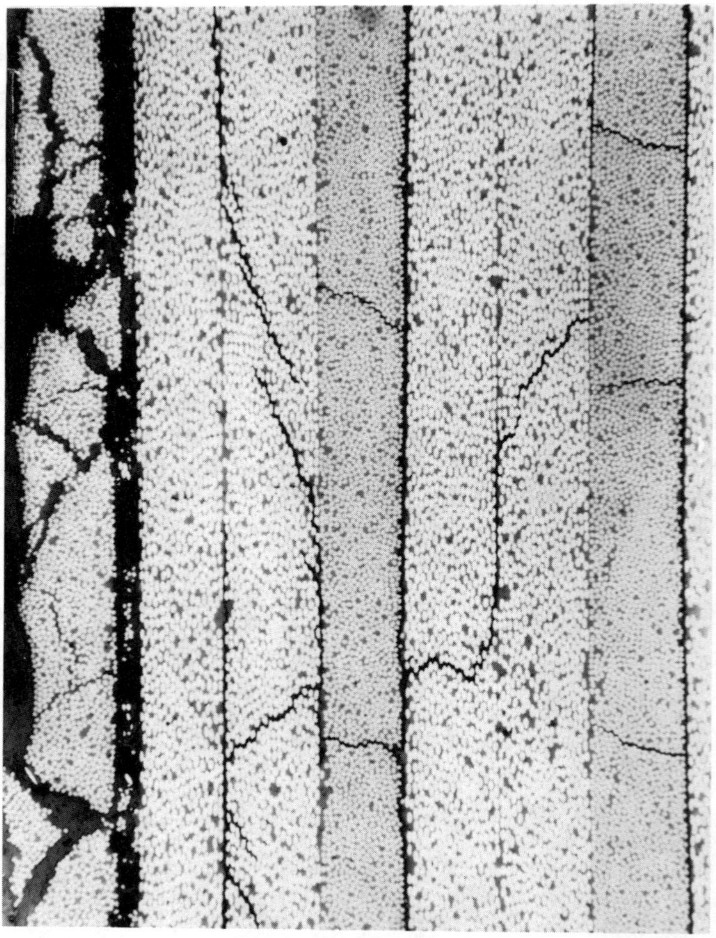

Fig. 9. Optical micrograph of typical fatigue damage in a multi-directional CFRP laminate with a $[(\pm 30°, 90°)_3]_s$ lay-up.[17]

5 NON-DESTRUCTIVE TESTING AND FATIGUE DAMAGE DEVELOPMENT

A key part of any investigative fatigue programme is the determination of fatigue damage development and failure processes of the type described in the previous section. This relies heavily upon non-destructive testing methods, applied to fatigue testing. This review will

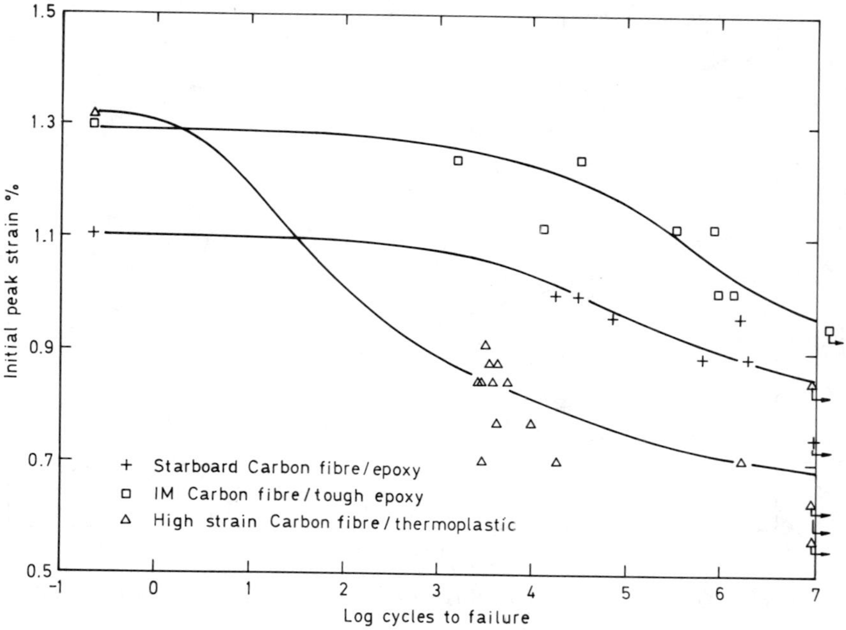

Fig. 10. Fatigue strain–life data for $[0°, 90°, 90°, 0°]_s$ carbon fibre/epoxy composites.[7]

not cover the details of the techniques but concentrate on how they can be applied to the fatigue testing of composites with some examples of the results obtainable.

5.1 Microscopy

Firstly, optical microscopy, which is available to most laboratories, can be most useful in examining damage at edges of coupons and inside holes. The usual technique is to polish the edges of the test coupon by conventional metallographic polishing routes and then carry out examinations at intervals during the test. Ideally the observations should be made without removal of the test coupon from the test machine, but this is not always feasible.

A slight variation of this technique, which obviates the need for specimen removal is the edge replication technique.[9] This relies upon applying a polymer solution to the coupon edge, allowing the solvent to

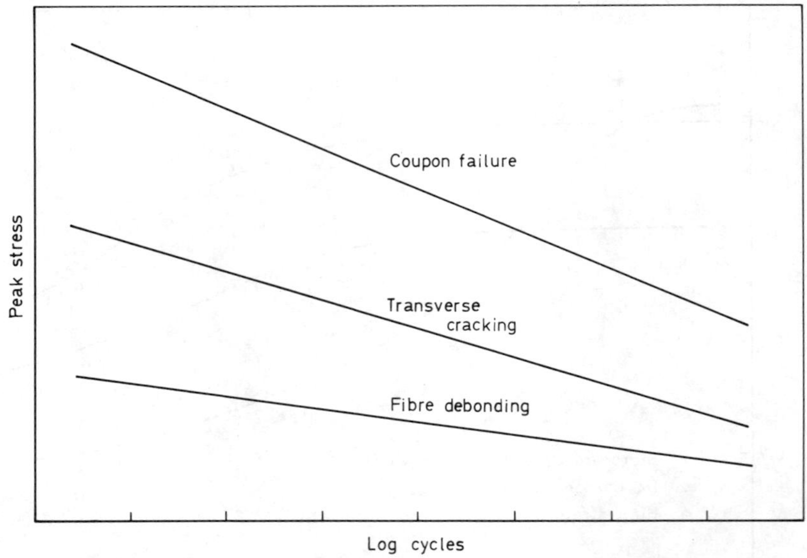

Fig. 11. Typical damage thresholds for a cross-plied composite tested in tensile fatigue.

evaporate, then peeling away the polymer film. This can then be examined conventionally in the optical microscope.

5.2 Ultrasonic C-Scanning

Another technique widely applied to composite materials in fatigue is ultrasonic C-scanning. It relies upon measuring the attenuation of an ultrasonic beam passed through the specimen and relating this to the damage present. The ultrasonic beam requires a transfer medium, which is usually water, and thus removal of the coupons from the test machine is almost inevitable to allow immersion in a tank of water. Techniques relying upon water jetted at the specimens are available, enabling coupons to remain on the test machine, but there are messy to use around complex testing equipment and also expensive, probably beyond the budget of most R and D laboratories. Ultrasonic inspection is particularly useful for detecting interlaminar damage, such as edge cracks and delamination, and studying the growth of this damage in fatigue loading. An example of the detection of damage growth by ultrasonic C-scan is shown in Fig. 12. The technique is less good for detecting in-plane cracks such as transverse cracks in 90° layers.

5.3 X-Radiography

X-ray inspection of composites during fatigue loading is less widely used, but nevertheless it is a useful technique. As applied to composites, it usually relies upon the introduction into the damage of a penetrant that is opaque to X-rays. The most commonly used penetrant is perhaps zinc iodide, introduced in an organic solvent known to be safe in contact with the matrix resin. The usual approach is to remove the fatigue coupon from the test machine for inspection, but equipment is

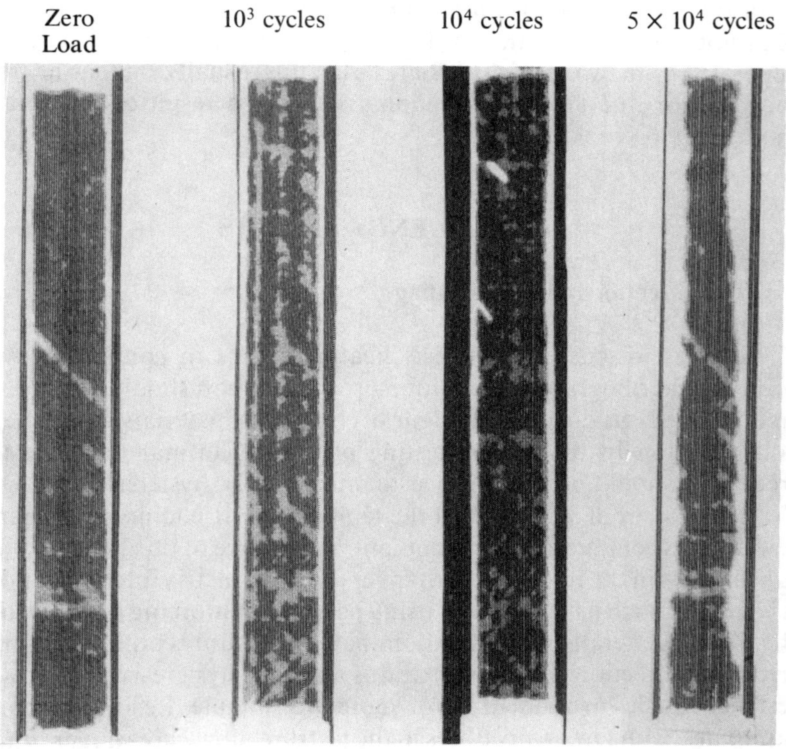

Fig. 12. Ultrasonic C-scans for a carbon fibre/epoxy quasi-isotropic laminate tested in compressive fatigue, as a function of cycles at the stress level giving a mean life of 100 000 cycles.[17]

available which permits inspection of the coupon within the test machine. The technique is particularly good for detecting in-plane damage such as transverse cracks in 90° layers and thus complements the ultrasonic technique well.

5.4 Infra-red Thermography

Finally a mention must be made of infra-red thermography, a slightly more exotic technique than the others but one particularly suitable for the study of damage development during the fatigue loading of composite materials. It has the advantage that the inspection requires no interruption of the fatigue test. The resolution obtainable depends upon the equipment, but is typically similar to that obtainable with ultrasonics (just under 1 mm) and rather less than can be achieved with X-radiography (better than 0·1 mm). The technique can detect heat generated from two sources: hysteresis heating, usually emanating from the resin or interface, and frictional heating, a result of differential movement at cracks.

6 FREQUENCY EFFECTS

6.1 The Effect of Induced Heating

Having just discussed hysteresis heating effects in connection with infra-red thermographic techniques, some mention should be made of frequency effects during the fatigue of composite materials. As a general consideration in the fatigue testing of composite materials, the test frequency should be chosen so as to minimise the hysteresis heating of the material, or at least to limit the temperature rise to no more than a few degrees centigrade above ambient. The source of this heating effect is hysteresis in the resin and perhaps at the fibre/matrix interface, and in a few cases, such as composites using polymer reinforcing fibres, also in the fibres. Generally laminates dominated by mainly continuous fibres in the test direction show lower strains and little hysteresis heating, and test frequencies around 10 Hz or more are suitable. Resin-dominated laminates, with few or any fibres in the test direction, show larger strains and marked hysteresis heating; as a guide, frequencies of 5 Hz or less are suitable.

Local heating at damage sites, the alternative source of heating, may still occur and could cause local overheating. Ideally the specimen temperature should be monitored during the test to ensure that overheating does not occur. This is difficult without expensive thermography equipment, although the strategic positioning of thermocouples, the use of hand-held temperature sensors or the application of temperature-sensitive coatings may be suitable alternatives, particularly when the site of the heating effect is known, such as when stress concentrators are present.

6.2 The Effect on Mechanical Properties

The effect of frequency on properties, i.e. the effect of fatigue loading rate, is negligible for most continuous fibre composites tested in the fibre direction, as long as hysteresis heating is also negligible. The main exception is GRP, in which there is a significant rate effect; the greater the rate of testing, the greater the strength. One group of workers[11] found a rate sensitivity for strength of over 100 MPa per decade rate (Fig. 13). The reason for this is not entirely clear, but it is believed to be due to the environmental sensitivity of the glass fibres rather than any visco-elastic effect[18] (see Chapter 1). Certainly the effect has been found to change when the environment surrounding the glass fibres changes. Testing composites with no fibres in the test direction, where the resin matrix has visco-elastic behaviour, will often also result in a significant rate effect.

When collecting fatigue data on composite materials, therefore, it is

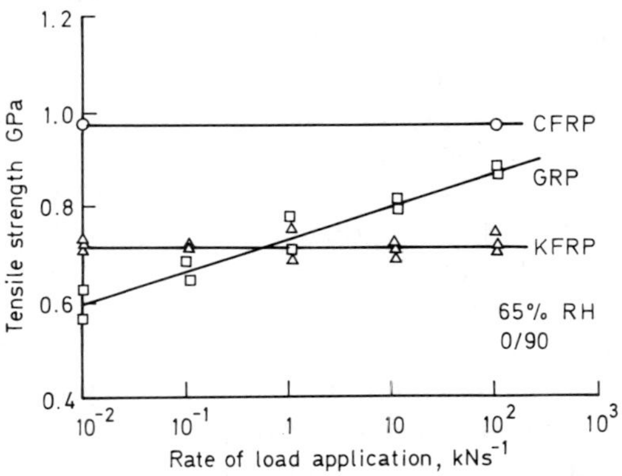

Fig. 13. Tensile strength versus loading rate for cross-plied composites.[11]

important to carry out the static tests at the same rate as the fatigue tests, or the static strengths may be artificially low. Indeed, the best policy with GRP, as well as with laminates showing marked hysteresis heating, is to carry out all fatigue tests at a constant rate of stressing.[12] Thus low load tests are performed at relatively high frequencies and high load tests at low frequencies.

7 EDGE EFFECTS

Edge-induced stresses can be a problem in many types of testing, but especially so in fatigue. Some tests, such as those investigating interlaminar effects, may aim to maximise edge effects (see Chapter 9), but in fatigue tests the policy is usually to try to minimise edge-induced stresses and the damage that inevitably develops as a result. The magnitude of edge stresses will change with both temperature, as layers have differing expansion coefficients, and also with moisture content, as layers expand to different extents on absorbing external moisture.[22] Layer stacking sequence is also a critical variable, the magnitude of edge stresses varying greatly with the relative positions of the layers. Figure 14 shows the distribution of interlaminar normal stress from the midplane to the specimen edge for four different lay-ups, all with the same percentages of 0°, 45° and 90° plies.[20,21] Two of the lay-ups (A and C) can be seen to develop large edge stresses at the specimen midplanes, the other two (B and D) lower stresses away from the midplane. In tests on these laminates all but lay-up B developed edge cracks before failure, those in lay-ups A and C being quite severe, which is in keeping with the magnitude of the edge stresses predicted.

Edge-induced damage, apparent in static loading, usually grows with increasing numbers of fatigue cycles, as shown in Fig. 12. In the worst cases the layers can become completely delaminated, as shown in Fig. 9, leading to potential environmental attack and certainly serious losses in compressive strength. In this coupon most of the layers have delaminated, as a result of both the large shear and normal edge stresses in the laminate and the extensive translaminar damage that developed.

There have been many treatments of edge-induced stresses in the literature,[19–21] some allowing the approximate magnitude of induced edge stresses to be calculated from elastic properties. Thus a laminate's susceptibility to edge effects may be determined before embarking on a fatigue test programme and laminates known to be relatively insensitive

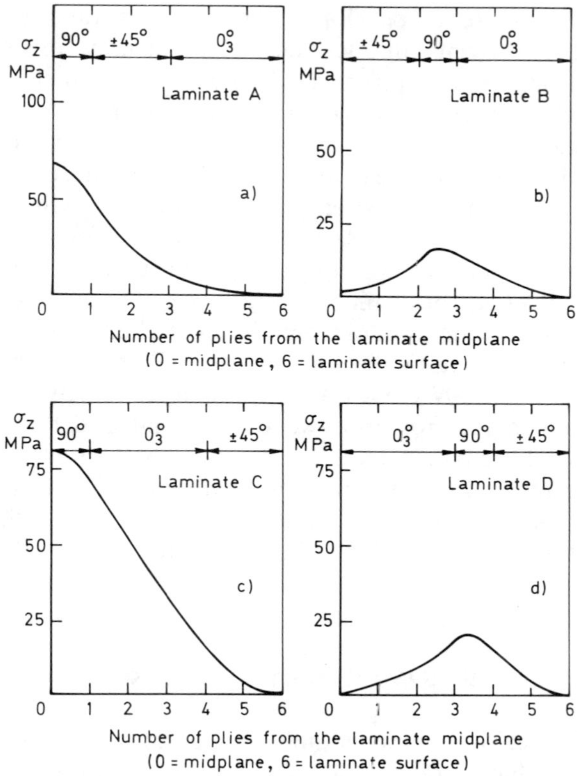

Fig. 14. Theoretical variation of interlaminar normal edge stress as a function of distance from the laminate midplane for four laminates with varying layer stacking sequences but the same percentage of 0°, 45° and 90° layers.[20]

to edge effects may be selected for the work. In general, laminates with thin, evenly distributed layers lead to the lowest edge stresses for both tensile and compressive externally applied loads.

8 THE EFFECT OF STRESS CONCENTRATORS

Stress concentrators such as notches, holes, fasteners, impact damage and other imperfections have less effect on fatigue strength than they do statically. Depending upon the laminate configuration, these stress concentrations can reduce static tensile strength by up to 50%. In fatigue, however, damage zones develop at stress concentrations which can serve to reduce their magnitude. These zones usually consist of cracks along the fibres within layers, and interlaminar cracking between layers, which, if they do not damage fibres, can lead to increased strength. Indeed if fatigue loading is applied by gradually increasing the applied load from zero to peak load over the course of several hundred cycles, then the apparent strength of composites with stress concentrators may actually increase, as shown in Fig. 15 for notched carbon fibre reinforced epoxy resin coupons.[23, 24] Further cycling results in some loss in strength, but typically fatigue strength, calculated on a net stress basis, approaches that of the plain unnotched material after long lifetimes, resulting in fairly flat *S–N* curves.[23, 26] This is also shown in Fig. 15, in which four *S–N* curves are plotted for two sets of composite materials, one made from non-woven material and the other from woven carbon fibre fabric. The static strengths of the notched coupons, both woven and non-woven, are significantly lower than those of the plain coupons, but after large numbers of fatigue cycles their fatigue strengths approach those of the plain coupons.

9 COMPRESSIVE LOADING

The Achilles heel of composite materials undoubtedly lies with their compressive behaviour. Even the mechanism of compressive failure in composites is not well understood.[7, 27] In compression, although the fibres remain the principal load-bearing elements, they must be supported from becoming locally unstable and undergoing a microbuckling type of failure.[23, 24, 28] This is the task of the matrix and the fibre/matrix interface, the integrity of both being of far greater importance in compressive loading than in tensile loading. Matrix and interfacial damage develops for much the same reasons as for tensile loading, but because of the greater demand on the matrix and the interface in compressive loading, compressive fatigue loading generally has a greater effect on the strength of composite materials than tensile loading.[29] In addition, local resin and interfacial damage leads to fibre instability in compressive loading which is more severe than the fibre isolation mode which occurs in tensile loading. Generally much less information is available on the compressive fatigue of composites, mainly because the compressive testing of these materials presents

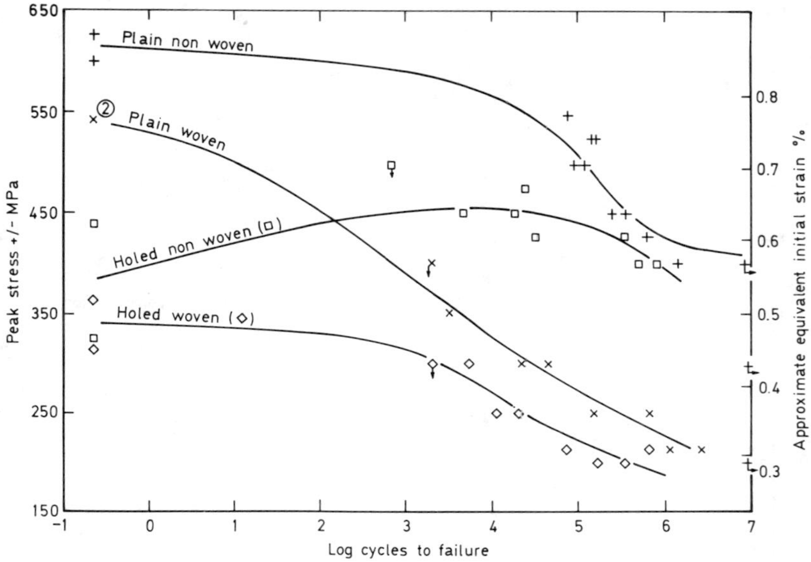

Fig. 15. *S–N* data for $[0°, 90°, 90°, 0°]_s$ woven and non-woven CFRP coupons with and without holes, tested in reversed axial fatigue loading.[24]

many problems, not the least of which is the need to support the specimens from undergoing global macro-buckling combined with limitations imposed on specimen geometry by the anisotropic nature of the materials.

Ultimately, the worst fatigue loading condition for composite materials is fully reversed axial fatigue, or tension–compression loading. The poorer behaviour of composite materials in reversed axial loading compared with tensile loading, depicted in Fig. 16, is because many of the laminate plies, without fibres in the test direction, develop intraply damage and this causes local layer delamination at relatively short lifetimes. In tensile loading this is not serious, since the layers containing fibres aligned along the test direction continue to support the majority of the applied load. In compression, however, tensile-induced damage of this type can lead to local layer instability and layer buckling, perhaps before resin and interfacial damage within the layers has initiated fibre micro-buckling. Thus fatigue lives in reversed axial loading are usually shorter than for zero-compression or zero-tension loading.

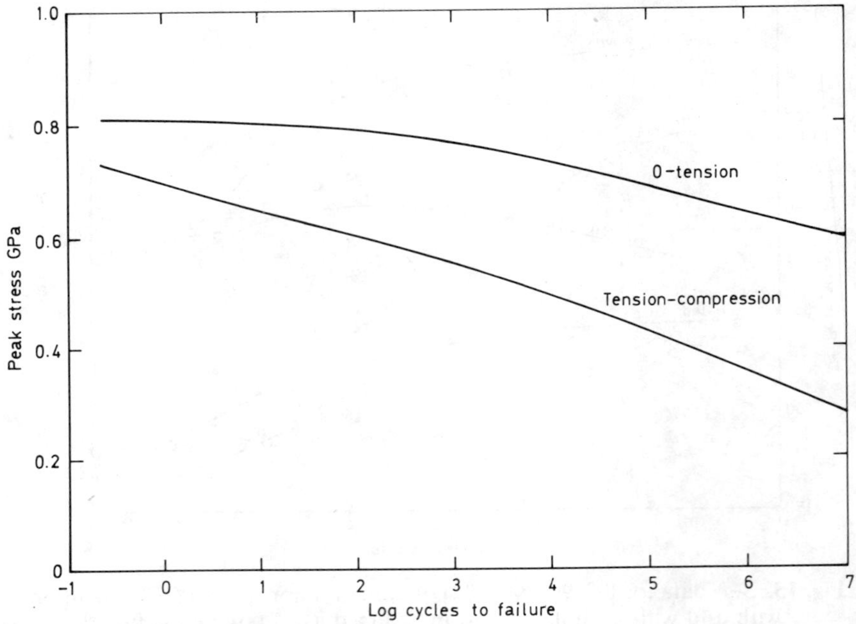

Fig. 16. A comparison of the tensile and reversed axial fatigue behaviour of $[\pm 45°, 0°_2]_s$ CFRP material.

10 FABRIC-BASED COMPOSITES

One of the main alternative forms of composite materials in use are those based on woven fibre reinforcement. These composites offer significant advantages in handling and fabrication over conventional non-woven materials. In addition, the use of woven composites can result in better containment of impact damage and improved residual properties after impact[30,31] (see Chapter 11). However, it is well known that laminate static stiffness and strength are reduced through the use of fabric reinforcement, mainly because of the inevitable distortion of the fibres in the weave, but also because fibre volume fractions of 60–65% found in non-woven continuous fibre material are not achievable in woven composites, a value of 50–55% being more typical.[24,30] This effect is also observed in fatigue, with stresses at short lifetimes being similarly reduced in the woven material. However, additional degradation processes lead to further reductions in strength (Fig. 15) which result in steeper S–N curves than for the non-woven material. This is due to the stress concentrating effect of fibre tow crossover points in the fabric, which become sites for fatigue-induced damage in the resin and fibre/resin interface. In the worst case, cross-plied CFRP, sustainable strains after one million cycles of reversed axial fatigue loading can be as low as 0·3%. Thus although woven composites have many processing advantages, their mechanical properties, particularly in fatigue loading, are generally poorer than those of the equivalent non-woven material.

11 SHORT FIBRE COMPOSITE MATERIALS

The use of short fibres usually leads to better handling and drapability of the fibre felts or pre-pregs used to make the composites, but the disadvantage is poorer static strength and stiffness due to the less efficient utilisation of the fibres and the stress concentrating effect of the fibre ends. The effect becomes even more greatly marked in fatigue, as additional damage develops at fibre ends and causes failure at shorter lifetimes than in continuous fibre materials. the effect may be worsened if the fibres are not aligned along the test direction, which will be the case for chopped strand mat composites and short fibre injection mouldings. Typical S–N curves for a short glass fibre reinforced thermoplastic injection moulded material are shown in Fig. 17. The figure demonstrates the much lower stresses sustainable by such materials, both statically and in fatigue, than in continuous fibre materials. In addition, coupons tested transverse to the general fibre alignment in the mould flow direction have poorer behaviour than those tested with the fibres roughly aligned along the test direction, as a result of the greater demand on the resin matrix, as expected.

12 FATIGUE OF HYBRID COMPOSITES

Another alternative form of composite material in general use is the hybrid composite. This term is usually reserved for composites fabricated with two different fibres in the same resin matrix, although strictly other hybrids are possible perhaps with different resins or with more than two types of fibres. The main reasons for hybridisation are improved toughness or stiffness matching. Some fibres, such as the polymer-based fibres, are particularly tough and when hybridised with

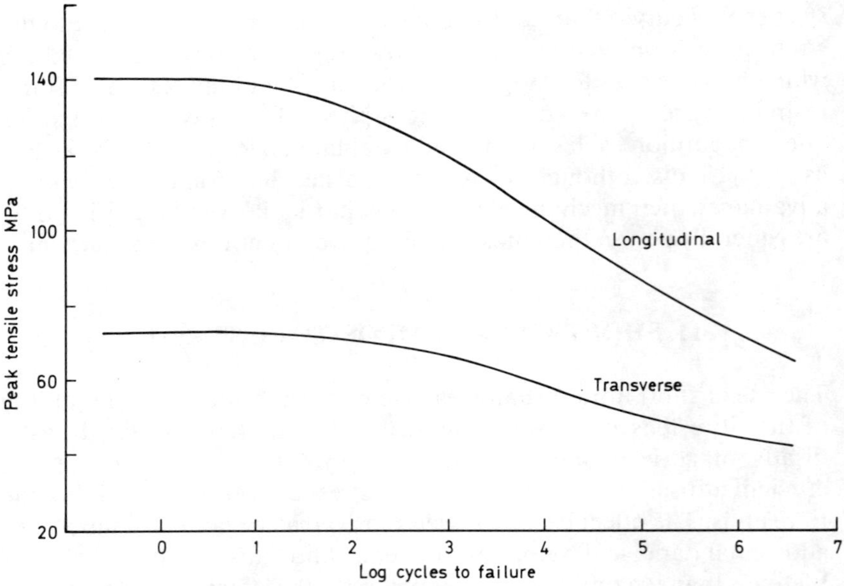

Fig. 17. Typical *S–N* data for an injection moulded, glass fibre reinforced thermoplastic tested longitudinal and transverse to the flow direction.

fibres such as carbon or glass can yield tough laminates without some of the disadvantages associated with polymer fibre composites. Alternatively some components, such as helicopter rotor blades, may require exact stiffness matching and this can more readily be achieved with hybrid composites, perhaps with mixtures of glass and carbon fibres.

The static mechanical performance of hybrid composites is dominated by the behaviour of the lower-strain phase, often carbon fibres. If this is present in only small proportions, then failure of this phase may be tolerable with the load carried being transferred to the higher-strain phase. Usually, however, failure of the low-strain phase triggers failure of the hybrid composite.

In fatigue, as has been shown earlier, the different fibres in composite materials result in very different composite fatigue behaviour. Since fatigue behaviour is primarily strain dependent, in hybrid composites the behaviour is usually dominated by the lower-strain phase. This is because this phase is usually the stiffer and as such carries the greater part of the load and is thus being worked the harder. The other phase is thus usually subjected to strains below normal and shows a smaller fatigue degradation than when tested in isolation. Figure 18 shows a series of *S–N* curves for carbon/Kevlar fibre hybrids with varying ratios of fibre type.[32] On increasing the Kevlar fibre content of the hybrid there is a gradual change from the flat *S–N* behaviour of the carbon fibre composite, to slightly steeper *S–N* plots. This change is not, however, linearly in proportion to the large difference in the fatigue behaviour observed between the carbon and Kevlar materials in isolation. This is because the behaviour of the hybrids with up to 50% Kevlar content is still dominated by the lower-strain carbon fibre phase, which has a relatively flat *S–N* response.

13 BIAXIAL FATIGUE TESTING

There are perhaps two main reasons for wanting to test composites in biaxial fatigue. Firstly, information may be gained on composite

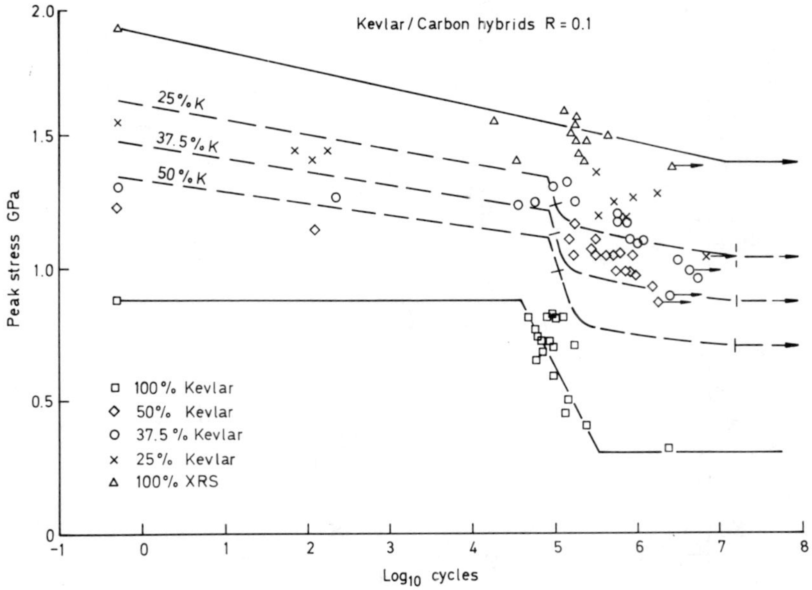

Fig. 18. *S–N* data for carbon/Kevlar fibre/epoxy resin hybrid materials tested in tensile fatigue.[32]

behaviour in a particular complex rather than uniaxial stress state, perhaps equibiaxial or pure shear, such as can be achieved by the application of biaxial loads. Secondly, the application of complex loads may allow service loads to be mimicked and the subsequent response of the material to be studied. In recent years there has been some limited success in the biaxial fatigue testing of composite materials using a cruciform-type specimen (Fig. 19).[33] By varying the ratio of stresses on the two loading arms, different stress states can be imposed. However, the specimen is prone to failure at the arms, especially when testing carbon fibre materials. More recent work has concentrated on specimens with stress concentrators to try to avoid failures out of the gauge area. Alternative means of applying complex fatigue stress fields

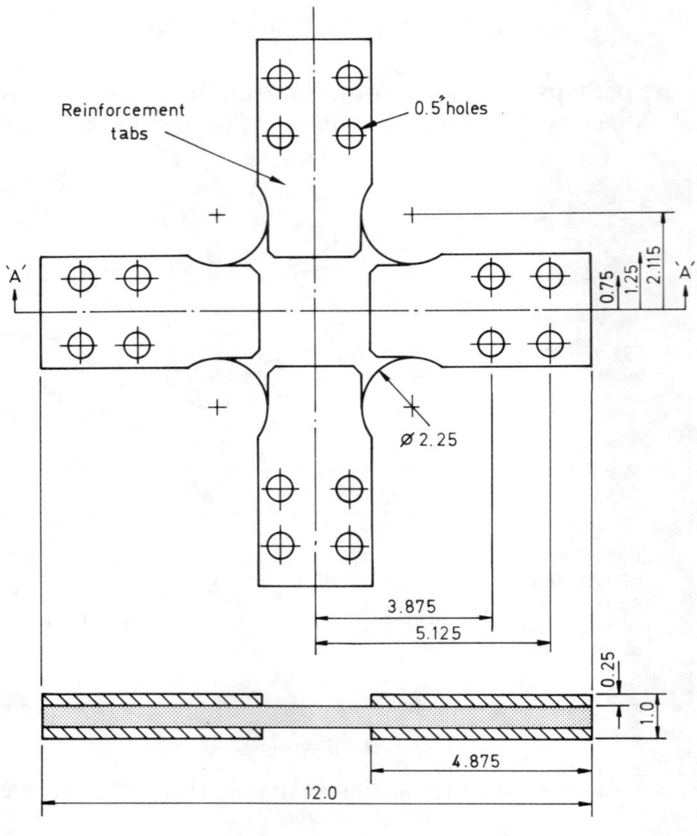

Fig. 19. Cruciform specimen for biaxial fatigue testing.[33]

to composite materials have been used, such as tubular specimens, but without much success.[34]

14 COMBINED FATIGUE/ENVIRONMENTAL EFFECTS

The first problem that must be tackled in this specialist area of composites fatigue work is to decide how to perform the test. Fatigue tests are inevitably medium- to long-term tests, sometimes lasting days or weeks; therefore the possibility of test coupons drying out during the period of the test must be considered. Tests performed at room temperature generally lead to little change in moisture content, since the times involved for most materials to absorb or desorb significant amounts of moisture are long (see Chapter 1). The problem of coupon drying is particularly acute, however, when fatigue testing takes place at elevated temperatures. Precautions must then be taken to preserve the moisture content of the specimen. One possibility is to test in a chamber in which the temperature and humidity are controlled. Although this method is used, it involves expensive equipment, perhaps beyond the budget of many laboratories. Alternative approaches that have been used include enclosing the specimen in a polyethylene bag in which a salt solution is used to maintain the required humidity, or sealing the specimen totally by encapsulation.

The effect of environmental exposure on the fatigue behaviour depends on the sensitivity of the laminate to matrix properties, since it is usually the matrix or fibre/matrix interface that is affected by the absorbed moisture, although some polymer fibres do also absorb moisture. Thus carbon fibre laminates, having a strong fibre/matrix interface, show little sensitivity to moisture content[35, 36] when tested at room temperature. This is shown clearly in Fig. 20, in which fatigue life is plotted versus moisture content for ($\pm 30°$, $90°$) coupons tested at a stress level that gave a life of 100 000 cycles when dry. It is clear that there is no change in life with increasing moisture content. In materials in which the fibre–matrix bond strength is less strong and susceptible to environmental attack, the laminate lay-up is also important, laminates with few fibres in the test direction being more susceptible than those with many fibres in the test direction. Even then, if tested at room temperature, the effect is small, as shown in Figure 21 for GRP $\pm 45°$ specimens tested in three different preconditioned states.[11] Only the pre-boiled specimens showed any noticeable reduction in the slope of the *S–N* curve. Data on materials tested at elevated temperature in fatigue are

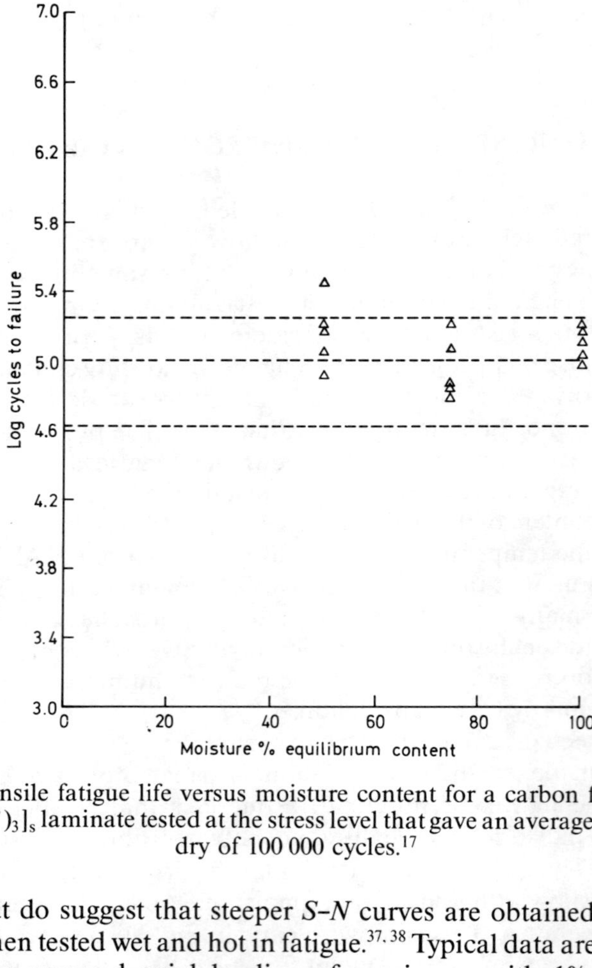

Fig. 20. Tensile fatigue life versus moisture content for a carbon fibre/epoxy $[(90°, \pm 30°)_3]_s$ laminate tested at the stress level that gave an average fatigue life dry of 100 000 cycles.[17]

scarce, but do suggest that steeper *S–N* curves are obtained, even for CFRP, when tested wet and hot in fatigue.[37,38] Typical data are shown in Fig. 22 for reversed axial loading of specimens with 1% absorbed moisture at 20 °C and 95 °C.[37] It is clear that the imposition of hot/wet conditions does result in reduced static strength and that this is carried over into the fatigue performance.

15 COMPARISON WITH METALS

Despite the references in previous sections to the poorer performance of composite materials in compressive fatigue loading compared with

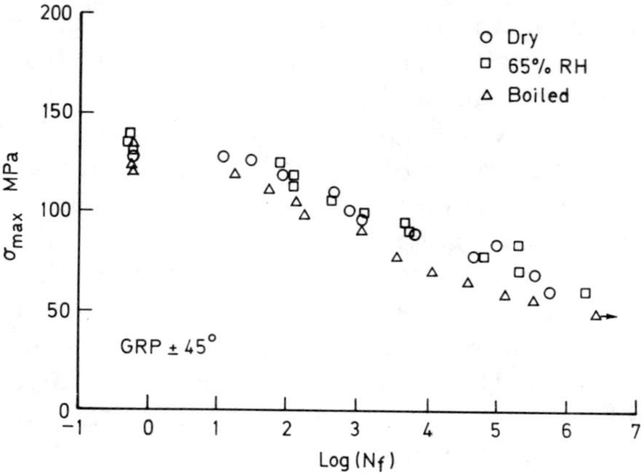

Fig. 21. *S–N* data for $[(\pm45, \mp45)_2]_s$ glass fibre/epoxy material subjected to different preconditioning treatments.[11]

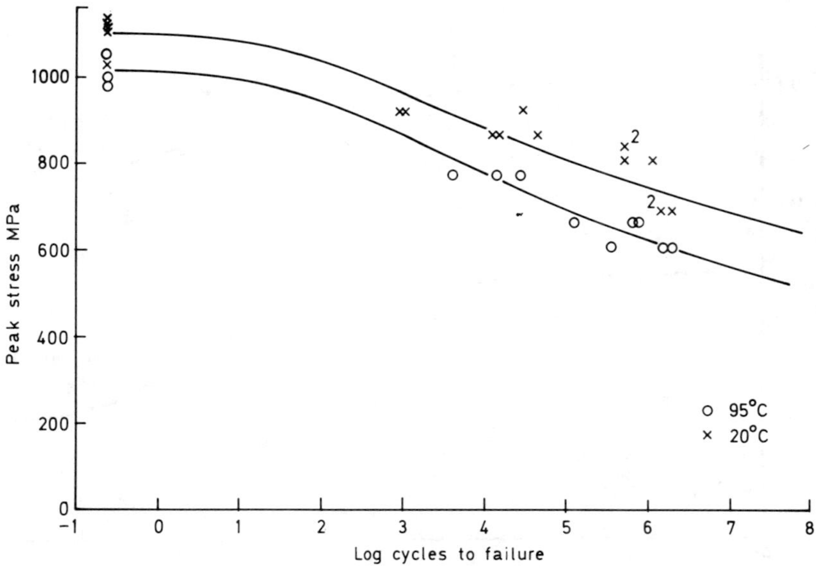

Fig. 22. Reversed axial fatigue data for carbon fibre epoxy with a [0°, ±45°] lay-up containing 1% moisture, tested at 20°C and 95°C.[38]

tensile loading, composite fatigue performance is generally very good when compared with that of metallic materials. Indeed, some of the uses of composite materials, such as in helicopter rotor blades, have been driven mainly by the materials' excellent fatigue resistance. In Fig. 23, typical fatigue *S–N* curves are presented for notched coupons of a standard laminated composite (suitable for aerospace use) and an aerospace aluminium–lithium alloy, under tensile loading.[39] The composite has a much larger ratio of fatigue stress at high life to static strength than the metal, such that at long lifetimes the difference in strength is considerable. Indeed notched composite materials have such flat *S–N* curves that airworthiness fatigue testing for clearance purposes can prove difficult. This is because testing to very large numbers of cycles would be necessary to generate the full distribution of fatigue lifetimes to enable the necessary probabilities of failure to be determined. The superior fatigue performance of composite materials is due to the very different damage mechanisms involved. Metals generally

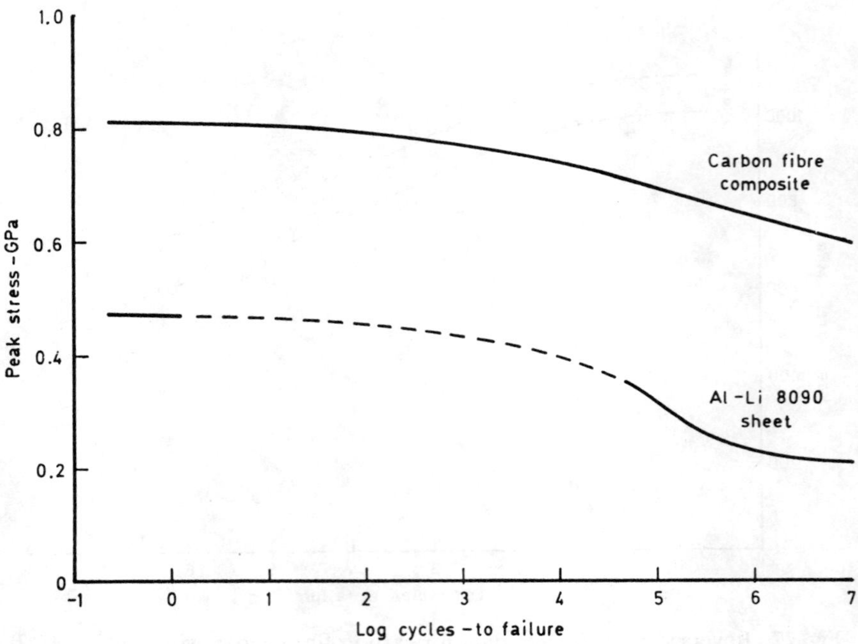

Fig. 23. Comparison of the tensile fatigue behaviour of notched CFRP and aluminium alloy.[39]

demonstrate a fatigue effect related to the progressive growth of a single flaw, and thus damage is very localised. Composites, however, generate complex damage zones at stress concentrations, such as notches, as shown in Fig. 24, and no single flaw growth is observed. These zones include damage in many directions, frequently parallel to fibres, both within and between the layers. Their effect is to blunt the stress concentrating effect of the notch on the load bearing fibres, as explained in Section 4, and results in good fatigue behaviour.

In summary, metals fatigue by the localised growth of a crack from a stress concentration, perpendicular to the local tensile stress, so they are more sensitive to tensile loading than compressive loading. In composites, the damage is more widespread and consists typically of interlayer damage and reduced support for the fibres with a consequent loss in compressive strength, so they are more sensitive to compressive loading than tensile loading.

16 FATIGUE LIFE PREDICTION OF COMPOSITE MATERIALS

The most widely applied technique for the life prediction of metallic components is linear elastic fracture mechanics. This technique is

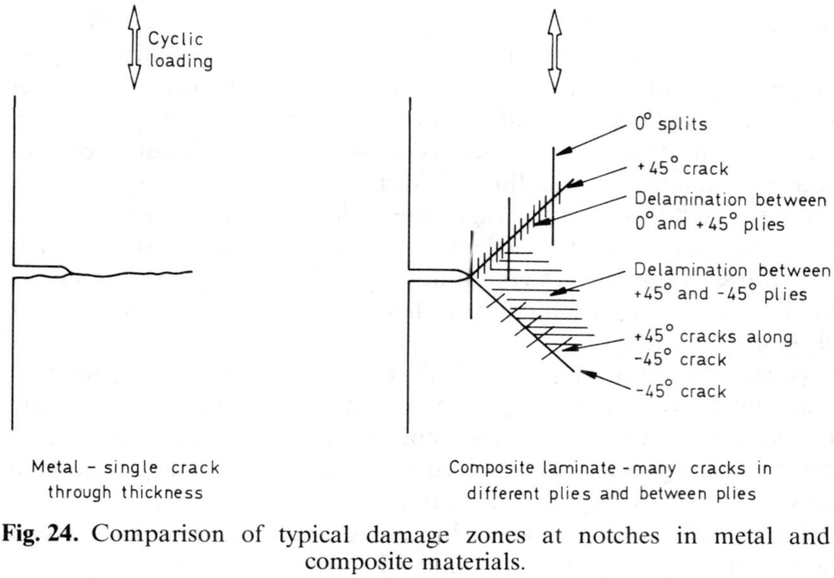

Fig. 24. Comparison of typical damage zones at notches in metal and composite materials.

based upon the growth of single flaws to failure, something not observed in composites (except perhaps for the growth of interlayer delamination). Thus the applicability of such a technique to composite materials must be considered at the very least to be very restricted.

The ideal life prediction technique for composite materials would enable fatigue life and residual strength to be predicted from the minimum amount of data, such as only static strength and perhaps fatigue data on a very few types of laminates. It should then be possible to extrapolate to any laminate or lay-up stacking sequence, loading mode, spectrum loading, environmental condition or notched state. The effect of changing failure modes as these parameters are varied must also be included. Unfortunately, in practice, current understanding is a long way from providing such a technique. Models that are available are usually restricted in their applicability to different laminates, loading modes and other conditions, and indeed do not provide very good estimates of fatigue behaviour. Useful reviews are provided in Refs 40 and 41.

The techniques that have been developed in attempts to predict the fatigue lives and residual strengths of composite materials fall into two major categories, empirical models[41–43] and wear-out models.[44,45] The former rely on the strength life equal rank assumption, which states that if a set of coupons could be tested both statically and in fatigue, their strengths in both tests would rank in the same order. The empirical models thus require large amounts of data, including static strength, fatigue life and residual strength distributions, in order to predict residual strengths at any number of cycles. Useful statistical means of manipulating data have been derived in these techniques, but agreement with experiment has been limited. More recent work linking scatter in static strength to scatter in fatigue life looks more promising.[9,41,46] The wear-out models are all based on the reduction of some property, such as stiffness or strength, and the ability to predict its decay during fatigue loading. The major problem with this approach is that the parameter chosen to describe the wear-out is inevitably a function of stress. This necessitates testing at many stress levels to describe its behaviour, reducing the usefulness of the technique in minimising data collection. In addition, such techniques have not been very successful at predicting composite fatigue behaviour. Indeed, no particularly successful technique currently exists for the life prediction of composite materials.

Historically, Miner's rule has been used extensively to calculate safe fatigue lives for metallic structures. Its applicability to composites, however, has been found to be rather limited, the main reason being that it provides unconservative estimates of fatigue lives.[29,47] There may be scope, however, for empirical modifications to be made to this approach to give better predictions of fatigue life under spectrum loading conditions.

17 STRUCTURAL DESIGN

Structures may be designed to strength, stiffness or minimum gauge requirements. The last two use relatively low stresses and should not give fatigue problems. Even strength designs use relatively low design stresses because of the scatter in static strengths, the reduction of compressive strength under hot/wet conditions and the reductions in static compressive strength caused by relatively small amounts of impact damage. These can result in design-allowable strains of 0·6% and design limit strains of only 0·4%. Even when notches are present, to reduce the applied failure strain from 1·4% to 0·7%, the subsequent *S–N* curves are so flat that fatigue strengths still exceed the static design allowables. Indeed, composite structures, such as rotor blades, have gone into service with infinite lifetimes, although the designs are fatigue-critical. One consequence of this is that when structures with both composite and metal components are fatigue tested, the metal parts almost always fail first.

It is possible that certain features in composite structures might be more susceptible to fatigue damage than the base composite material. Ply drop-offs, used for tapering sections, bonded or bolted joints, and the complex stressing around attachment points could present problems if not considered adequately in the design. Also new materials aimed at raising the static design allowables and with steeper *S–N* curves could result in fatigue being of more prominent concern in design.

18 CONCLUSIONS

Different fibres and resins behave differently in fatigue, the resin being particularly important. In general, if fibres are aligned with the main loads, then the *S–N* curves are very flat. Stress concentrations can produce marked reductions in static strength but result in even flatter *S–N* curves, the fatigue strength not being greatly affected.

Because of relatively low static design allowables, fatigue strength is not usually a serious design criterion for composite structures. Composite materials and structures are still developing relatively quickly and if static design allowables are increased, fatigue behaviour could become more important in future.

REFERENCES

1. Harris, B., *Composites,* **8**(4) (1977) 214.
2. Curtis, P. T. & Dorey, G., Fatigue of composite materials. In *Proc. Int. Conf. Fatigue of Structural Materials and Structures, Sheffield, 1986.* Mechanical Engineering Publications Ltd London, 1986, p. 297.
3. Sturgeon, J. B., Fatigue testing of carbon fibre reinforced plastics. RAE TR75135, 1975.
4. Curtis, P. T. & Moore, B. B., A comparison of plain and double waisted coupons for the static and fatigue tensile testing of unidirectional GRP and CFRP. RAE TR82031, 1982.
5. Curtis, P. T., CRAG test methods for the measurement of the engineering properties of fibre reinforced plastics. RAE TR85099, 1985.
6. Butler, R. J., Barnard, P. M. & Curtis, P. T. In *Composite Materials: Testing and Design* series, ASTM STP972, 1988.
7. Curtis, P. T., An investigation of the mechanical properties of improved carbon fibre composite materials. RAE TR86021, 1986.
8. Talreja, R., *Proc. Roy. Soc. London, A,* **378** (1981) 461.
9. Jamison, R. D., Schulte, K., Reifsnider, K. L. & Stinchcomb, W. W. In ASTM STP836, 1983.
10. Barnard, P., Butler, R. J. & Young, J. Cranfield College of Aeronautics interim report under MOD agreement 2028–0134 XR/MAT, 1985.
11. Jones, C. J., *et al., Proc. Roy. Soc. London, A,* **396** (1984) 315.
12. Sims, G. & Gladman, D. G. NPL report DMA(A) 59, 1982.
13. Boeing specification XBMS 8–276, 1983.
14. Bader, M. G. & Boniface, L., Damage development during quasi-static and cyclic loading in GRP and CFRP laminates containing 90 degree plies. *Proc. ICCM-V*, San Diego. AIME, 1985.
15. Ogin, S. L., Smith, P. A. & Beaumont, P. W. R., Cambridge University Engineering Dept report TR105, 1984.
16. Reifsnider, K. L. & Jamison, R. D., *Int. J. Fatigue* (Oct. 1982) 187.
17. Curtis, P. T. & Moore, B. B., The effect of environmental exposure on the fatigue behaviour of CFRP laminates. RAE TR84027, March 1984.
18. Metcalf, A. G. & Schmitz, G. R., *Glass Technol.,* **13** (1972) 5–16.
19. Pagano, N. J. & Pipes, R. B., *J. Compos. Mater.,* **5** (1971) 50–7.
20. Curtis, P. T., The effect of edge stresses on the failure of (0°, 45°, 90°) CFRP laminates. RAE TR80054, 1980.
21. Curtis, P. T., *J. Mater. Sci.,* **19** (1984) 167–82.
22. Curtis, P. T., Residual strains and the effects of moisture in fibre reinforced laminates. RAE TR80045, 1980.

23. Curtis, P. T. & Moore, B. B., A comparison of the fatigue performance of woven and non-woven CFRP laminates. In *Proc. 5th Int. Conf. Composite Materials*, San Diego, 1985.
24. Curtis, P. T. & Moore, B. B., A comparison of the fatigue performance of woven and non-woven CFRP laminates. RAE TR85059, 1985.
25. Schutz, D., Gerharz, J. J. & Alschweig, A. In ASTM STP723, 1981, p. 31.
26. Dorey, G., Sigety, P., Stellbrink, K. & t'Hart, W. G. J., Impact damage tolerance of a carbon fibre composite laminate. RAE TR84049, 1984.
27. Piggott, M. R., *J. Mater. Sci.,* **16** (1981) 2837.
28. Purslow, D., *Composites,* **12**(4) (1981) 241.
29. Rosenfeld, M. S. & Gause, L. W. In ASTM STP723, 1981, p. 174.
30. Bishop, S. M. & Curtis, P. T., An assessment of woven carbon fibre reinforced plastics for aerospace use. RAE TR83010, 1983.
31. Curtis, P. T. & Bishop, S. M., *Composites,* **15** (1984) 259–65.
32. Adam, T., Dickson, R. F., Fernando, R., Harris, B. & Reiter, H. In *Proc. Int. Conf. Fatigue of Engineering Materials and Structures, Sheffield, 1986.* Mechanical Engineering Publications Ltd, London, 1986.
33. Tutton, P. A. & Pascoe, K. J., Cambridge University Engineering Dept. report under MOD agreement 2029/0153 XR/MAT, Nov. 1982.
34. Sturgeon, J. B., The use of tubular test specimens for material characterisation of carbon fibre composites. RAE TR75117, 1975.
35. Curtis, P. T. & Moore, B. B., The effect of environmental exposure on the fatigue behaviour of CFRP laminates. RAE TR84027, 1984.
36. Curtis, P. T. & Moore, B. B., *Composites,* **14** (1983) 294–300.
37. Rouchon, M., Centre D'Essais Aéronautique de Toulouse report CEAT MO542600, Toulouse, France, 1983.
38. Webb, J. M. Progress note issued under MOD contract A93b/458 with BAe Warton Division, 1985/86.
39. McDarmaid, D. S., Fox, S. & Flower, H. M., Paper presented at European Materials Research Soc. Conf. Advanced Materials Research and Development for Transport, Strasbourg, Nov. 1985.
40. Gordon, B. & Whitehead, R. S., BAe report SON(P) 179, Feb. 1979.
41. Barnard, P. M. & Young, J. B., Final Report under MOD agreement 2028/0134 XR/MAT, Nov. 1986.
42. Hahn, H. T., USAF Wright Patterson Laboratories report AFML-TR-78-45, 1978.
43. Yang, J. N. & Du, S., *J. Compos. Mater.,* **17** (1983) 511.
44. Pipes, R. B., Kulkarni, S. V. & McLaughlin, P. V., *Mater. Sci. Engng,* **30** (1977) 113.
45. Poursatip, A., Ashby, M. F. & Beaumont, P. W. R., Damage accumulation during fatigue of composites. In *Proc. Int. Conf. Composite Materials IV*, Tokyo, 1982.
46. Barnard, P. M., Butler, R. J. & Curtis, P. T., Fatigue scatter of UD glass/epoxy — a fact or fiction? In *Proc. 3rd Int. Conf. Composite Structures*, Paisley, 1985, pp. 69–82.
47. Heath-Smith, J. R., Fatigue of structural elements in carbon fibre composite — present indications and future research. RAE TR79085, 1979.

Chapter 11

Damage Tolerance and Damage Assessment in Advanced Composites

Graham Dorey

Materials and Structures Department, Royal Aerospace Establishment, Farnborough, Hampshire, UK

1 INTRODUCTION

Aircraft structures have to perform satisfactorily under normal service conditions for long periods of time. With current advanced composite materials, structural loads cause few problems but accidental damage can cause significant reductions in performance. Although glass and aramid fibres are used for their damage tolerance, carbon fibre composites (CFRP) can be susceptible to impact damage[1-3] and can cause design limits to be set below 0·5% strain, less than one-third of the fibre failure strain. This limits the efficient use of advanced composites. There is a need to detect damage, to predict the effects on structural performance, and to develop more damage-tolerant materials and structures.

The various fracture modes of composite materials affect the impact behaviour of fibrous laminates and have important implications for the design of composite structures. A better understanding of the micromechanics of the failure processes should increase confidence in the performance of carbon fibre composites and allow more efficient use of the fibre strength.

The fracture characteristics of fibrous composites are reviewed briefly and it is shown how toughness depends on preferential splitting parallel to the fibres. The extent of this splitting depends on the properties of the fibres and of the matrix, on the interfacial bond strength, and on interactions between the plies in the laminate. Under impact loading, the damage produced shows similar trends to material properties, but also depends on the velocity, energy and properties of the impacting projectile, and in particular on the dynamic response of the composite specimen. Testing at different velocities can produce strain-rate effects or changes in failure mode, and this will affect the design of standard impact tests.

The various types of impact damage affect different mechanical properties, mainly producing significant reductions in static strengths but also possibly affecting fatigue properties and environmental resistance. For structural applications, the design requirements should be related to the possible incidence of impact damage and its effects on performance.

Improvements in impact performance are being made through changes to the basic materials. Recent improvements in both the fibre and matrix properties have further increased the useful performance of carbon fibre composites and will widen the range of possible applications (see for example Fig. 1). Further improvements should come from combining different composites in hybrid forms and through modifications in structural designs.

2 FRACTURE BEHAVIOUR

2.1 Unidirectional Laminates

Unidirectional fibre composites have high strength and stiffness, imparted by the fibres, in the fibre direction or longitudinal direction. In the transverse direction the strength is much lower (about one-thirtieth of the longitudinal strength). In a notched composite, subjected to a longitudinal tensile stress, there is a complex stress field near the notch tip. The greatest stress concentration is at the notch tip in the longitudinal stress, which is in the direction of maximum strength parallel to the fibres. However, smaller local stresses, such as shear stresses at the notch tip or transverse tensile stresses ahead of the notch, can exceed the low transverse strengths and cause fibre/matrix splitting parallel to the fibres, as shown in Fig. 2.

This splitting effectively blunts the notch and removes the stress concentration at the notch tip, and thus unidirectional composites are

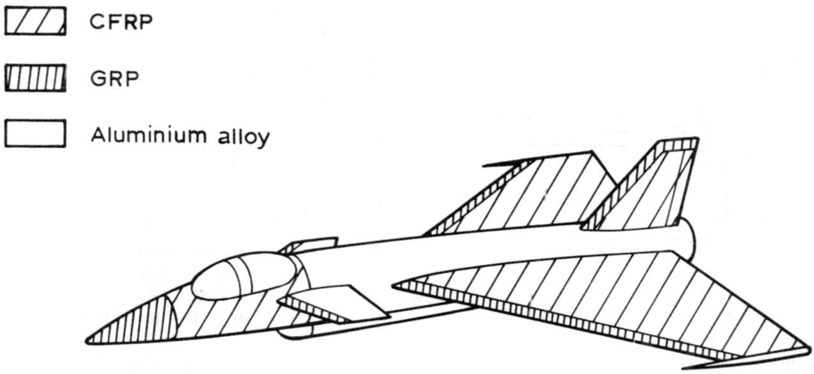

Fig. 1. Material distribution typical of advanced design combat aircraft.

usually 'notch-insensitive'. The splitting parallel to the fibres is a relatively low-energy fracture process. For carbon fibre/epoxy composites the initial fracture energy is about 100 J/m^2, similar to that for unreinforced epoxy resins, but as the split progresses misaligned fibres form a tied zone behind the crack front and the fracture energy can increase as these fibres have to be broken for the crack to advance (Fig. 3). Fracture across the fibres requires much more energy, of the order of 50–100 kJ/m^2 depending on the amount of associated splitting.

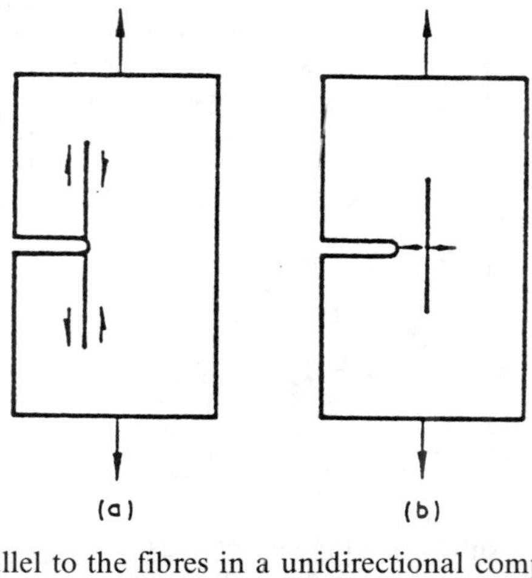

Fig. 2. Splits parallel to the fibres in a unidirectional composite due to local stresses: (a) shear, and (b) transverse tension.

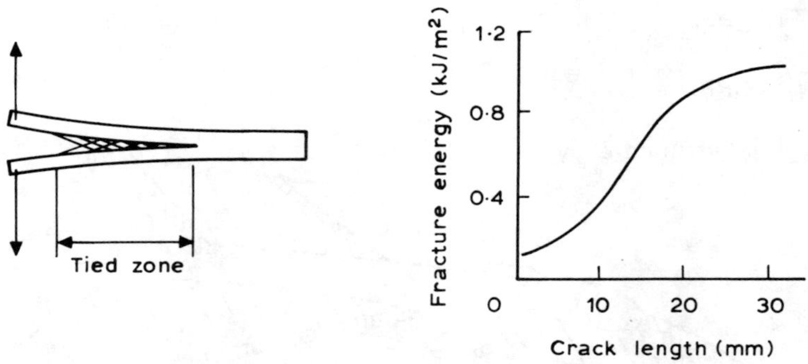

Fig. 3. Misaligned fibres forming a tied zone behind the crack front in a unidirectional CFRP composite, and effect of the tied zone on the Mode I fracture energy.

2.2 Multidirectional Laminates

Multidirectional laminates are usually notch-sensitive, as illustrated for CFRP under tensile loading in Fig. 4. The failure stress is significantly below the line expected from the net section strength for a notch-insensitive material. In this particular laminate, sharp notches were found to be more detrimental than round holes, but for smaller notches

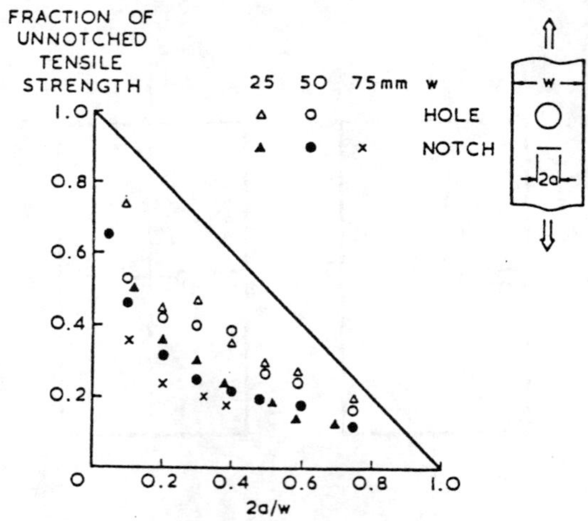

Fig. 4. The notch sensitivity of a $[0°, 90°, 0°, \pm45°, 0°]_s$ CFRP laminate under tensile loading.

and tougher materials this distinction gets less marked. However, the failure stresses are generally greater than would be expected for brittle materials.

As with the unidirectional composites, the complex stresses near the notch tip cause splitting parallel to the fibres within the individual plies, and delamination occurs between the plies, forming a damage zone at the notch tip,[4] as shown in Fig. 5. This damage zone absorbs energy, increasing the fracture energy, but most important are the splits parallel to the longitudinal 0° fibres because they are the most effective in blunting the notch and reducing the stress concentration in the 0° fibres which carry most of the load. The size of this damage zone depends on fibre, matrix and interface properties but also on ply thicknesses and on proportions of plies in the various directions.[5]

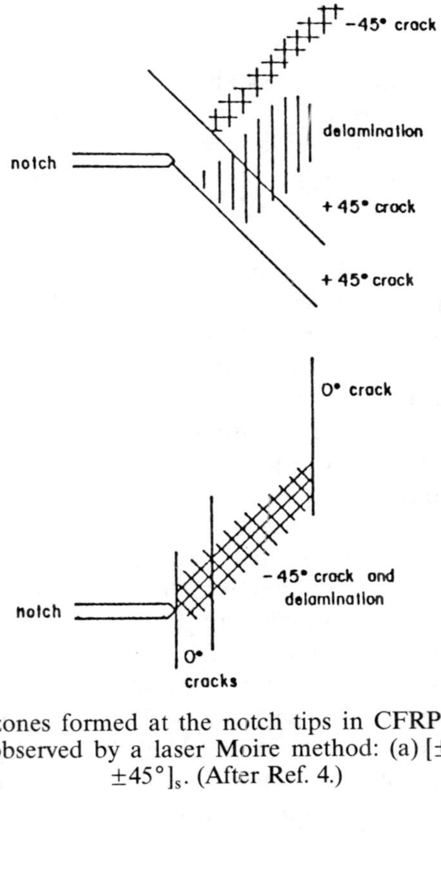

Fig. 5. Damage zones formed at the notch tips in CFRP laminates under tensile loading, observed by a laser Moire method: (a) $[\pm 45°, 0°]_s$, (b) $[0°, \pm 45°]_s$. (After Ref. 4.)

Similar crack blunting mechanisms have been observed under compression loading,[6,7] even though failure is usually associated with some fibre buckling instability, as shown in Fig. 6. Under room temperature, dry conditions, final failure often occurs by in-plane shear buckling of the 0° fibres along a 45° ply, with a net section stress similar to that for unnotched material. However, under hot/wet conditions the shear modulus of the resin can be reduced, so that there is less resistance to fibre buckling, and at relatively low loads the local stresses can precipitate an out-of-plane shear failure in the 0° fibres along a 90° direction.

3 FIBRE/MATRIX INTERFACE

The reinforcement efficiency of a fibre/matrix system, and the degree to which the material splits during failure processes, depends on the adhesion at the fibre/matrix interface. The only reason that brittle fibres in a brittle matrix can have any appreciable toughness at all, is that cracks get diverted along the fibre/matrix interface, like greenstick fractures observed with wood and bone. If there is too weak a fibre/matrix bond, the material will not support loads in shear or compression; too strong a bond and the material will be brittle. This can be affected by the fibre structure, the fibre surface condition, the size or surface finish, and the matrix structure both at the interface and away from the interface.

Glass fibres have a silane finish applied, both to protect the glass surface from degradation and to increase the adhesion to resin matrices. Polymeric fibres can have their surfaces etched to increase adhesion, but high-modulus polymer fibres such as aramid tend to defibrillate at internal interfaces so there is a limit to which the fibre/matrix bond strength is effective. In both glass fibre and aramid fibre composite laminates the failure processes are dominated by the relatively weak interfaces.

In carbon fibre composites the interface bond strength is increased by means of an oxidative treatment to the surface of the carbon fibres.[8] This increased bond strength can be seen in Fig. 7, where the interlaminar shear strength (ILSS) increased with fibre surface treatment up to a certain value; for greater levels of surface treatment the ILSS remained reasonably constant and this may be because shear failures were occurring in the resin rather than at the interface.

Other properties of unidirectional CFRP such as transverse tensile strength, longitudinal compression strength and flexural strength

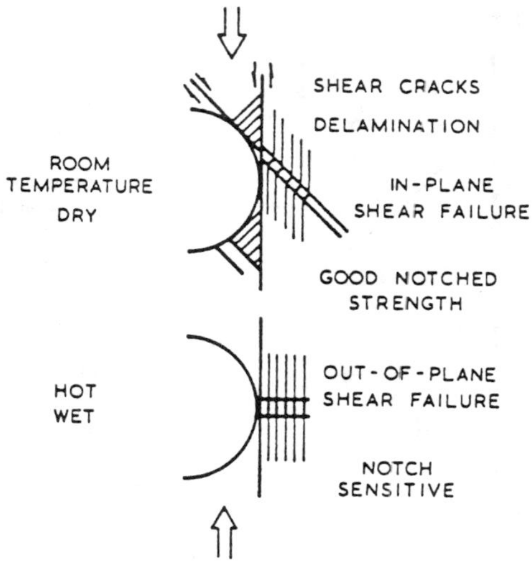

Fig. 6. Changes in damage zones and failure modes with environmental conditions in a notched [+45°, 0°, −45°, 0°]$_{3s}$ CFRP laminate, under compressive loading. (After Ref. 7.)

showed similar trends. Multidirectional laminates behaved differently and the notched tensile strength continued to decrease, as the fibre surface treatment was increased, even in the range where the ILSS was not changing. The fibre surface treatment clearly affects the intraply splitting at the notch tip and the formation of the damage zones.

The mechanisms of the bonding between carbon fibres and resin matrices are still receiving much attention with attempts to correlate the surface topography and chemical activity with composite failure modes and mechanical properties. The important technological point is that, unlike glass or aramid fibre composites, carbon fibre composites can have too great a bond strength between the fibre and the resin, resulting in brittle materials. The fibre surface treatment has to be optimised for the fibre/resin system and it has to be carefully controlled.

4 IMPACT DAMAGE AND DETECTION

4.1 Impact Testing

Traditionally, impact tests such as Izod and Charpy tests have been used to indicate the impact toughness of isotropic materials, and they

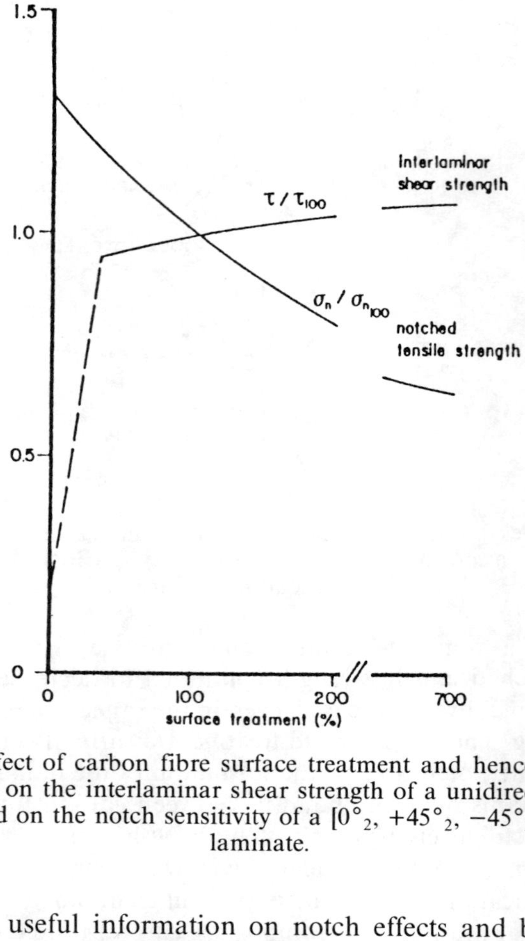

Fig. 7. The effect of carbon fibre surface treatment and hence fibre/matrix bond strength on the interlaminar shear strength of a unidirectional CFRP composite, and on the notch sensitivity of a $[0°_2, +45°_2, -45°_2, 0°_2]_s$ CFRP laminate.

have yielded useful information on notch effects and brittle–tough transition temperatures. However, because of the complexity of composite failure processes, these tests are only of limited value with composite materials and they give no indication of residual properties after impact. Dropweight tests (simulating dropped tools) and ballistic tests (simulating runway stones) have proved more useful.

The various forms of impact damage observed are shown schematically in Fig. 8, together with expressions for the energy needed to cause them, typical values for 2 mm thick CFRP and the area of damage caused. It can be seen, from the effects of geometry, that delaminations are more likely with short spans, thick laminates or laminates with low

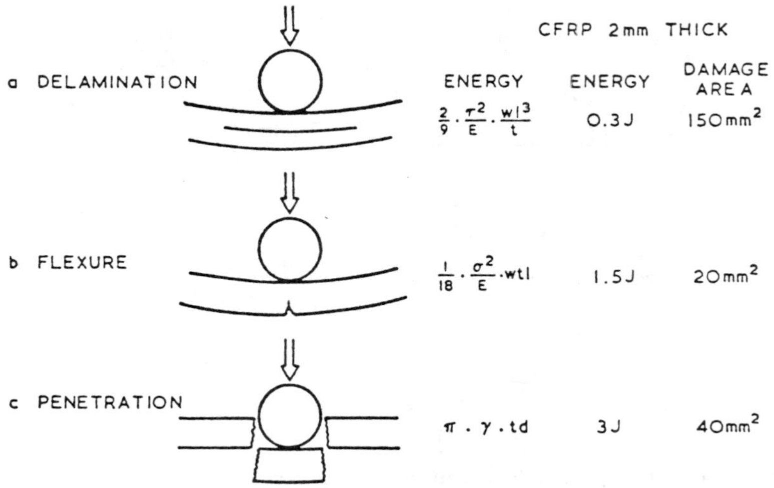

Fig. 8. The primary failure modes of composite laminates under impact loading, the elastic strain energy at the point of failure and typical values of threshold energy and corresponding areas of damage for CFRP; w, l and t are the width, unsupported length and thickness of the specimen, d is the diameter of the ball, τ is the ILSS, σ is the flexural strength, E is the flexural modulus, and γ is the fracture energy.

interlaminar shear strength. Flexural failures are more likely with large spans or thin skins. Penetration is most likely for small projectiles moving at a sufficiently high velocity that the laminate cannot respond quickly enough in flexure and high stresses are generated closer to the point of impact. If high velocity particles have insufficient energy for full penetration they can still cause local fractures and erode the surface.

In dropweight tests, the supports may be instrumented with strain gauges to give a load–time trace during the impact, as illustrated in Fig. 9. The area under this trace is the impulse between the projectile and the specimen and is equal to the change in momentum of the projectile. When there was no failure, the peak load occurred when the projectile came to rest, its kinetic energy being converted into elastic strain energy within the composite specimen, and the second half of the curve corresponds to the projectile rebounding. When a flexural failure occurred the projectile did not rebound, the elastic strain energy causing damage to the composite. The shear failure resulted in a change of stiffness of the laminate, but it could still carry load and the projectile

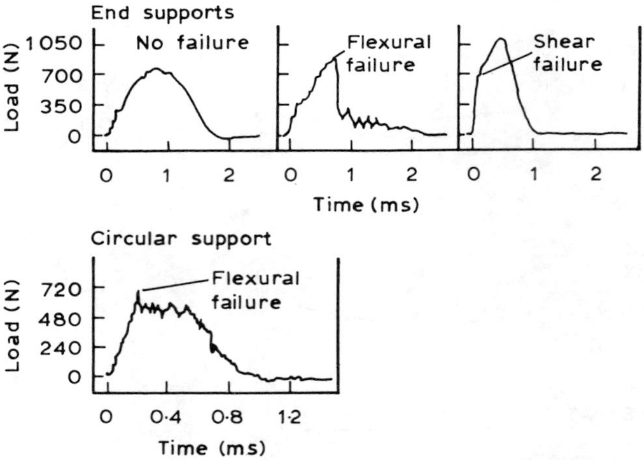

Fig. 9. Load–time traces during impact on CFRP laminates, using either end supports or a circular support.

rebounded but with a reduced momentum. With the circular support, the specimen was supported all round its edge, unlike the first three cases with just end supports. This made the specimen stiffer and critical loads were reached sooner and for lower energy. The fracture process continued longer at high loads because cracks were being driven into regions of lower stress nearer the support, and the low stored elastic energy meant that further input of energy was required from the projectile to continue the fracture process.

Increasing use is being made of instrumented impact testing. This is usually done on dropweight machines or pendulum impact machines, where the striker is instrumented to measure the applied load. The impacting energy is usually much greater than that needed to break or penetrate the specimen and displacement can usually be related to time. However, some machines have independent means of measuring displacement. Thus traces of load–time and load–displacement can be obtained and these can be converted to give energy–time and energy–displacement traces, so that features such as peak load and absorbed energy can be related to fracture processes occurring in the material. Examples are shown in Fig. 10 for a $[0°, 90°]_{5s}$ laminate,[9] and in Fig. 11 for a sandwich panel comprising aramid skins on a Nomex honeycomb core;[10] the peak loads associated with the penetration of the two skins can be clearly distinguished.

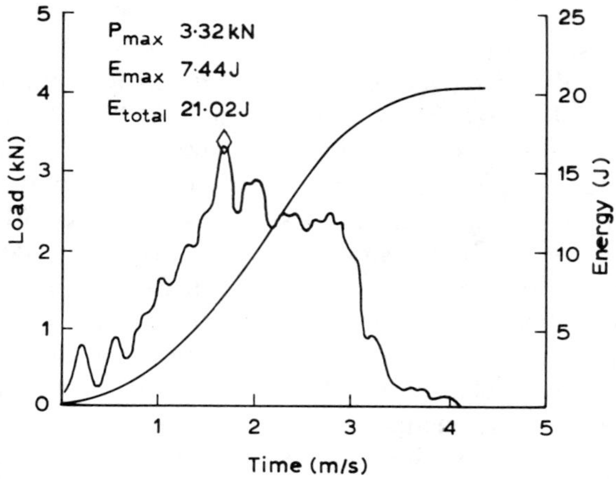

Fig. 10. Load–time and energy–time traces from an instrumented dropweight machine testing a $[0°, 90°]_{5s}$ CFRP laminate. (After Ref. 9.)

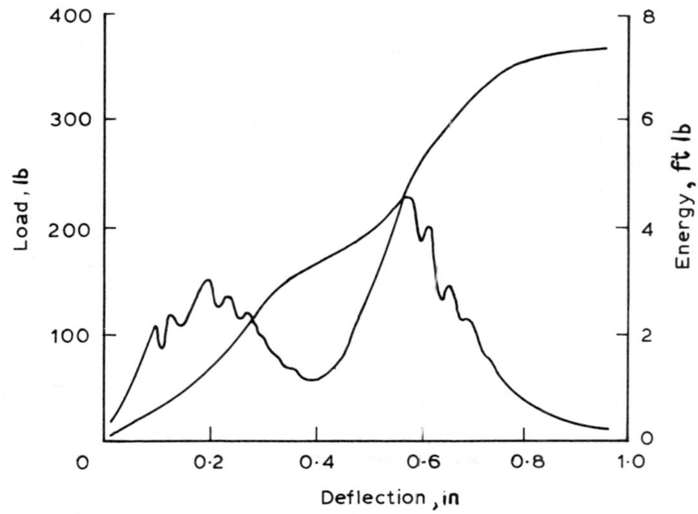

Fig. 11. Load–deflection and energy–deflection traces during dropweight impact on an aramid faced sandwich panel. (After Ref. 10.)

4.2 Strain Rate Effects

Also of interest in impact testing are effects due to strain rate. Figure 12 shows the results of tensile tests on various CFRP laminates, $[0°_8]$, $[0°, 90°, 0°, \pm45°, 0°]_s$ plain and notched and $[(\pm45°)_3]_s$ plain and notched, over a range of strain rates from 0·001/s to 100/s; there was no effect on failure stress. This has been confirmed[11] at higher strain rates using a Hopkinson bar technique as shown in Fig. 13. Although the CFRP showed no significant strain rate effect up to over 1000/s, the glass fibre composite specimens showed a significant increase in failure strain. Similar effects have been seen when comparing strain rates used in static and fatigue tests; composites using carbon fibres or aramid fibres showed no effect of strain rate whereas GRP did show an increase in failure load with increasing strain rate. Although strain rate may have little effect in any given failure mode, projectiles travelling at different velocities can cause the failure mode to change because of the dynamic response of the specimen.

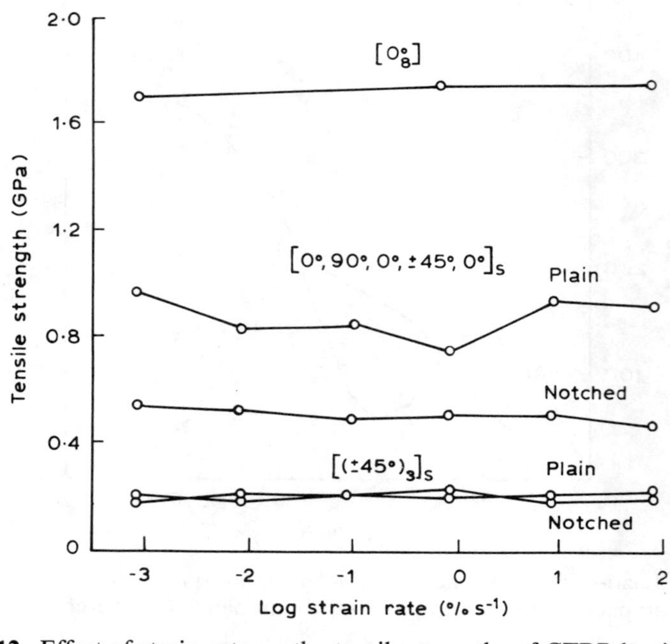

Fig. 12. Effect of strain rate on the tensile strengths of CFRP laminates.

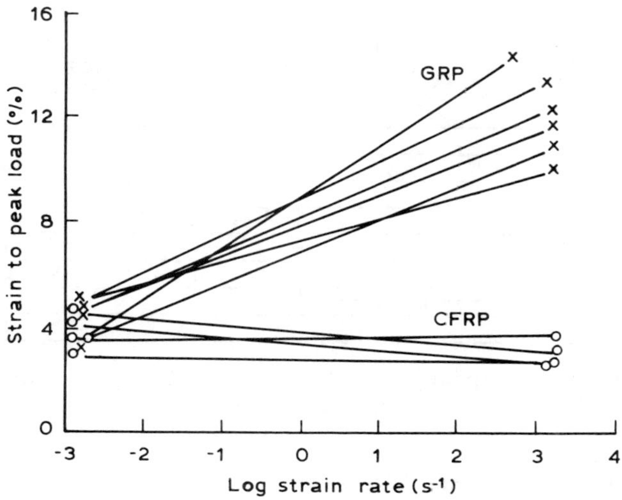

Fig. 13. Effect of strain rate on the impact strain capability of various glass fibre and carbon fibre composites. (After Ref. 11.)

4.3 Damage (Detection)

If the impact results in visible damage, this can be clearly seen on inspection and if it occurred on a composite structure the appropriate remedial action could be taken. Delaminations however are not visible even though there may be significant internal damage to the laminate as shown in Fig. 14. The damage is often worse towards the back of the laminate, which makes detection more difficult. Delaminations can be detected by ultrasonic techniques. C-scans give a good silhouette of the damaged areas and A- or B-scans give useful information on the depth of the damage; computer graphics can be used to give a three-dimensional representation of the damage. Useful information on the damage, especially through thickness cracks, can also be obtained using X-rays but this requires the use of radio-opaque dye to penetrate the damaged region. Holography combined with heating can be used to show local disturbances in thermal conductivity caused by damage. Changes in vibration mode have also been proposed as indications of damage. There are many possible inspection techniques that work well in the laboratory on coupon specimens,[12] but they are usually difficult to apply to large complex structures in the field.

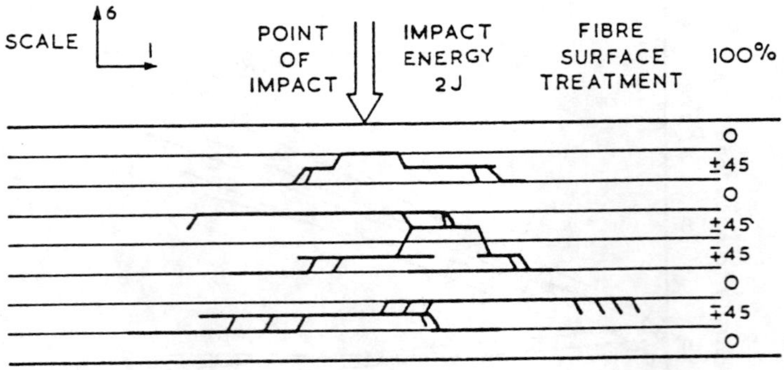

Fig. 14. Multiple delaminations in a $[(0°_2, \pm45°)_2]_s$ CFRP laminate caused by dropweight impact.

4.4 Damage in Composite Structures

Thicker laminates cannot always respond by flexing and impact causes front surface damage, rather like erosion damage by small particles. However, if the projectile has sufficient energy to penetrate the thick laminate, then it may result in front face erosion for approximately half the thickness, and then delamination and the usual flexural response can occur in the rear half of the laminate,[13] as shown in Fig. 15.

Similar work has been done on composite structures incorporating features such as stiffeners. It was found that impact in the middle of the bay required 30–50 J of energy to cause barely visible impact damage (BVID), because of the relatively large volume of material that was being deformed, whereas near a stiffener where the structure was stiffer, only a few joules were needed. Also, the stiffeners altered the local response of the panel so that damage was sometimes observed to one

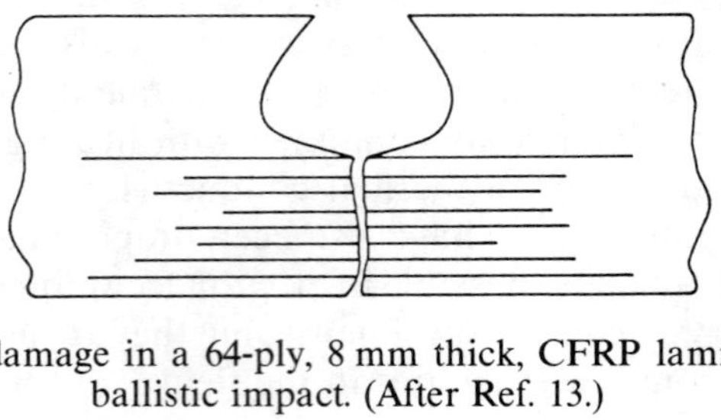

Fig. 15. Impact damage in a 64-ply, 8 mm thick, CFRP laminate caused by ballistic impact. (After Ref. 13.)

side of the point of impact or even away from the point of impact. Clearly the more complex the impact conditions the more closely must the test condition simulate the practical situation.

5 POST-IMPACT PERFORMANCE

The different forms of impact damage affect the various laminate properties to different degrees, and for this reason it is difficult to have a single impact parameter to describe the 'impact resistance'. Localised damage has little effect on global properties such as stiffness, although in helicopter rotor blades small changes in stiffness can cause detectable vibrations before there are significant reductions in strength.

Damage usually has a greater effect on residual strength. Delamination causes reductions in interlaminar shear strength (Fig. 16) and the residual ILSS can be predicted from the area of delamination using a fracture mechanics model and an interlaminar fracture toughness.[14] The damage observed for both dropweight and ballgun impact depended on the energy of the projectile, even though the velocities and momenta were very different.

Larger spans resulted in flexural damage, with reductions in flexural and tensile strength, as illustrated in Fig. 17. Again the reductions in

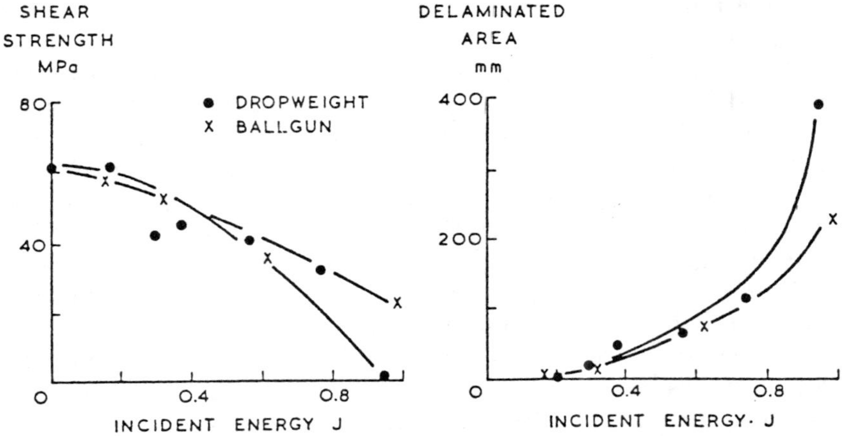

Fig. 16. Effect of impact damage on the residual ILSS of a $[(0°, 90°)_3]_s$ CRFP laminate.

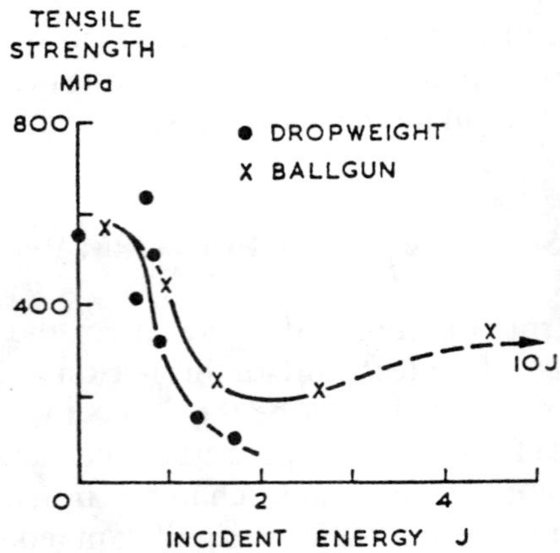

Fig. 17. Effect of impact damage on the residual tensile strength of a $[(0°, 90°)_3]_s$ CFRP laminate. The broken line represents penetration by the ball.

strength occurred at similar incident energies for both the dropweight and ballgun impact, although in this case the threshold energy for damage was significantly higher than for the smaller span used for the residual shear strength measurements. For the ballgun impact, the maximum amount of damage was done at incident energies just sufficient to cause penetration by the 6-mm ball. At higher velocities, the ball penetrated the laminate relatively cleanly, losing little energy, causing little cracking away from the hole and resulting in smaller reductions in strength.

The threshold energy for flexural failures can be affected by the material properties and the ply stacking sequence, as shown in Fig. 18. Increasing the resin toughness increased the flexural strength (σ), and therefore increased the energy ($C\sigma^2/E$), needed to cause failure. Putting the 45° plies on the outside reduced the flexural strength but it also reduced the flexural modulus (E), so that the threshold energy ($C\sigma^2/E$) was increased.

One of the main energy absorbing mechanisms in composite materials is the strain energy to failure of the fibres, or the area under the fibre stress–strain curve. Lower modulus fibres such as glass fibres and aramid fibres have a larger energy absorbing capacity. It is therefore possible to make hybrid laminates incorporating glass or aramid fibres to improve the impact resistance of CFRP. It is preferable to put the glass or aramid fibres on the surface[14] because the strains are greater

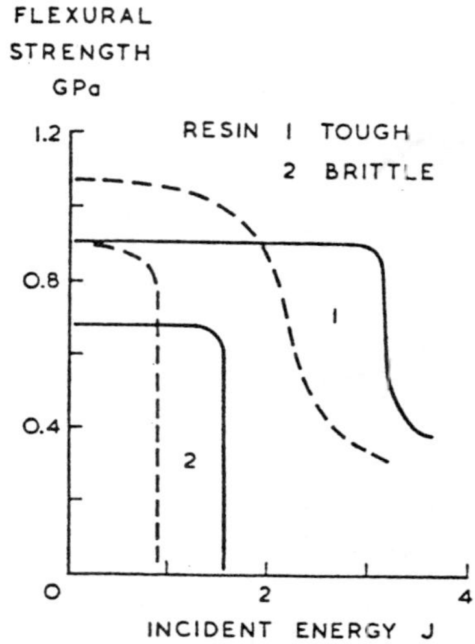

Fig. 18. Effect of resin toughness and ply stacking sequence on the threshold dropweight energy for flexural failure in CFRP laminates; ——, $[(\pm45°)_2, 0°_4]_s$; - - - -, $[0°_2, +45°_2, -45°_2, 0°_2]_s$.

there in flexure, and these surface plies can prevent fretting of the carbon fibres or fibre break-out on the back face.

Residual compression strength after impact is particularly important. In order to have a sufficiently large gauge section to accommodate the damage it is necessary to have anti-buckling guides to support the specimen, but these should not prevent local instabilities in the damaged region. Delaminations, under in-plane compression loading, tend to grow transverse to the applied load because this gives the greatest increase in compliance or the greatest strain energy release rate. Typical results are shown in Fig. 19. There was a marked reduction in compression strength for delamination damage (up to 2 J), that had little effect on tensile strength. The tensile strength did not decrease until there was significant fibre damage. Lower values of fibre surface treatment gave tougher laminates in tension especially in the region of fibre damage, as explained for notches in Section 3. But the residual compression strengths were lower with less fibre surface treatment, because of larger areas of delamination and less resistance to delamination growth. Some improvements can be made by the judicious use of stitching through the thickness of the laminate before

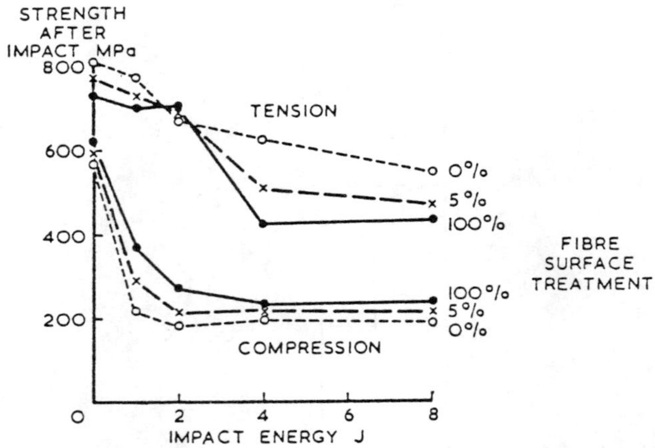

Fig. 19. Effect of fibre surface treatment on the residual tensile and compressive strengths of $[(0°_2, \pm 45°)_2]_s$ CFRP laminates after dropweight impact.

moulding, to reduce the incidence and growth of delamination (Chapter 9).

A factor that might affect the damage and hence the residual strength is the static load that a component is carrying when it is impacted. It might be expected to be less capable of deforming elastically under impact and fail for a lower incident impact energy. Figure 20 shows that in tension there was a slight reduction in residual strength with preload,

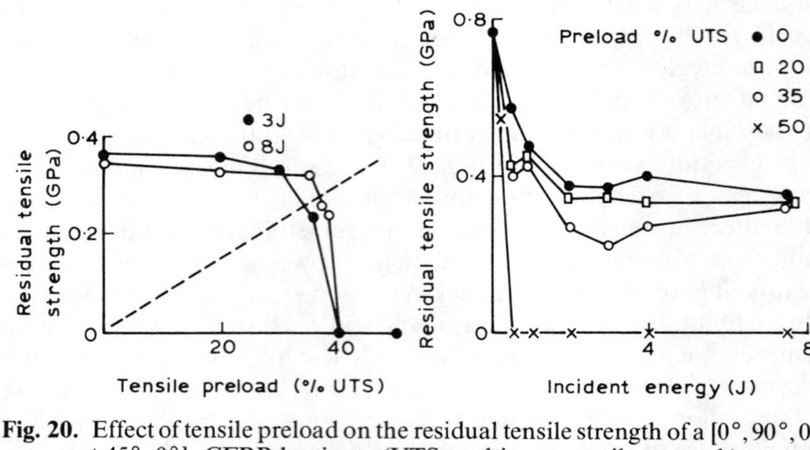

Fig. 20. Effect of tensile preload on the residual tensile strength of a $[0°, 90°, 0°, \pm 45°, 0°]_s$ CFRP laminate (UTS = ultimate tensile strength).

especially at intermediate impact energies where the maximum damage was done, but the main effect was a complete failure on impact if the tensile preload exceeded the expected residual tensile strength.

Similar results have been obtained with compression preloads,[15] as shown in Fig. 21. At high incident energies the transition from no-failure to failure-on-impact coincided with the residual strength for specimens that had not failed on impact. At intermediate incident energies the residual strengths were significantly higher than this transition preload.

Of practical concern for service structures is the possible growth of small sub-critical areas of delamination under fatigue loading, to the point of becoming critical. Figure 22 shows the effect of a 3 J impact on a CFRP laminate, causing internal delaminations and a reduction in compression strength. Figure 23 shows *S–N* curves for the same laminate, both undamaged and containing the delaminations caused by the 3 J impact, under fully reversed cyclic loading. It can be seen that the damage significantly reduced the static and low cycle fatigue strengths but the *S–N* curve for the damaged laminate was very flat and after one million cycles the fatigue strengths of the damaged and undamaged specimens were similar. It seems that loads that are not

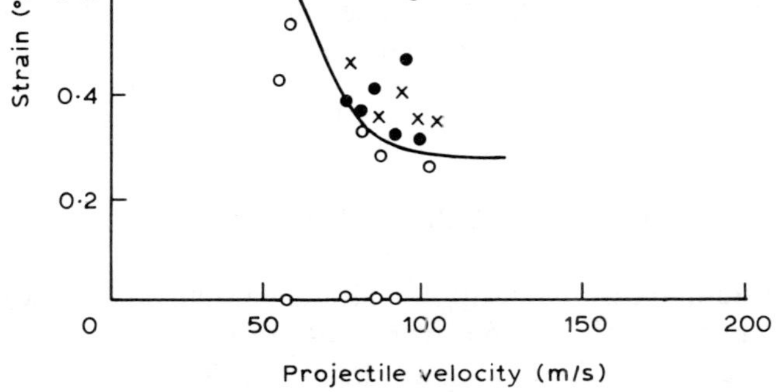

Fig. 21. Effect of compressive preload or prestrain on the impact behaviour of a $[(\pm45°, 0°_2)_2, \pm45°, 0°, 90°]_{2s}$ CFRP laminate. (After Ref. 15.)

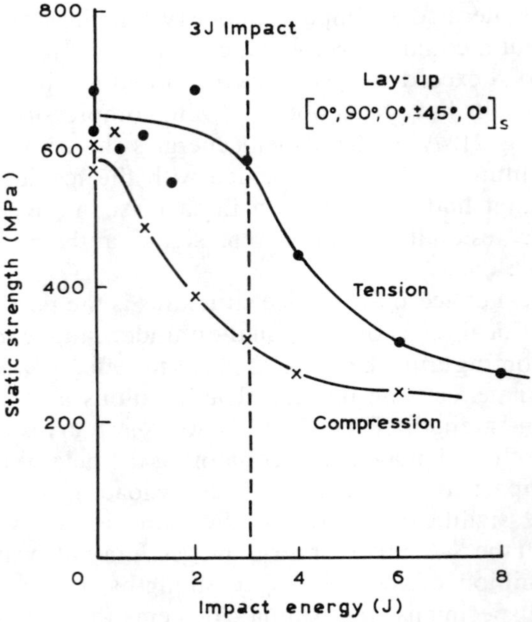

Fig. 22. Effect of dropweight impact damage on the tensile and compressive strengths of a [0°, 90°, 0°, ±45°, 0°]$_s$ CFRP laminate.

great enough to cause unstable damage growth and static failure are not likely to cause damage growth under fatigue loading. This has been found also with machined notches in CFRP laminates made from woven and non-woven plies.[16] If design limits are set to take care of impact damage and machined notches then it seems that fatigue will not be a problem (see also Chapter 10).

6 STRUCTURAL IMPLICATIONS

In the traditional safe life designs of aircraft structures there was a need to define the curve of probability of failure as a function of the number of load cycles, and in particular the first part of the curve where probabilities of failure were very low. With the move to damage-tolerant designs there is more emphasis on inspections, on predictions of

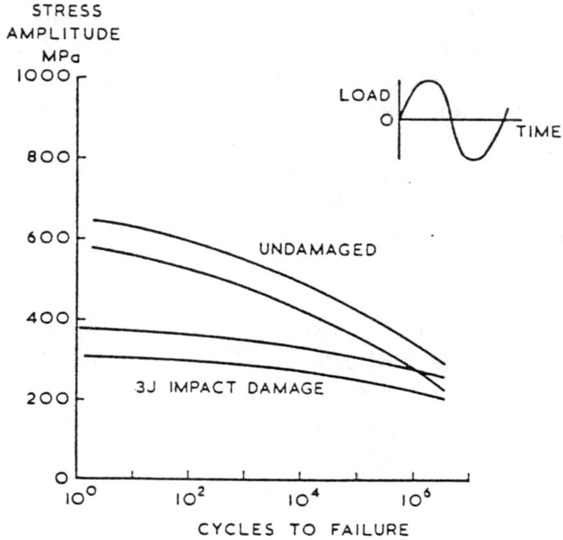

Fig. 23. Post-impact fatigue, under fully reversed cyclic loading, for damaged and undamaged [0°, 90°, 0°, ±45°, 0°]$_s$ CFRP laminates.

damage growth and on definition of critical damage conditions. Damage-tolerant materials have slow damage growth rates and can tolerate relatively large amounts of damage that can be readily detected. In metals, this has been concerned with fatigue crack growth, and for example in some helicopter rotor blades the inspection period was as short as 4 h. With composites, fatigue has not been so important and rotor blades have been designed with infinite lifetimes.

However, accidental damage can be significant and there is still a need to inspect for damage and to predict residual strengths, so that decisions can be made on whether to monitor the damage or to repair or replace the component. Because accidental damage can occur at any time, inspections need to be frequent and therefore to be quick and easy. There have been schemes proposed for large area inspection, based on infra-red irradiation or vibration characteristics, but they have not proved feasible and visual inspection is still relied on. Thus, invisible or barely visible damage might be missed and for this reason the design allowables assume that the components contain such damage.

There is still a need to be able to predict the static strengths for notched laminates under tensile, compressive or shear loading and for delaminated laminates under compressive or shear loading. There are various semi-empirical models to predict the strengths of notched laminates, and finite element techniques are being developed to predict the compressive strengths of delaminated panels, but we are still some way from being able to predict the true effects of service damage. If damage could be reduced to an idealised notch and/or delamination then it might be possible to relate these to local stress fields in a realistic structure and therefore be able to predict structural failure loads.

7 DAMAGE TOLERANCE ASPECTS OF THERMOSETTING AND THERMOPLASTIC MATRICES

Most of the applications for CFRP have so far used epoxy resin matrices and in some instances polyimides for higher temperature capability. High temperature, solvent resistant thermoplastic matrix composites are now available (Chapter 2).

Comparisons between carbon fibre/PEEK and carbon fibre/epoxy composites[17] showed that fibre-dominated properties were similar, there being good translation of fibre properties to the carbon fibre/PEEK composites. However, the shear modulus of the PEEK was lower than that of the epoxy, resulting in slightly lower compression strengths in the carbon fibre/PEEK. The PEEK matrix was tougher than epoxy resins and it showed some ability to flow under shear stresses. Thus it did not give a shear failure under three point bend, and interlaminar fracture toughness measurements gave values significantly higher than for carbon fibre/epoxy laminates. The increased toughness and shear strain capabilities resulted in much greater strains in tension tests on [$\pm 45°$] laminates: there were as many incipient cracks, indicating that the interface bond strength was similar, but these did not readily grow into resin cracks and delaminations.

On impact at low energies, there was an indentation on the front face with associated radial compression cracks perpendicular to the fibres in the surface ply. Sections through these compression lines showed them to be compression failures with characteristic shear bands. Ultrasonic C-scans and X-rays of the impacted laminates showed less delamination with carbon fibre/PEEK and less fibre damage at the point of impact, but there was some 0° splitting not seen with epoxy matrices. It could be that the fibre/matrix bond could still fail in transverse tension, but that this did not cause the interlaminar shear cracking as with epoxies and the stress was relieved by transverse tensile cracks within the plies. Plots of delaminated areas against impact energy (Fig. 24) showed that with carbon fibre/epoxy the delamination could extend a considerable distance, but that with carbon fibre/PEEK the damage was confined near the point of impact. Figure 25 shows polished cross-sections through the point of impact for an incident energy of 7 J. With the PEEK matrix there were fewer delaminations through the thickness and they were restricted in extent, whereas the laminate with the epoxy matrix showed delamination at many more interfaces and they extended much further. The carbon fibre/PEEK cross-section shows the front face indentation and the compression cracking.

Because of the restriction of damage and the 0° splitting which tended to blunt any cracks produced by impact, the residual tensile strengths (Fig. 26) were greater for the carbon fibre/PEEK laminates.[18] In compression too (Fig. 27) residual strengths were significantly greater in carbon fibre/PEEK because of the reduced damage and the greater resistance to damage growth. The greater residual compression strengths after impact, combined with the damage being visible on the front face at lower impact energies, give carbon fibre/PEEK a potential for raising design stresses when they are associated with BVID (Fig. 28).

8 IMPROVED COMPOSITE MATERIALS

Advanced composite materials are still relatively young and improvements are continually being made. By removing strength-limiting flaws, carbon fibre strengths have been increased significantly over the past few years. The manufacturers have produced two main classes of high strength carbon fibre, as shown in Fig. 29. High strain carbon fibres have been produced by keeping the modulus the same and increasing the strength to about 5 GPa and the strain approaching 2%. Intermediate modulus fibres have maintained similar working strains but have increased the strength to about 5 GPa (Chapter 3). In either case the strain energy to failure of the fibres has increased and this should restrict the damage and improve the impact performance.

Resin matrices are also being improved and significant increases can be made in the interlaminar fracture toughness (Chapter 4), but this has usually been at the expense of hot/wet compression strength. With toughened epoxy matrices, the areas of delamination on impact are reduced but the residual tensile strengths may be no greater because the

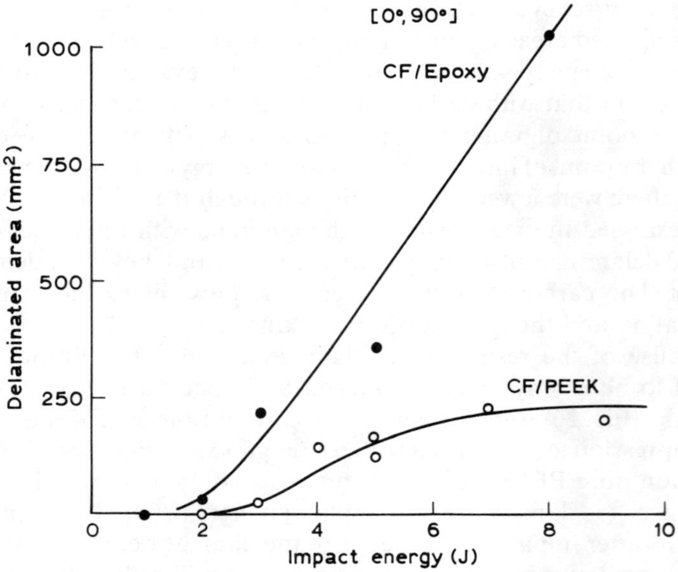

Fig. 24. Delaminated areas measured by ultrasonic C-scan, due to dropweight impact on [0°, 90°] laminates of carbon fibre/PEEK and carbon fibre/epoxy.

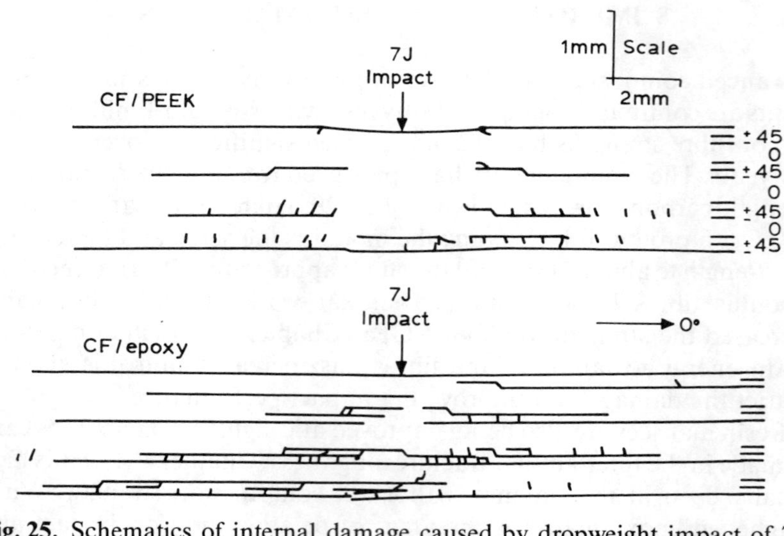

Fig. 25. Schematics of internal damage caused by dropweight impact of 7 J energy on $[\pm45°, 0°_3, \pm45°, 0°_2]_s$ laminates of carbon fibre/PEEK and carbon fibre/epoxy.

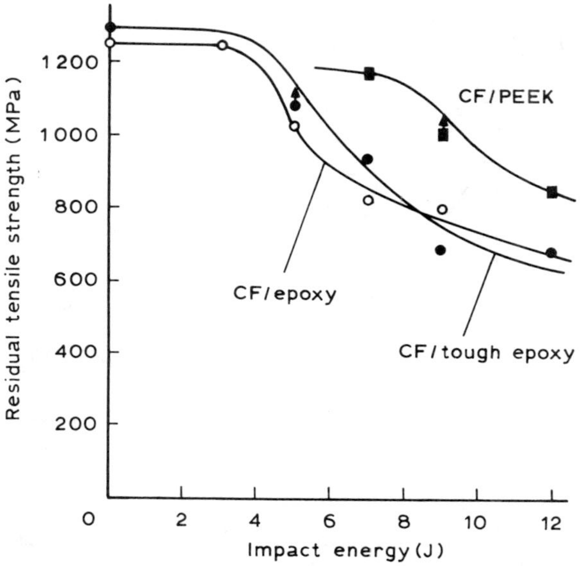

Fig. 26. Effect of dropweight impact on the tensile strengths of [±45°, $0°_3$, ±45°, $0°_2$]$_s$ laminates of carbon fibre/PEEK, carbon fibre/epoxy and carbon fibre/tough epoxy. (After Ref. 18.)

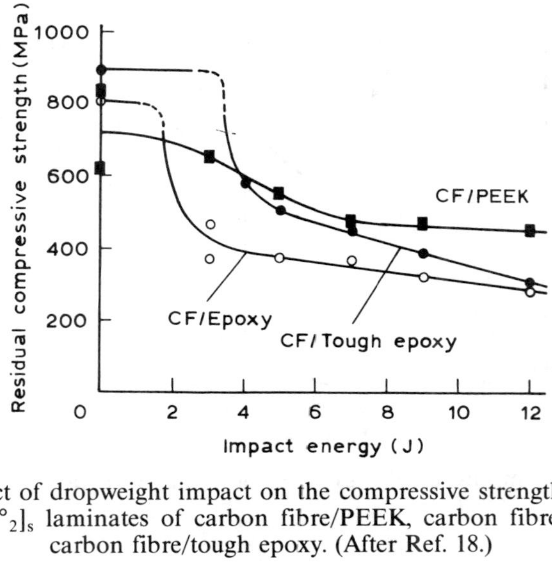

Fig. 27. Effect of dropweight impact on the compressive strengths of [±45°, $0°_3$, ±45°, $0°_2$]$_s$ laminates of carbon fibre/PEEK, carbon fibre/epoxy and carbon fibre/tough epoxy. (After Ref. 18.)

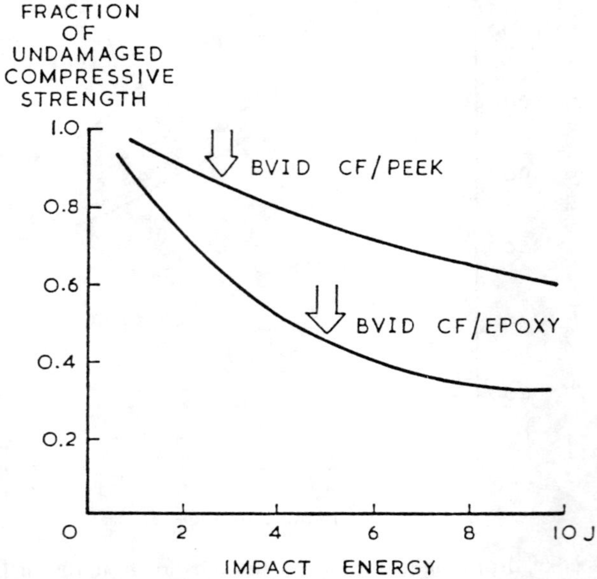

Fig. 28. The effect of barely visible impact damage (BVID) on the design allowable compression stress for carbon fibre/PEEK and for carbon fibre/epoxy composite structures.

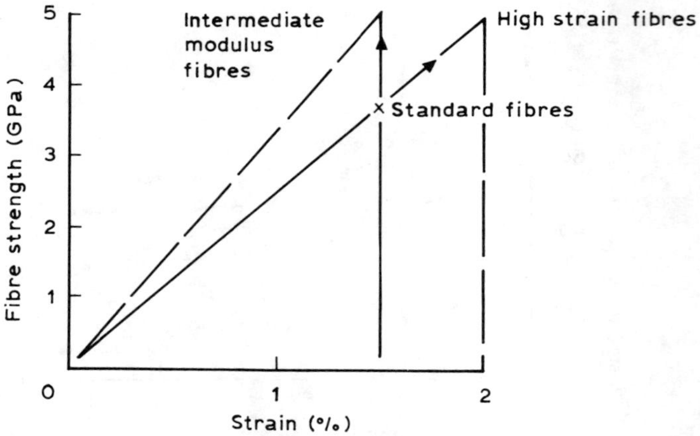

Fig. 29. Improvements in the tensile strengths of carbon fibres.

restriction of crack blunting mechanisms increases the notch sensitivity. However, the residual compression strengths are significantly better with toughened epoxy matrices, and critical strains for the impact of compression loaded laminates are significantly greater,[15] as shown in Fig. 30. The moduli of the toughened resins[15] were slightly lower but the tensile strains to failure had been increased from about 2% for the standard epoxy to about 6% for the toughened resins. The toughened laminates were able to withstand 'standard' impacts when preloaded in compression to a strain of 0·5%, whereas the standard epoxy system could not, although Fig. 30 shows that the selection of a 'standard' impact could be misleading and one really needs a range of impact energies.

The tougher matrix systems, both thermoplastic[17, 18] and thermoset,[15] tend to produce compression failure modes comprising fibre or ply buckling instabilities, rather than the usual delamination growth associated with compression failures of impact-damaged CFRP laminates, as shown in Fig. 31. These changes in failure mode need to be carefully considered in case apparent improvements in some properties produce degradations in others.

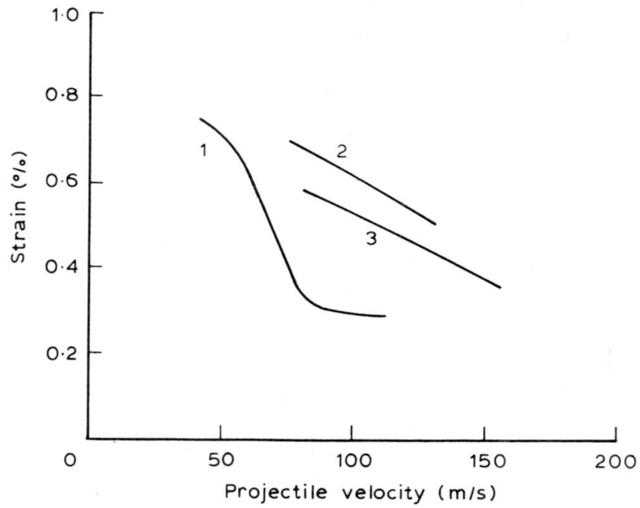

Fig. 30. Effect of ballistic impact on the critical compression strain for preloaded $[(\pm 45°, 0°_2)_2, \pm 45°, 0°, 90°]_{2s}$ CFRP laminates; 1, with a standard epoxy matrix; 2 and 3, with toughened matrices. (After Ref. 15.)

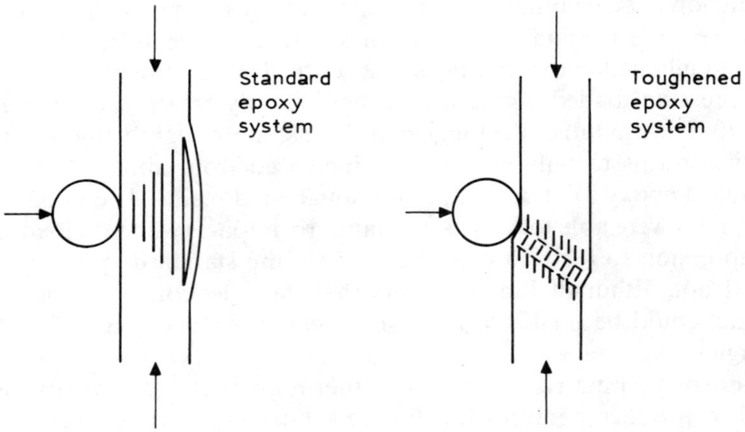

Fig. 31. Failure modes associated with the impact of compression loaded CFRP laminates. (After Ref. 15.)

It is becoming evident that the requirements for composite components for civil aircraft and for military aircraft are diverging. Civil aircraft require hot/wet compression strength at 90 °C and impact resistance, usually measured by compressive performance after impact, as shown in Fig. 32. For military aircraft the hot/wet compression strength has to

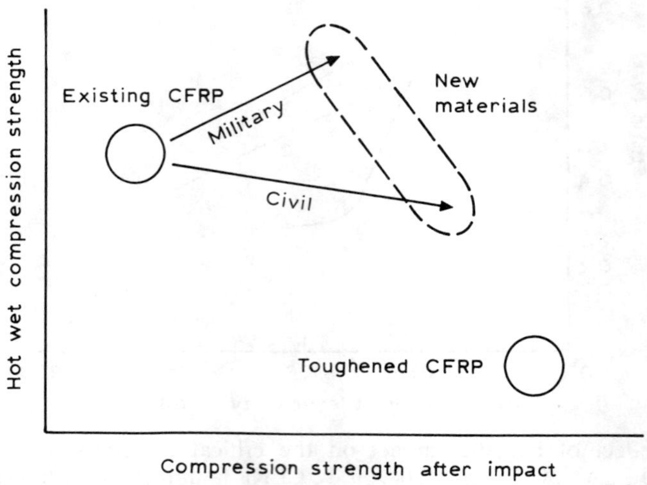

Fig. 32. The development of improved CFRP materials combining critical properties.

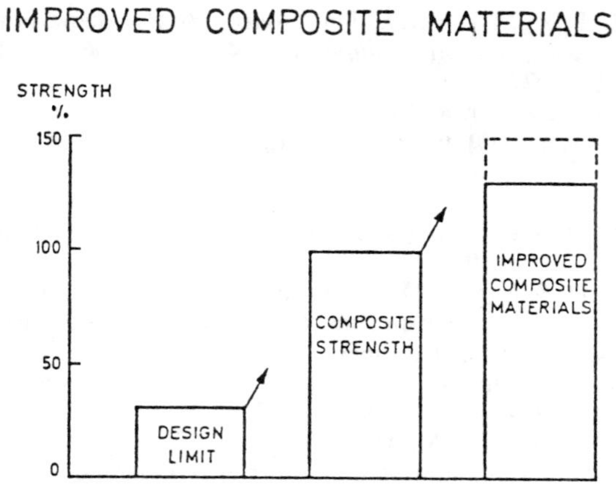

Fig. 33. Potential improvements in the design allowables for components made from advanced composite materials.

be demonstrated at about 130 °C. Higher-T_g epoxies and bismaleimides are being developed for this but they tend to be less tough and could have impact problems (Chapter 1).

In summary, the present generation of carbon fibre composites are being severely limited by poor impact resistance and poor hot/wet compression strengths. This makes design limits low compared with the materials' capability (Fig. 33), but significant mass savings can still be made. Improved understanding of the failure processes might improve the ratio of design stress to material strength, but greater advances are likely to come from material developments and improvements, and imaginative combinations of materials in hybrid structures.

REFERENCES

1. *Foreign Object Impact Damage to Composites*, ASTM STP568, 1975.
2. Effect of service environment on composite materials. AGARD *Conf. Proc. No. 288, Athens, 1980.*
3. Dorey, G., Impact and crashworthiness of composite structures. In *Structural Impact and Crashworthiness*, Vol. 1, ed. G. A. O. Davies. Elsevier Applied Science, London, 1984.
4. Bishop, S. M., Deformation near notches in angle plied carbon-fibre composites. RAE Technical Report TR 77093, 1977.
5. Dorey, G., Fracture of composites and damage tolerance. In *Practical Considerations of Design, Fabrication and Tests for Composite Materials.* AGARD Lecture Series No. 124, 1982.

6. Potter, R. T. & Purslow, D., The environmental degradation of notched CFRP in compression. *Composites,* **14** (1983) 206–25; RAE Technical Report TR 83029, 1983.
7. Purslow, D. & Potter, R. T., The effect of environment on the compression strength of notched CFRP — a fractographic investigation. *Composites,* **15** (1984) 112–20.
8. Dunford, D. V., Harvey, J., Hutchings, J. & Judge, C. H., The effect of surface treatment of type 2 carbon fibres on CFRP properties. RAE Technical Report TR 81096, 1981.
9. Winkel, J. D. & Adams, D. F., Instrumented drop weight impact testing of cross-ply and fabric composites, *Composites,* **16** (1985) 268–78.
10. Wardle, M. W., Impact damage tolerance of composites reinforced with Kevlar aramid fibres. In *Proc. ICCM-IV on Progress in Science and Engineering of Composites, Tokyo, 1982.*
11. Kawata, K., Hondo, A., Hashimoto, S., Takeda, N. & Chung, H. L., Dynamic behaviour analysis of composite materials. In *Proc. Japan–US Conf. on Composite Materials, Tokyo, 1981.*
12. Stone, D. E. W. & Clarke, B., Non-destructive evaluation of composite structures — an overview. In *Proc. ICCM-VI and ECCM-II, London, 1987*, Elsevier Applied Science Publishers, London, 1987.
13. Cantwell, W. J., Impact damage in carbon fibre composites. PhD Thesis, Imperial College, University of London, 1985.
14. Dorey, G., Sidey, G. R. & Hutchings, J., Impact properties of carbon fibre/Kevlar 49 fibre hybrid composites. *Composites,* **9** (1978) 25–32.
15. Williams, J. G. & Rhodes, M. D., Effect of resin on the impact damage tolerance of graphite/epoxy laminates. In *Composite Materials: Testing and Design (6th Conf.).* ASTM STP787, 1982.
16. Curtis, P. T. & Dorey, G., Fatigue of composite materials. In *Proc. Inst. Mech. Eng. Conference on Fatigue of Engineering Materials and Structures, Sheffield, 1986.*
17. Dorey, G., Bishop, S. M. & Curtis, P. T., On the impact performance of carbon fibre laminates with epoxy and PEEK matrices. *Compos. Sci. Technol.,* **23** (1985) 221–37.
18. Bishop, S. M., The mechanical and impact performance of advanced carbon-fibre reinforced plastics. In *Proc. ECCM-1, Bordeaux, 1985.*

Chapter 12

Certification of Civil Aircraft Composite Structures

John W. Bristow

Civil Aviation Authority, South Area — Gatwick Airport, Gatwick, West Sussex, UK

1 INTRODUCTION

The period from 1980 to 1985 saw the evolution of a set of working requirements and interpretations covering the airworthiness of composite materials used in the primary structural components of civil aircraft. This work was in conjunction with European and USA airworthiness authorities. To achieve this a number of different aspects had to be considered and they are to be discussed in this chapter. Firstly the situation existing in the early 1980s is reviewed. Then the supply and control of materials is discussed, the importance of manufacturing processes control is outlined, key issues in structural design are raised, and finally, the revised requirement basis is discussed.

The term composite can be interpreted in many ways; in the context of this chapter the materials primarily considered are thermosetting matrices, chiefly epoxy, reinforced with glass, carbon or aramid continuous fibres, although the possibility of both thermoplastic matrices and shorter fibres is not excluded.

2 THE EXISTING SITUATION

The range of civil aircraft projects involving composite materials submitted for airworthiness consideration has been wide and varied. By

1985 projects under consideration or already certificated ranged from airship gondola skins to vertical tail fin structures. Table 1 gives a more detailed list of typical components considered in the United Kingdom.

For the UK the structural requirements relevant to civil applications are given by British Civil Aviation Requirements (BCAR) and the European code, Joint Airworthiness Requirements (JAR) in a number of sections which are listed as Table 2.

In the early 1980s composite materials were not generally addressed by these requirements, and the guidance given in supplementary material was normally in respect of metallic structures. Federal Aviation Regulations, the USA requirements, were in a similar position except that an initial form of guidance material on composites was issued in 1978.[1]

TABLE 1

Range of Projects Considered for Certification in the UK

Wing flap structures on several aircraft in carbon/epoxy.
Control rods for two helicopters and one transport aircraft in carbon/epoxy.
Flying control surfaces on five twin jet transport aircraft in carbon/epoxy.
Transmission drive shafts of a large helicopter in carbon/epoxy.
Rotor blades (main and tail) on 10 helicopters mainly in glass/epoxy.
Empennage structure of three helicopters in carbon and glass/epoxy.
Fin and tailplane of a large transport aircraft in carbon/epoxy.
Airship gondolas in aramid/epoxy.
Four types of propeller blade in glass, carbon aramid/epoxy.
Fuselage pressure bulkhead in carbon/epoxy.
Seven-seat executive transport all in carbon/epoxy.
Engine mounting structure of a twin engined transport in carbon/epoxy.
Wing structure of a business jet in carbon/epoxy.
Two-seater trainer aircraft all in glass/epoxy.
Tailfin substructure for airship in glass, aramid honeycomb/epoxy.
Four-seat twin jet light aircraft primarily glass/epoxy.
A range of home-built and microlight aircraft in glass/epoxy.
Many applications of glass, carbon and aramid honeycomb/epoxy in galley structures, floor panels, access doors, fairings, undercarriage doors and engine cowlings.

TABLE 2
Applicable Civil Aircraft Requirements in the UK

Large aircraft	JAR 25 (was BCAR Section D)
Light aircraft	BCAR Section K
Airships	BCAR Section Q
Gliders	JAR 22 (was BCAR Section E)
Helicopters	BCAR Section G
Microlights	BCAR Section S
Propellers	BCAR Section C 5

The following sections outline the considerations that lead up to the formulation of additional guidance material to cover composites structure airworthiness requirements.

The reasons for special consideration of composites lie in their essential differences from metallic materials which they may replace in aircraft structures. These differences are:

— Composite materials generally have small strains to failure and do not exhibit yielding and plastic behaviour.
— The formed laminate of composite material has properties which are very highly anisotropic.
— The composite, with its attendant properties, is created at the same time as the component achieves its final form.

3 SUPPLY AND CONTROL OF MATERIALS

Although the range of composite material is vast, in practice for aircraft structural applications the range of fibres and matrices that need to be considered is limited.

The fibres in frequent use are glass, aramid or carbon in continuous fibre form as tow, tape or woven fabric. The major resin system in use is epoxy (by far), with some applications in polyester, phenolic, polyimide and bismaleimide. Developments using thermoplastic systems are envisaged.

Within one resin system different combinations of catalyst and hardeners can modify the cure times and temperatures over a wide range to suit particular applications.

The materials supply route is a complex one and it can involve a number of organisations, often on an international basis, before reaching the component manufacturer. It is not unknown for material to pass through a fibre maker, a weaver, a resin manufacturer, and a pre-impregnator before distribution to an aircraft constructor.

Obviously the number of combinations of the above factors are legion and it is necessary to have precise definition of the system and traceability at all stages, to ensure a consistent, reproducible product. (The long-term, 20–30-year, effects of small changes are as yet unknown.) Such a goal is achieved by adequate specifications and quality control to those specifications.

Although a very small number of national specifications exist in the UK, generally speaking there are not any national or international systems of specification covering these materials for aerospace use. In the industry procurement and supply are to a multiplicity of house specifications; for example one material supplier has to meet over 16 different specifications when producing one particular unidirectional carbon fibre epoxy pre-impregnated tape. This is further complicated because precise definitions of resin systems are not usually available to the user.

In principle there is no reason why composite materials supply should not be controlled in the same way as that of metallic material. In the long term, a system of nationally or internationally based composite material specifications including adequate characterisation of resin formulations and fibre treatments still needs to be established. However, for the present, civil airworthiness authorities seek a system of 'in house' manufacturers' material specifications and insist upon 'no change' clauses with adequate on-receipt inspection and testing to ensure that production materials supplied are to the same standard as that used for design evaluation. Table 3 gives a suggested minimum contents list for the infrastructure of material and process specifications. Any proposed changes to an existing material or to a new material have to be supported by a comparability study between old and new including any necessary testing of differences. Table 4 gives an example check list for carrying out a comparability study for material changes.

4 PROCESS AND MANUFACTURING CONTROL

One of the major appeals of composite materials is the wide range of ways in which they can be manufactured into their final form. Such methods have been documented in many other publications and are not

TABLE 3
Typical Contents List for Material and Process Specifications

Basic fibre properties
Types, numbers and frequencies of tests
Basic matrix properties and chemical characterisation
Types, numbers and frequencies of tests
Basic composite properties
Types, numbers and frequencies of tests
Condition of fabrication
Method, lay-up, tooling, workshop environment
Significant parameters of cure cycle
Limits, ranges and recording methods
Cured component properties
Types, numbers and frequencies of tests
Inspection criteria at each stage
Storage and handling conditions throughout the process

covered here in any detail; hand lay-up, filament winding and tape laying are the most common methods employed (see Chapter 7). With each it is necessary to ensure that there is no moisture or dirt contamination of the subsequent bonding process and that the correct fibres are laid in the correct numbers, in the correct direction, in the correct location.

Subsequent moulding and curing is done by a variety of techniques (e.g. autoclave, pultrusion, elastomeric tooling and platen press). In order to achieve the correct pressure–temperature–time sequence for resin cure, adequate thermal calibration of tooling and individual component quality control test pieces, preferably on the same lay-up, are required, particularly in the early stages of production.

Faults such as voids, porosity, fibre misalignment, etc., can be attributed to improper cure cycle and tooling. Such defects can significantly affect properties of the composite and not all can truly be assessed by subsequent inspection without component sectioning. Thus, the process control is of prime importance in achieving not only geometrically acceptable parts but also in forming components of adequate strength and stiffness — a situation not unlike the production of premium quality forgings.

Inspection following manufacture involves increased use of NDI methods to supplement visual inspection. Inevitably defects will be found and there is a need for adequate procedures involving design,

TABLE 4
Composite Material Change Check List

Likely effect on ± / *Material change* ↓	*Quality assurance tests*										*Basic properties*										*Structural properties*							*Process properties*				
	Flexural strength	Flexural modulus	ILSS	Shear modulus	Peel	Lap shear	45° Tension	Flatwise tension	etc.	etc.	Tensile strength	Tensile modulus	Compressive strength	Compressive modulus	In-plane shear	Fatigue	Creep	Impact	etc.	etc.	Stress concentration	Ply ramp	Thickness	Hole	Joint	etc.	etc.	Tack	Flow	Gel	etc.	etc.
Fibre																																
Form																																
Size																																
Treatment																																
etc.																																

Matrix																																
Type																																
Hardener																																
Catalyst																																
etc.																																
Interface																																
Process																																
Environment																																

production and inspection staff to assess the significance of a particular defect in a particular location.

As the production process can have a significant effect on the design properties of a component, the civil airworthiness authorities require that once a production route has been established and associated with the necessary certification testing, it will need to remain unchanged. Any proposal for change will have to be supported by a comparability study of the old and new methods along with any necessary testing of differences.

5 ISSUES IN STRUCTURAL DESIGN

5.1 Static Loads and Safety Factors

The static structural loading cases and the minimum ultimate factor of safety to be considered are the same as for conventional metallic aircraft structures, i.e. the design ultimate load is $1{\cdot}5 \times$ the design limit load (limit load equates to the maximum load expected in service). Consideration of variability and environmental effects may lead to supplementary factors in certain cases.

5.2 Static Strength Substantiation

Composite structures can exhibit a number of different failure modes in fibre, matrix or interface,[2] which may be affected by the environment. Furthermore with their elastic and highly anisotropic behaviour they can be particularly sensitive to detail design features. For these reasons civil airworthiness authorities normally require a combination of structural analysis and test evidence to cover each significant structural feature of a design.

For analysis, allowable strength values are established in a statistical manner, usually in a similar way to metallic materials:

- For failsafe/redundant design to a 'B' basis, the value above which at least 90% of the population of values is expected to fall with a confidence of 95%.
- For single load path design to an 'A' basis, the value above which at least 99% of the population of values is expected to fall with a confidence of 95%.

Such allowable values are currently established by each constructor, but work is in hand in the UK to produce a composite materials data handbook, like the *MIL-Handbook 17* in the USA.

Due in part to the multistage production process, composite components can exhibit significant strength variation[3] when compared with their metallic counterparts. This variability can change with the type of design feature, related to the lay-up geometry, etc. The use of statistically based allowables covers this effect in analysis. However, with test substantiation any additional variability, when compared with metals, needs to be considered.

Humidity, temperature, solar radiation and cyclic loading can reduce strength of composites significantly.[4] As an example, Fig. 1 shows the synergistic effect of temperatures and absorbed moisture on carbon/epoxy laminate compressive strength. Consideration must be given to the effects relevant to the composite system in use, related to the expected service environments, when deriving allowables and establishing test conditions.

An acceptable route for static strength substantiation is set out below. For all relevant design cases:

(a) carry out a structural analysis and identify the significant structural design features,
(b) subsequently confirm the structural analysis by loading a fully instrumented structure,
(c) establish allowable values for each significant structural feature from relevant test evidence with allowance for variability and environmental effects,
(d) test the complete airframe structure to a minimum of design ultimate loads and check for each design case that allowables are not exceeded at design ultimate loads.

In such a procedure, consideration should be given to the possible interactive effects of different loading cases that could occur in service. For example, consider a wing box (see Fig. 2) where a transverse load (from a wing flap track) produces a local delamination at a ply runout without loss in tension strength, but significantly reducing the compressive strength under wing bending loads. Also, it is strongly recommended that the structure should be taken to failure to establish the failure mode in at least one case in confirmation of analytical assumptions.

Major static tests may be carried out on new structures or after cyclic

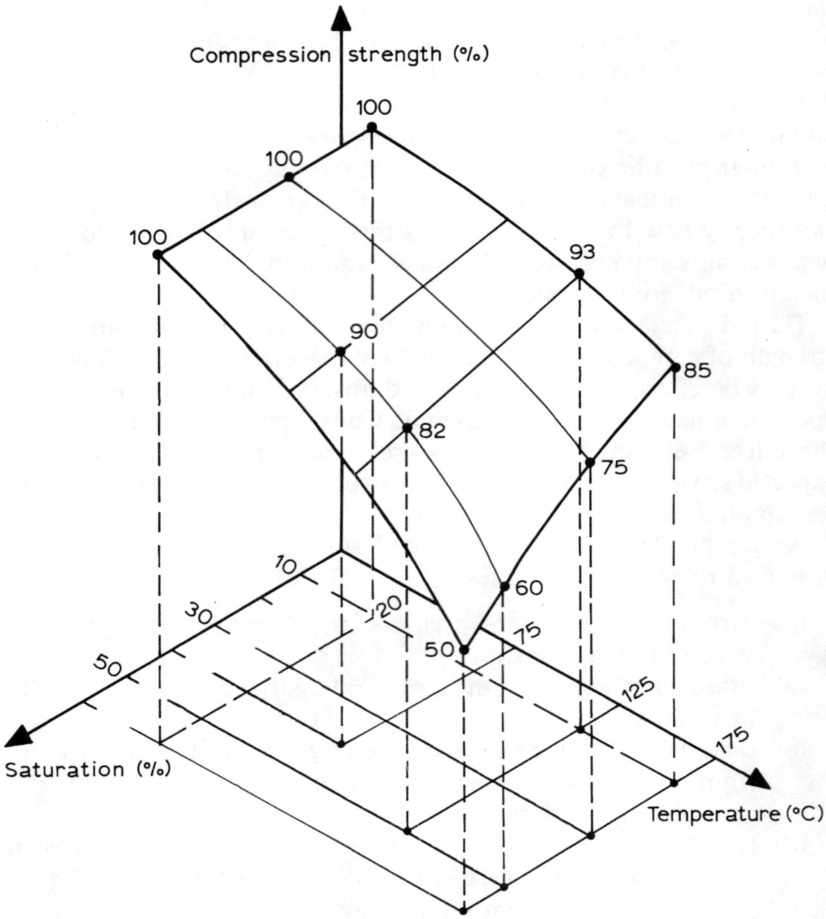

Fig. 1. Effect of temperature and moisture on CFC strength. After Ref. 5.

loading if it is the more critical case, as established by tests on specific details of the structure.

5.3 Cyclic Loading

There are two approaches to the effects of cyclic loading (fatigue) on aircraft structures.

(a) *Safe life.* Where a 'life' is declared below which no significant damage is likely to occur, a component is not permitted to operate beyond this life.
(b) *Failsafe (or damage tolerant).* By means of multiple load-paths, arrest features or slow flaw growth, a component can remain in service indefinitely provided that it is inspected with a method and frequency that will detect any flaw before it affects the safety of the aircraft.

The relevant requirements determine which approach may be

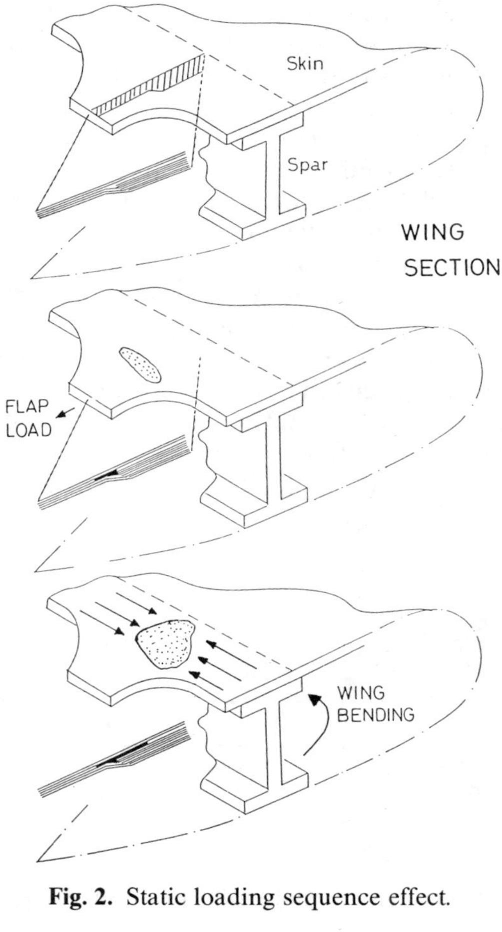

Fig. 2. Static loading sequence effect.

adopted for different aircraft. Typically, large transport aircraft are failsafe (damage tolerant) whereas helicopters have a large number of safe life components.

Structures in composite materials often have good intrinsic fatigue properties in comparison with metals[6] and are therefore good candidates for either approach. The data and analytical tools for failsafe evaluation are not at such a developed state as those for crack growth and residual strength estimation in metals. Currently, therefore, acceptable failsafe substantiations for composites will be largely based on test work with induced realistic flaws in actual components.

In some particular configurations of composites, large readily detectable flaws have good residual strength capabilities and can be demonstrated not to grow under cyclic loading (Chapters 10, 11). Figure 3 shows a schematic comparison of damage growth and residual strength of metallic structures with a 'no growth' composite flaw. This 'no-growth' approach is being increasingly adopted in the industry[7] and is acceptable to airworthiness authorities provided there is adequate demonstration of the 'non'-growth and an inspection programme is provided to detect any such flaw should it occur.

In areas where such failsafe substantiation is not possible an authority will require adequate safe life demonstration. Associated with the good intrinsic properties is a large scatter in fatigue life terms[8] which leads to large reduction factors to establish a safe life. However, consideration can be given to a factor in strength terms, analogous to the way many helicopter components are tested, in the substantiation for safe life or flaw growth intervals in composites.

Due to the many and varied failure modes possible in a composite

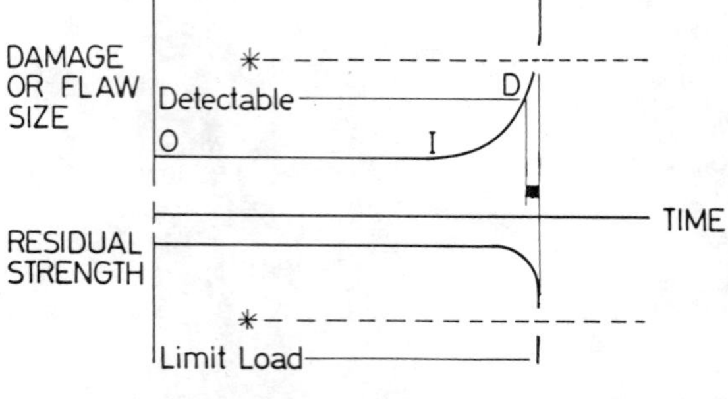

Fig. 3. Flaw growth and residual strength comparison.

structure it may be necessary to substantiate part as failsafe and part as safe life within one component.

5.4 Impact

Aircraft structures may experience impacts in any of four categories:

(1) Low-energy impact.
(2) Bird strikes.
(3) Discrete source damage, e.g. turbine disc burst.
(4) High-energy crash impact.

For composites the first and last cases are of special interest. At low energy levels of impact, particular composites such as carbon epoxy can experience significant compression strength reductions without the impact damage being visible.[9] A typical behaviour is shown schematically in Fig. 4. Therefore, for the range of energies up to the maximum likely to be experienced, it is necessary to demonstrate that the structure can withstand ultimate loads for damage up to the point where it becomes detectable. Furthermore, detectable impacts should not reduce the strength below limit value (two-thirds of ultimate) (see the 'requirement boundary' on Fig. 4).

The second and third cases are treated in a similar way to metallic structures. Direct testing of bird strikes is required to be carried out and also adequate strength after significant structural cutting has to be shown in the case of disc failure.

Due to the elastic behaviour up to failure, energy absorption

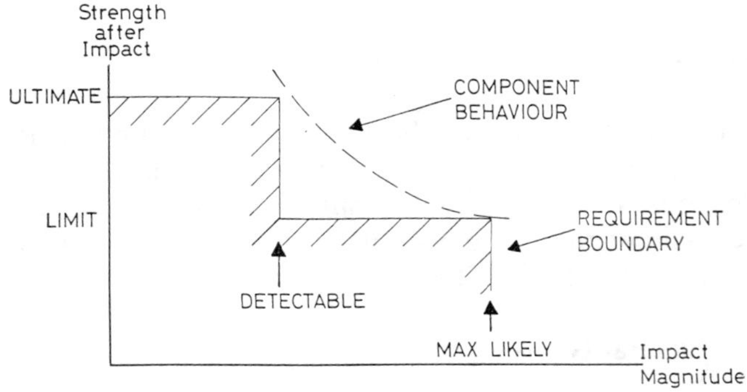

Fig. 4. Required strength versus impact damage.

characteristics of composites are likely to be different from those of metallic materials. In the crash case the objective is to establish a structure with a survival capability at least equal to current metallic structures.

5.5 Repairs

During the design of a composite aircraft structure it is necessary to establish the capability of the structure to be repaired, the techniques to be used, and in particular any allowances that have to be made for the lower standard of process control that may be expected in the field. Particularly detailed guidance instructions are required for repairs allowed in the field.

5.6 Erosion

Some composite materials are liable to rain, sand or dust erosion to a greater extent than metals. Erosion damage can lead to moisture ingress and delamination of the composite. Allowance must be made in design for this effect in vulnerable areas by the use of suitable materials for protective coatings or shields.

5.7 Stiffness

In addition to strength considerations it is necessary to check that the variations of mass and stiffness possible in production and the changes in stiffness due to damage progression in service do not invalidate the deflection and aeroelastic assumptions made in the design stage.

5.8 Inspection

Inspection of composite materials in service can be complicated by the lack of commonly accepted visual clues such as surface deformation and reflections. Great care and detailed instruction are needed for adequate maintenance of the aircraft in service with an increased reliance on extensive NDI likely.

5.9 Non-structural Issues

Alongside the direct structural issues there are a number of topics that must be addressed such as lightning strike, flammability, toxicity, smoke emission and electromagnetic effects. These are mentioned as part of the overall picture in assessing a composite design but are not the subject of this present work.

6 THE NEW REQUIREMENTS

Early in 1982 European structural specialists, from both airworthiness authorities and industry, commenced a series of meetings to review requirements for composite materials in the structure of large transport aircraft. Eighteen months later, after several meetings, a jointly agreed working paper was produced, at its sixth draft, on the certification of composite aircraft structure. In 1983 on the other side of the Atlantic the FAA (Federal Aviation Administration) was meeting with their industry to rewrite the existing advisory circular on Composite Aircraft Structures. By mid-1983 that paper had reached its tenth modification stage. It became possible to bring the European and USA groups together and hold three joint meetings. These meetings resulted in a text (now at the 16th modification) mutually agreed at the working level by the end of 1983. In airworthiness requirements terms, the achievement of this agreed technical wording before its issue between the Europeans and USA was something of a first and it is helping to achieve a common approach on both sides of the Atlantic.

The outcome was that the airworthiness requirements needed no significant change; however, the need was for advisory literature on how to comply with the existing regulations when using composite materials.

The overall objective for such literature is to achieve a safety level equivalent to that of current conventionally constructed aircraft with the underlying principles that the structure should be designed to retain ultimate load capability at any time during its service life, including the effect of repeated load and environmental exposure.

The material was published in 1986 by JAR[10] and the full text is given in the Appendix. Similar material has been issued by the Federal Aviation Administration.[11] The topics covered are:

Material and Fabrication Development;
Proof of Structure — Static — Fatigue/Damage Tolerant — Flutter;
Impact Dynamics;
Protection of Structure;
Quality Control;
Production Specifications;

Inspection and Maintenance;
Substantiation of Repairs.

The following paragraphs give details of the guidance material produced and will review some of the experience gained in the practical application to the airworthiness certification for a range of civil aircraft structures, as seen by an airworthiness authority. The subject will be dealt with in sections stating the guidance material in smaller type taken from Ref. 10, followed by a commentary on the practical experience to date.

> '4 Material and fabrication development
>
> 4.1 To provide an adequate design data base, environmental effects on the design properties of the material system should be established.
>
> 4.2 Environmental design criteria should be developed that identify the most critical environmental exposures, including humidity and temperature, to which the material in the application under evaluation may be exposed. . . . .' (See Appendix for full text.)

The concept of a data base is not new but when it involves a new material the question that is often asked is what should be contained in a data base? The structural substantiation of composites is based on a pyramid shape culminating in a full-scale structural test at the top, supported by a series of tests on major subcomponents (e.g. a tailplane-to-fuselage joint). That is in turn supported by a range of detailed tests on simple structural features (e.g. ply changes, fastener holes, etc.) and in turn that is based on a larger number of material property tests on basic laminates. It is the lowest two tiers of this pyramid that form the data base (see Fig. 5). Its extent will vary with the project under consideration, but it should encompass all the relevant design properties in the range of environments expected in service. It has been shown in results of recent full-scale tests that it would be wise to include total thickness, stacking sequence and impact damage in the range of variables to be considered in such a data base.

Taking two examples from the environmental investigations carried out to date, the maximum temperature to be considered is focused around 70–80 °C for subsonic transport operations and around 50–55 °C for gliders and light aircraft with reflective paint schemes.

Figure 6 is an example of how a composite structure can cool during take-off before the most significant flight loads can occur. The different behaviour of internal and external structure can be clearly seen. Figure

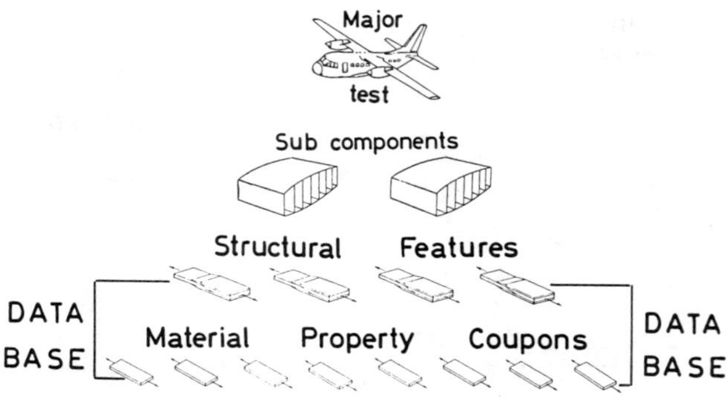

Fig. 5. Data base for CFC.

7 shows how structural temperature may vary with external paint scheme colour under solar radiation at two ambient temperatures.

In the earliest days of carbon fibre composites, boiling water immersion was considered as an appropriate definition of the environmental effect. When this was found to be too severe, a 1% by weight uptake of moisture was assumed. However there is a danger that this 1% figure will be read across to other composite systems and also the definition of moisture uptake is a particularly tricky one in defining the zero uptake position. For these reasons the currently favoured approach is to expose the material to a relative humidity appropriate to worldwide exposure (say 84% from Ref. 12) until saturation is reached. Then the properties in that material condition are evaluated at the appropriate temperature.

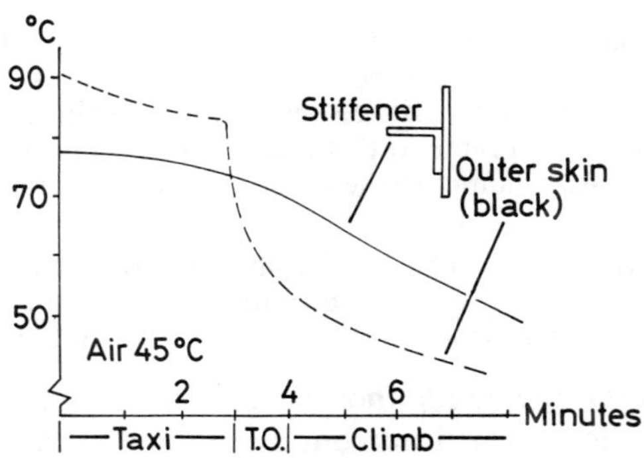

Fig. 6. Example of maximum temperatures for CFC structures.

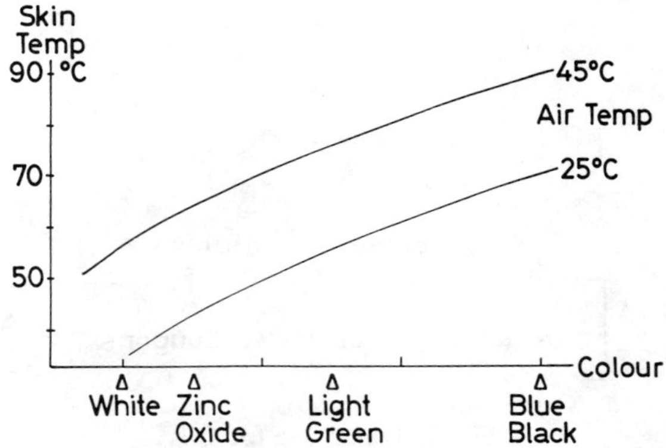

Fig. 7. Structural temperatures related to surface colour.

In establishing the design value for a structure the same statistical basis may be used for composite material as for metallic materials (see Ref. 13), although, in the early days of a particular material, small sample sizes are allowed provided that the statistics are adjusted for this reduced sample size.

The establishment of design values of specific material configurations by test is an important approach, either in its own right or used to validate analytical approaches that may be used. The significant anisotropy, the elastic behaviour and the variety of failure modes in composites make it extremely difficult at present to analyse complex structures reliably for all loading cases to be considered using laminate material property data alone.

> '5 Proof of Structure — Static
>
> 5.1 The static strength of the composite design should be demonstrated through a programme of component ultimate load tests in the appropriate environment, unless experience with similar designs, material systems and loadings is available to demonstrate the adequacy of the analysis supported by subcomponent tests, or component tests to agreed lower levels.
>
> 5.2 The effects of repeated loading and environmental exposure which may result in material property degradation should be addressed in the static strength evaluation . . .' (See Appendix for full text.)

The key point of the guidelines is a static test to ultimate load considering the effects of environment, cyclic loading and manufacturing variabilities. Aircraft constructors have chosen a variety of ways to meet this requirement. The simplest way of testing is at room temperature without environmental conditions, using a test factor on loads to cover the environment, cycling and variability. These factors, which are specific to the construction used, can vary. Some examples are:

— For the effect of variability in carbon fibre epoxy laminates compared with conventional aluminium alloy plate, a factor of 1·06.
— For aramid/phenolic honeycomb sandwich under warm moist conditions, a factor of 1·7.
— For the hot wet effect in carbon fibre epoxy laminates a factor of 1·31.

As the necessary additional factors are greatest for the matrix dominated properties, such as compression strengths, a room temperature test with large factors necessitates over-design of tension elements. Additionally, any metallic component in the structure will also have to be designed to cope with the increased test loading.

For this reason some constructors elect to carry out structural testing at elevated temperature on a moisture conditioned test specimen. This is a more expensive approach, but eliminates the need for additional structural size (i.e. weight) to cover the increased test load in lieu of environmental exposure.

Another approach to the problem is to establish by extensive subcomponent test and analysis the complete range of allowable strains for the material system in the environmentally degraded state, with a statistical allowance for variability, then to carry out a fully instrumented structural test up to ultimate load at room temperature without environmental conditions. At design ultimate load, if any strain measured in the structure is indicated as above the degraded statistical allowable, previously derived, the test is not passed.

Major static tests may be carried out on the fatigue test specimen, as the elastic behaviour of typical composites may not lead to the strongly misleading interactive effects seen in conventional metallic structures statically loaded before fatigue testing. It needs to be established, however, on small-scale specimens, whether prior cyclic loading can have a beneficial effect on the subsequent static strength.

The need for full-scale testing to ultimate load has been underlined by the results of a significant number of full-scale composite airframe tests in recent years. The majority of these have failed below ultimate load at the first attempt. There have been a variety of reasons for the failures: inadequate allowance for environmental effects; omission of relevant loading cases in the design; the effect of impact; and certain design features. Such testing has enabled the structures to be modified accordingly with subsequent testing proving satisfactory.

> '6 Proof of Structure — Fatigue/Damage Tolerance
>
> 6.1 The evaluation of composite structure should be based on the applicable requirements. The nature and extent of analysis or tests on complete structures and/or portions of the primary structure will depend upon applicable previous fatigue/damage tolerant designs, construction, tests, and service experience on similar structures. In the absence of experience with similar designs, structural development of components, subcomponents, and elements should be performed. The following considerations are unique to the use of composite material systems and should be observed for the method of substantiation selected by the applicant. When selecting the damage tolerance or safe life approach, attention should be given to geometry, inspectability, good design practice, and the type of damage/degradation of the structure under consideration.
>
> 6.2 Damage Tolerance (Fail-Safe) Evaluation
>
> 6.2.1 Structural details, elements, and subcomponents of critical structural areas should be tested under repeated loads to define the sensitivity of the structure to damage growth . . .' (See Appendix for full text.)

A number of issues have arisen in the approaches outlined. The most significant one is something of a dilemma. Some composites, although they have a good fatigue strength, have a very flat *S–N* curve which leads to high scatter in life. Thus, larger factors than the currently accepted ones for metallic materials would be needed for adequate demonstration, say between 10 and 50. However, the factor on stress, such as is typically used in helicopter fatigue testing, can cover the point, but problems may arise if metallic components are also present in the structure and would in consequence be over-stressed. The author is aware of two approaches to this dilemma at present. One is to use separate test procedures to validate the composite component and the metallic component independently. The other is to use a combined life and stress test factor, so that the test does not go on for too long, nor that the structural loads become unrepresentatively high.

Associated with a good fatigue strength, the margin of static strength over fatigue strength can be considerably reduced (see Fig. 8). This may influence the nature of substantiation testing, particularly for helicopter components where substantiation for fatigue has taken precedence over static testing for many years, due to a large implied margin.

Environmental interaction with fatigue performance is another issue.

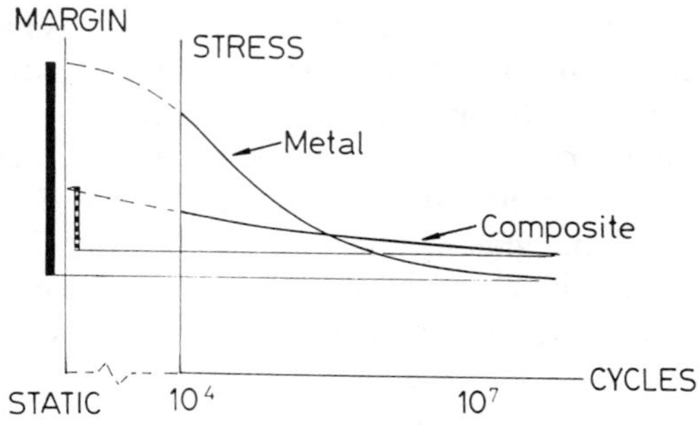

Fig. 8. Static and fatigue strength margins.

Helicopter rotor blades are among the longest serving primary structural components and here the problem is addressed in two ways, using real time component exposure tests and also re-testing of service components. In one approach the constructor has assumed a degradation in performance by increasing the fatigue test loads leading to fixed reduced life. In the other, the results of the long-term and component recall tests are used to reassess the fatigue life periodically. In general the results on these thick section glass/epoxy structures have been shown to be conservative.

The no-growth concept, referred to above, is that for some configurations of composite structures, large readily detectable flaws have good residual strength capability and can be demonstrated not to grow under cyclic loading. There is however a need to have an adequate inspection programme of sufficient frequency, such that if the flaw does occur the aircraft is not exposed to extended periods of operation with the large flaw and hence reduced strength capability.

Impact damage must be considered both in the damage tolerant, and in the safe life, approach. Impact is discussed in Section 5.4 of this chapter (see also Chapter 11).

> '7 Proof Structure — Flutter. The effects of repeated loading and environmental exposure on stiffness, mass and damping properties should be considered in the verification of integrity against flutter and other aeroelastic mechanisms. These effects may be determined by analysis supported by test evidence, or by tests at the coupon, element or subcomponent level.'

No fundamental issues in the flutter area have arisen specifically because of the use of composites, rather than metallic material.

However, two examples have come up on the effect of changing the type of composite materials providing the skins to control surfaces. In one of these cases the minor change revealed a mild flutter situation that was inherent in the aircraft but had escaped detection previously, so the composite material could hardly be claimed to be the culprit. In the other a change from glass to aramid fabric, associated with other changes to primer, filler and paint, led to water passing through the skins and remaining trapped in the surface, thus affecting the balance of the control surface.

> '8.1 Impact Dynamics. The present approach in airframe design is to assure that occupants have every reasonable chance of escaping serious injury under realistic and survivable impact conditions. Evaluation may be by test or by analysis supported by test evidence. Test evidence includes but is not limited to element or subcomponent tests and service experience. Analytical comparison to conventional structure may be used where shown to be applicable.'

As mentioned previously, aircraft structures may experience impact in any of four categories: (1) low-energy impact. (2) bird strike; (3) discrete source damage, e.g. turbine disc burst; and (4) high-energy crash impact.

In practice it has been shown that not only is the impact energy important, but so is the shape of the impacting object. Sharp objects render damage visible at lower energies than rounded ones. Thus for airworthiness clearances purposes, impact with hemispherical objects is requested to establish maximum undetectable damage sizes.

The definition of maximum likely impact is obviously an important issue and will vary from application to application. Significant energies can be involved.[14] To take a simple example, a full executive briefcase (weight 5 kg) dropped from an overhead locker (height 2 m) can impart (98 J) to a composite seat structure below.

In the fourth category, as stated previously, the elastic behaviour of a composite structure up to failure would indicate that the energy absorption characteristics are likely to be different from a metallic structure (see the typical line in Fig. 9), showing little energy absorption. However, by suitable composite configuration significant energy absorption can be achieved by inducing progressive collapse (see the optimum line on the same figure). Furthermore, in reality wing leading edge structures have been successfully designed for bird impact in aramid epoxy laminates. Also glass carbon epoxy propeller blades have shown a gentle shredding behaviour during impact after undercarriage collapse.

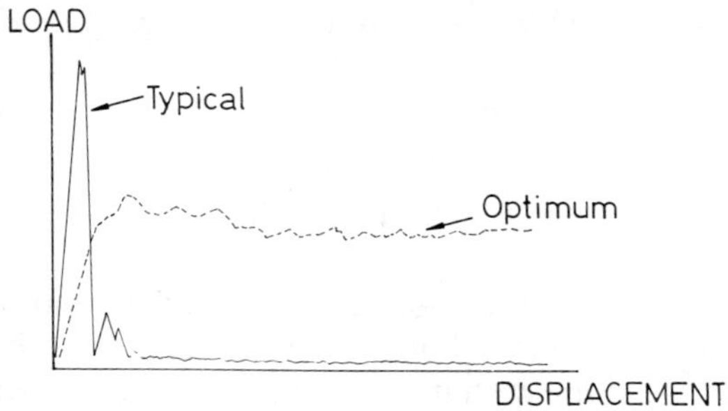

Fig. 9. Load–displacement curves for composite tubes. After Ref. 15.

'Additional considerations
8.2 Flammability
8.3 Lightning Protection
8.4 Protection of Structure
8.5 Quality Control
8.6 Production Specifications
8.7 Inspection and Maintenance
8.8 Substantiation of Repairs.' (See Appendix for full text.)

The important issue of fire, smoke, toxicity and electrostatic/electromagnetic effects must be addressed in considering the overall airworthiness of composites in aircraft. They are not relevant, however, in this structural chapter.

The composites typically in use are demonstrating good resistance to most commonly used aeronautical fluids, apart from water, as previously discussed. Erosion shields are important on rotating components, as are reflective coatings to prevent ultraviolet degradation of surfaces exposed to sunlight. There is, however, one practice in parts of the industry which is a little difficult to understand. That is the use of conventional metallic primers on composite components 'to make them look like metal'. This practice cannot help any maintenance task, nor can any use for a chromate primer be seen on a non-metallic structure.

Quality control, specifications, inspections, maintenance, and repair have been highlighted in the early parts of this chapter. They are topics of fundamental importance to the production and operation of airworthy composite structures. A good dialogue between design, engineering, production quality and maintenance disciplines is essential if any composite project is to be successful. Further requirement development work on a European–USA collaborative basis on these topics began in 1988.

ACKNOWLEDGEMENT

The views expressed in this work are those of the author personally. However, he would like to acknowledge with gratitude the many useful discussions on the subject that have taken place with colleagues in the industry and the permission of the CAA to present the work.

REFERENCES

1. Composite Aircraft Structure. Advisory Circular 20–107. Federal Aviation Administration, July 1978.
2. Schwartz, M. M., *Composite Materials Handbook*. McGraw–Hill, New York, 1983.
3. Gerharz, J. J. & Schutz, D., Literature Research on the Mechanical Properties of Fibre Composite Materials. RAE Translation 2045, Aug. 1980.
4. Imperial College London, Environmental effects on fibre-reinforced plastics. *Composites,* **14**(3) (July 1983).
5. Bergmann, H. W. & Nitsch, P. In *AGARD Conf. Proc.,* **288** (August 1980), Paper 3–1.
6. Cardrick, A. W., A strategy for the fatigue substantiation of helicopter dynamic components. Paper presented at R Ae. Soc. Symposium, Dec. 1984.
7. McCarty, J. E. *et al.*, 737 graphite–epoxy horizontal stabiliser certification. AIAA Paper 82–0745, 1982.
8. Hahn, H. T., Fatigue behaviour and life prediction of composite laminates. ASTM STP674, 1979, pp. 383–417.
9. Guynn, E. G. & O'Brien, T. K., The influence of lay-up and thickness on composite impact damage and compression strength. AIAA Paper 85–0646, 1985.
10. Joint Airworthiness Requirements JAR 25, ACJ 25–603, 16 June 1986.
11. Composite Aircraft Structure. Advisory Circular 20–107A, Federal Aviation Administration, April 1984.
12. Collins, T. A., The effect of observed climatic conditions on the moisture equilibrium level of fibre reinforced plastics. *Composites,* **17**(1) (Jan. 1986).
13. Ministry of Defence Standard, *Metallic Materials Data Handbook*, DEF STAN 00–932, August 1985.
14. McCarty, J. E., Damage tolerance of composites. ICAS (International Conference of Aeronautical Sciences) Paper, London, Sept. 1986.

APPENDIX: JAR GUIDANCE DOCUMENT ON COMPOSITE AIRCRAFT STRUCTURE

ACJ 25.603

Composite Aircraft Structure (Acceptable Means of Compliance) See JAR 25.603

1 Purpose. This ACJ sets forth an acceptable means, but not the only means, of showing compliance with the provisions of JAR-25 regarding airworthiness type certification requirements for composite aircraft structures, involving fibre-reinforced materials, e.g. carbon (graphite), boron, aramid (Kevlar), and glass-reinforced plastics. Guidance information is also presented on associated quality control and repair aspects.

This ACJ material is identical, apart from minor editing, to the structural content of FAA Advisory Circular AC 20.107A, dated 25 April 1984.

The individual JAR paragraphs applicable to each ACJ paragraph are listed in Table 1 of this ACJ.

2 Definitions

2.1 Design values. Material, structural element, and structural detail properties that have been determined from test data and chosen to assure a high degree of confidence in the integrity of the completed structure (see JAR 25.613(b)).

2.2 Allowables. Material values that are determined from test data at the laminate or lamina level on a probability basis (e.g. A or B base values) (see JAR 25.615(a)).

2.3 Laminate level design values or allowables. Established from multi-ply laminate test data and/or from test data at the lamina level and then established at the laminate level by test validated analytical methods.

2.4 Lamina level material properties. Established from test data for a single-ply or multi-ply single-direction oriented lamina layup.

2.5 Point design. An element or detail of a specific design which is not considered generically applicable to other structures for the purpose of substantiation (e.g. lugs and major joints). Such a design element or detail can be qualified by test or by a combination of test and analysis.

2.6 Environment. External, non-accidental conditions (excluding mechanical loading), separately or in combination, that can be expected in service and which may affect the structure (e.g. temperature, moisture, UV radiation, and fuel).

2.7 Degradation. The alteration of material properties (e.g. strength, modulus, coefficient of expansion) which may result from deviations in manufacturing or from repeated loading and/or environmental exposure.

TABLE 1
ACJ Paragraphs and Related JAR Texts

ACJ Paragraphs	JAR-25 Paragraphs
1 Purpose	No relevant JAR Paragraph
2 Definitions	No relevant JAR Paragraph
3 General	No relevant JAR Paragraph
4 Material and Fabrication Development	25.603 25.605 25.613 25.615 25.619
5 Proof of Structure — Static	25.305 25.307(a)
6 Proof of Structure — Fatigue/Damage Tolerance	25.571
7 Proof of Structure — Flutter	25.629
8 Additional Considerations	
8.1 Impact Dynamics	25.561 25.601 25.721 25.783(c) & (g) 25.785 25.787(a) & (b) 25.789 25.801 25.809 25.963(d) & (e)
8.2 Flammability	25.609(a) 25.853 25.855 25.859 25.863 25.865 25.867 25.903(c)(2) 25.967(e) 25.1121(c) 25.1181 25.1182 25.1183 25.1185 25.1189(a)(2) 25.1191 25.1193(c), (d) & (e)

TABLE 1—*contd.*

ACJ Paragraphs		JAR-25 Paragraphs
8.3	Lightning Protection	25.581 25.609 25.899 25.954
8.4	Protection of Structure	25.609 25.1529
8.5	Quality Control	**
8.6	Production Specifications	25.603 25.605

**Guidance material on quality control for composites is under consideration.

2.8 Discrepancy. A manufacturing anomaly allowed and detected by the planned inspection procedure. They can be created by processing, fabrication or assembly procedures.

2.9 Flaw. A manufacturing anomaly created by processing, fabrication or assembly procedures.

2.10 Damage. A structural anomaly caused by manufacturing (processing, fabrication, assembly or handling) or service usage. Usually caused by trimming, fastener installation or foreign object contact.

2.11 Impact damage. A structural anomaly created by foreign object impact.

2.12 Coupon. A small test specimen (e.g. usually a flat laminate) for evaluation of basic lamina or laminate properties or properties of generic structural features (e.g. bonded or mechanically fastened joints).

2.13 Element. A generic element of a more complex structural member (e.g. skin, stringers, shear panels, sandwich panels, joints, or splices).

2.14 Detail. A non-generic structural element of a more complex structural member (e.g. specific design configured joints, splices, stringers, stringer runouts, or major access holes).

2.15 Subcomponent. A major three-dimensional structure which can provide complete structural representation of a section of the full structure (e.g. stub-box, section of a spar, wing panel, wing rib, body panel, or frames).

2.16 Component. A major section of the airframe structure (e.g. wing, body, fin, horizontal stabilizer) which can be tested as a complete unit to qualify the structure.

3 General

3.1 This ACJ is published to aid the evaluation of certification programmes for composite applications and reflects the current status of composite technology. It is expected that this ACJ will be modified periodically to reflect technology advances.

3.2 The extent of testing and/or analysis and the degree of environmental accountability required will differ for each structure depending upon the expected service usage, the material selected, the design margins, the failure criteria, the data base and experience with similar structures, and on other factors affecting a particular structure. It is expected that these factors will be considered when interpreting this ACJ for use on a specific application.

4 Material and Fabrication Development

4.1 To provide an adequate design data base, environmental effects on the design properties of the material system should be established.

4.2 Environmental design criteria should be developed that identify the most critical environmental exposures, including humidity and temperature, to which the material in the application under evaluation may be exposed. This is not required where existing data demonstrate that no significant environmental effects, including the effects of temperature and moisture, exist for the material system and construction details, within the bounds of environmental exposure being considered. Experimental evidence should be provided to demonstrate that the material design values or allowables are attained with a high degree of confidence in the appropriate critical environmental exposures to be expected in service. The effect of the service environment on static strength, fatigue and stiffness properties should be determined for the material system through tests (e.g. accelerated environmental tests, or from applicable service data). The effects of environmental cycling (i.e. moisture and temperature) should be evaluated. Existing test data may be used where they can be shown directly applicable to the material system.

4.3 The material system design values or allowables should be established on the laminate level either by test of the laminate or by test of the lamina in conjunction with a test-validated analytical method.

4.4 For a specific structural configuration of an individual component (point design), design values may be established which include the effects of appropriate design features (holes, joints, etc.).

4.5 Impact damage is generally accommodated by limiting the design strain level.

5 Proof of Structure — Static

5.1 The static strength of the composite design should be demonstrated through a programme of component ultimate load tests in the appropriate environment, unless experience with similar designs, material systems and loadings is available to demonstrate the adequacy of the analysis supported by subcomponent tests, or component tests to agreed lower levels.

5.2 The effects of repeated loading and environmental exposure which may result in material property degradation should be addressed in the static strength evaluation. This can be shown by analysis supported by test evidence, by tests at the coupon, element or subcomponent level, or alternatively by relevant existing data.

5.3 Static strength structural substantiation tests should be conducted on new structures unless the critical load conditions are associated with structure that has been subjected to repeated loading and environmental exposure. In this case either:-

a. the static test should be conducted on a structure with prior repeated loading and environmental exposure, or

b. coupon/element/subcomponent test data should be provided to assess the possible degradation of static strength after application of repeated loading and environmental exposure, and this degradation accounted for in the static test or in the analysis of the results of the static test of the new structure.

5.4 The component static test may be performed in an ambient atmosphere if the effects of the environment are reliably predicted by subcomponent and/or coupon tests and are accounted for in the static test or in the analysis of the results of the static test.

5.5 The static test articles should be fabricated and assembled in accordance with production specifications and processes so that the test articles are representative of production structure.

5.6 When the material and processing variability of the composite structure is greater than the variability of current metallic structures, the difference should be considered in the static strength substantiation by:-

a. deriving proper allowables or design values for use in the analysis, and the analysis of the results of supporting tests, or

b. accounting for it in the static test when static proof of structure is accomplished by component test.

5.7 Composite structures that have high static margins of safety may be substantiated by analysis supported by subcomponent, element and/or coupon testing.

5.8 It should be shown that impact damage that can be realistically expected from manufacturing and service, but not more than the established threshold of detectability for the selected inspection procedure, will not reduce the structural strength below ultimate load capability. This can be shown by analysis supported by test evidence, or by tests at the coupon, element or subcomponent level.

6 Proof of Structure — Fatigue/Damage Tolerance

6.1 The evaluation of composite structure should be based on the applicable requirements of JAR 25.571. The nature and extent of analysis or tests on complete structures and/or portions of the primary structure will depend upon applicable previous fatigue/damage tolerant designs, construction, tests, and service experience on similar structures. In the absence of experience with similar designs, approved structural development tests of components, subcomponents, and elements should be performed. The following considerations are unique to the use of composite material systems and should be observed for the method of substantiation selected by the applicant. When selecting the damage tolerance or safe life approach, attention should be given to geometry, inspectability, good design practice, and the type of damage/degradation of the structure under consideration.

6.2 Damage Tolerance (Fail-Safe) Evaluation

6.2.1 Structural details, elements, and subcomponents of critical structural areas should be tested under repeated loads to define the sensitivity of the structure to damage growth. This testing can form the basis for validating a no-growth approach to the damage tolerance requirements. The testing should assess the effect of the environment on the flaw growth characteristics and the no-growth validation. The environment used should be appropriate to the expected service usage. The repeated loading should be representative of anticipated service usage. The repeated load testing should include damage levels (including impact damage) typical of those that may occur during fabrication, assembly, and in service, consistent with the inspection techniques employed. The damage tolerance test articles should be fabricated and assembled in accordance with production specifications and processes so that the test articles are representative of production structure.

6.2.2 The extent of initially detectable damage should be established and be consistent with the inspection techniques employed during manufacture and in service. Flaw/damage growth data should be obtained by repeated load cycling of intrinsic flaws or mechanically introduced damage. The number of cycles applied to validate a no-growth concept should be statistically significant, and may be determined by load and/or life considerations. The growth or no-growth evaluation should be performed by analysis supported by test evidence, or by tests at the coupon, element or subcomponent level.

6.2.3 The extent of damage for residual strength assessments should be established. Residual strength evaluation by component or subcomponent testing or by analysis supported by test evidence should be performed considering that damage. The evaluation should demonstrate that the residual strength of the structure is equal to or greater than the strength required for the specified design loads (considered as ultimate). It should be shown that stiffness properties have not changed beyond acceptable levels. For the no-growth concept, residual strength testing should be performed after repeated load cycling.

6.2.4 An inspection programme should be developed consisting of frequency, extent, and methods of inspection for inclusion in the maintenance plan. Inspection intervals should be established such that the damage will be detected between the time it initially becomes detectable and the time at which the extent of damage reaches the limits for required residual strength capability. For the case of no-growth design concept, inspection intervals should be established as part of the maintenance programme. In selecting such intervals the residual strength level associated with the assumed damage should be considered.

6.2.5 The structure should be able to withstand static loads (considered as ultimate loads) which are reasonably expected during a completion of the flight on which damage resulting from obvious discrete sources occurs (i.e. uncontained engine failures, etc.). The extent of damage should be based on a rational assessment of service mission and potential damage relating to each discrete source.

6.2.6 The effects of temperature, humidity, and other environmental factors which may result in material property degradation should be addressed in the damage tolerance evaluation.

6.3 Fatigue (Safe-Life) Evaluation. Fatigue substantiation should be accomplished by component fatigue tests or by analysis supported by test evidence, accounting for the effects of the appropriate environment. The test articles should be fabricated and assembled in accordance with production specifications and processes so that the test articles are representative of production structure. Sufficient component, subcomponent, element or coupon tests should be performed to establish the fatigue scatter and the environmental effects. Component, subcomponent and/or element tests may be used to evaluate the fatigue response of structure with impact damage levels typical of those that may occur during fabrication, assembly, and in service, consistent with the inspection procedures employed. The component fatigue test may be performed with an as-manufactured test article if the effects of impact damage are reliably predicted by subcomponent and/or element tests and are accounted for in the fatigue test or in analysis of the results of the fatigue test. It should be demonstrated during the fatigue tests that the stiffness properties have not changed beyond acceptable levels. Replacement lives should be established based on the test results. An appropriate inspection programme should be provided.

7 Proof of Structure — Flutter. The effects of repeated loading and environmental exposure on stiffness, mass and damping properties should be considered in the verification of integrity against flutter and other aeroelastic mechanisms. These effects may be determined by analysis supported by test evidence, or by tests of the coupon, element or subcomponent level.

8 Additional Considerations

8.1 Impact Dynamics. The present approach in airframe design is to assure that occupants have every reasonable chance of escaping serious injury under realistic and survivable impact conditions. Evaluations may be by test or by analysis supported by test evidence. Test evidence includes, but is not limited to, element or subcomponent tests and service experience. Analytical comparison to conventional structure may be used where shown to be applicable.

8.2 Flammability. (See appropriate JAR requirements in Table 1 of this ACJ)

8.3 Lightning Protection. (See appropriate JAR requirements in Table 1 of this ACJ)

8.4 Protection of Structure. Weathering, abrasion, erosion, ultraviolet radiation, and chemical environment (glycol, hydraulic fluid, fuel, cleaning agents, etc.) may cause deterioration in a composite structure. Suitable protection against and/or consideration of degradation in material properties should be provided for and demonstrated by test.

8.5 Quality Control. An overall plan should be established and should involve all relevant disciplines (i.e. engineering, manufacturing and quality control). This quality control plan should be responsive to special engineering requirements that arise in individual parts or areas as a result of potential failure modes, damage tolerance and flaw growth requirements, loadings, inspectability, and local sensitivities to manufacture and assembly.

8.6 Production Specifications. Specifications covering material, material processing, and fabrication procedures should be developed to ensure a basis for fabricating reproducible and reliable structure. The discrepancies permitted by the specifications should be substantiated by analysis supported by test evidence, or tests at the coupon, element or subcomponent level.

8.7 Inspection and Maintenance. Maintenance manuals developed by manufacturers should include appropriate inspection, maintenance and repair procedures for composite structures.

8.8 Substantiation of Repair. When repair procedures are provided in maintenance documentation, it should be demonstrated by analysis and/or test that methods and techniques of repair will restore the structure to an airworthy condition.

Index

Ablative resistance, 239
Acrylic adhesives, 275–6
Acrylic modified urethanes, resistance of, 176
Adhesive bonding, thermoplastic composites, 94–5
Adhesives
  applications of, 291–5
  chemically reactive, 274–6
  classification, 273–7
  engineering design and, 282–91
  high-temperature, 276
  non-reactive, 277
  requirements of, 271–2
  selection of, 272–3
  test methods for, 277–82
  types of, 274–7
  see also Bonded-structure design *and* specific types of
Adhesive test methods, 278–80
Advanced composite fabrication, 251–68
Aircraft, 371
  anti-corrosion, 166
  civil, *see* Civil aircraft, composite structure certification
  cowl doors, 295
  cyclic loading, 408–11
  impact and, 411–12
  propeller blades, bonded joints, 291–2
Aircraft—*contd.*
  thermoplastic composites, 99–101
Alloys
  eutectic, 223–5
  light, CVD-reinforced, 229–44
  light, inorganic, 226–8
Anti-corrosion and reinforced plastics
  aircraft, 166
  applications, 164–8
  aramid, 186
  carbon fibres, 185
  chemical plant, 164–5
  civil engineering, 166
  design of, 192–4
  failure, 168
  fibre reinforcement, 182–6
  glass fibre, 183–5
  laminate construction, 190–2
  mechanism, 186–90
  properties, 168
  reinforced plastics, 163–95
  resins, 169–82
  underground tanks, 165
  vehicles, 165
Anti-corrosion design of reinforced plastics, 192–4
  fire and, 193–4
  joints, 192
  operating conditions, 193
  plastic laminate cure, 193

Anti-corrosion design of reinforced plastics—*contd.*
  static charge and, 193
  thermal mismatching, 192
Aramid, 134–7, 142
  anti-corrosion, 186
  price of, 139–41
  structure of, 135
Arcan fixture, 321–2
Areas data reduction, 312–15
Aromatic polyamide fibres, *see* Aramid
Asbestos, 186
Assembly technique, for thermoplastics, 94–5
Automated laminating, 264–5

Bagging materials, thermoplastic composites, 94
Biaxial fatigue testing, 357–9
Biocompatibility, and inorganic matrix composites, 241–2
Bismaleimide resins, 17, 19, 21
  future of, 36–7
  properties of, 22
Bisphenol resins, 169–71
Blended resins, 28–35
  elastomers, 32–5
  low molecular weight modifiers, 30–2
  monomer, 29–30
  thermoplastic modifiers, 32–5
BMI, *see* Bismaleimide
Bonded joints, and adhesive application, 290–5
  collapsible tube mast, 294–5
  cowl doors, 295
  launch vehicle attachment ring, 292–4
  propeller blades, 291–2
  racing cars, 290–1
Bonded-structure design, 269–301
Boron filaments, 123–6, 142
  deposition process, 125
  price of, 140–1
Broadgoods
  enhancements, 264–6
  methods, 253–4
Butt torsion test, 280–1

Cantilever beam enclosed notch test, 319
Carbon–carbon composites, and impregnation, 217
Carbon–carbon materials, 238–42
  ablation, 239
  biocompatibility, 242
  oxidation resistance, 239
  properties, 239
Carbon fibres, 127–34, 374
  anti-corrosion and, 185
  internal structure, 127–8
  mechanical properties, 133
  PAN, 128–32
  pitch precursor, 129–32
  precursor types, 128–32
  price, 140–1
  range available, 132, 134
  wettability, 211
Carboxyl terminated butadiene-acrylonitrile, 147–8, 152–3
  Young's modulus and, 150
Casting
  low pressure, at, 214
  squeeze, 214
Ceramic–ceramic composites, and impregnation, 217–18
Ceramic precursors, 218
Ceramics
  fibre-reinforced, 242–4
  whisker reinforced, 228–9
  see also Inorganic matrix composites
Ceramic tooling, 262
Chassis, bonded joints, 290–1
Chemical plant anti-corrosion, 164–5
Chemical vapour infiltration, *see* Vapour infiltration
Civil aircraft composite structure certification, 399–430
  manufacture control, 402–6
  materials control/supply, 401–2
  new requirements, 413–26
  outline of guidelines, 423–30
  present, at, 399–401

Civil aircraft structure, 399–430
Civil engineering anti-corrosion, 166
Classical analysis, *see* Closed form analysis
Cleaning, tooling, 262
Closed form analysis, 283–6
Coated fibres, 152–3
Co-lamination, high speed, 209
Collapsible tube mast, bonded joints, 294–5
Co-mingling, *see* Hybridisation
Compliance data reduction, 312–15
Compliance moulding thermoplastic composite fabrication, 90–2
Compocasting, 204–6
Composites
  carbon–carbon, 217
  ceramic–ceramic, 217–18
  civil aircraft structures, 399–430
  continuous fibre-reinforced thermoplastic matrix, 43–109
  damage tolerance, 369–98
  fabrication, advanced, 251–68
  improved, and damage, 391–7
  inorganic matrix, 197–250
  in situ inorganic, 223–6
  organic, fatigue, 331–67
  overview of, 1–5
Composite tooling, 258–9
Compression testing, 332, 334–5
Compressive loading, 352–3
Continuous fibre pre-impregnated technology, 252–7
Continuous fibre reinforced thermoplastic matrix composites, 43–109
  advantages, 46
  applications, 99–101
  assembly, 94–5
  background to, 44–8
  co-weaving, 60–1
  durability properties, 84–6
  economics of, 97–8
  environmental resistance, 79–84
  fabrication science, 95–6
  fabrication techniques, 86–97
  flow mechanisms, 96–7
  future, 97–101
Continuous fibre reinforced thermoplastic matrix composites—*contd.*
  history of, 44–6
  mechanical properties, 69–74
  melt, 57
  polymer availability, 50–1
  polymer properties, 51–6
  pre-impregnation techniques, 57–8
  processing, 63–9
  properties, 69–86
  pseudo-, pre-impregnated, 66–7
  reasons for using, 46–7
  solvent aided, 58
  stacking, film, 59–60
  temperature and, 78–9
  toughness of, 74–8
  trade names, 62
  true, pre-impregnated, 63–8
Corrosion mechanisms, 186–90
  fibre-resin debonding, 189–90
  leaching, 189
  osmotic cracking, 189
  polymer swelling, 187
  solubility, 187–9
  stress cracking, 187
  see also Anti-corrosion
Coupon profiles, 332
Co-weaving, 60–1
Cowl doors, bonded structure, 295
Cracked lap shear delamination test, 320–1
Crystalline polymers, 188
Cure, *see* Laminate cure
CVI, *see* Vapour infiltration
Cyclic loading, 408–11

Damage
  assessment, 369–98
  detection, 381–3
  tolerance, 369–98
Data reduction procedures, 312–15
DDS, 10–13
Delamination, 303–29
  fracture mechanisms approach, 309–10
  interlaminar stress, 304–9
  testing, *see* Delamination testing

Delamination testing, 310–24
  cracked lap shear, 320–1
  DCB, 311–15
  EDT, 319–20
  ENF, 315–17
  mixed mode, 319–23
  mode I, 311–15
  mode II, 315–19
  parameters, 323
Derakane, 175
DGEBA, 150, 153
Diacryl, 175
4,4′-Diaminodiphenylsulphone, *see* DDS
Diaphragm forming, thermoplastics, 91–2
Dicyandiamide (Dicy), 11
Diffusion bonding, inorganic matrix composites, 207–10
Diglycidyl ether, bisphenol A, *see* DGEBA
Directional solidification, 223–5
Discontinuous reinforced composites, 226–9
Double cantilever beam testing, 311–15
Dowty rotol composite propeller blades, 291–2
DRAMC (discontinuous reinforced aluminium matrix composites), *see* Discontinuous reinforced composites

Edge delamination tension test, 319–21
Edge-induced stress, 350–1
E-glass, 182–3
  see also Fibre glass
Elastomers, 32–5
Electroform tooling, 259
End notch flexure test, 315–17
Enhancement, and advanced composite fabrication, 264–6
  Broadgoods, 264–6
  hand lay-up, 264
Epikote 155, 12
Epikote 1031, 12
Epoxides, and anti-corrosion, 172
Epoxy adhesives, 275
Epoxy resins, 10–14, 172
  anti-corrosion and, 166–7
  multifunctional, properties, 15
  tetrafunctional, 10–12
  trifunctional, 12–14
Eutectic alloys, 223–5

Fabricated metal tooling, 258
Failure criteria, 287–90
  implementation, 289
  interfacial, 289
  optimization design, 289–90
  specific materials, 287–8
Fatigue life prediction, 363–5
Fatigue, organic matrix composites, 301–67
Fatigue testing methods, organic matrix composites, 331–67
Fiberite-934, 11
Fibredux-914, 33
Fibredux-6376, 33
Fibre/matrix interface, 374–5
Fibre-matrix liquid compounds, 204–6
Fibre-reinforced thermoplastic composites, *see* Continuous fibre reinforced thermoplastic matrix composites
Fibre-resin debonded corrosion, 189–90
Fibres, *see* specific forms of *and* High-performance fibres
Fibre yarns, *see* Yarns
Filament winding, thermoplastic, 93
Film stacking, 59–60
Finite element analysis, 286–7
Fire
  anti-corrosion, and, 193–4
  PEEK, 181
  thermoplastics and, 86–7
Firing, of organic/organo-metallic precursors, 216–18
Flexural testing, 335–6
Foils, composites, 213

Foundary techniques, and impregnation, 213
Fracture behaviour, 370–4
Fracture mechanisms, and delamination, 309–10
FRP, *see* Reinforced plastics
Furan resins, anti-corrosion and, 174, 176

Gerber knife cutter, 253–5
G-glass, 183
  see also Fibre glass
Glass–ceramic matrices, 214–16
Glass fibre, 116–23, 374
  anti-corrosion and, 183–5
  compositions of, 184
  PEEK and, 184–5
  pre-peg forms of, 118, 122–3
  price of, 139–41
Glyamine-120, 11
Graphite, *see* Carbon fibre
Graphite tooling, 259, 262

Hand lay-up
  enhancement, 264
  tape parts, 252–3
Heat, and organic matrix composite fatigue, 348–9
Heavy-metal reinforcement, 238
Hercules-8551-7, 33
High-performance fibres, 111–44
  aramid, 134–7
  boron filament, 123–6
  carbon, 127–34
  comparative, 140–3
  glass, 116–23
  history of, 111–14
  mechanical properties of, 119–21
  performance of, 114–16
  polyethylene, 137–9
  price of, 139–40
  silicon carbide, 126
High-speed co-lamination, 209
High-temperature adhesives, 276
High-temperature thermosetting matrix resins, 7–41
  background to, 7–8
High-temperature thermosetting matrix resins—*contd.*
  blends of, 28–35
  epoxy, 10–14
  new types of, 23–8
  phenolics, 9–10
  polyimide, 14–21
  polystyrylpyridine, 21–3
  progress in, 36–7
  review of, 8–9
  types of, 9–23
  see also specific chemical forms of
Hot compression technique, 207–9
Hot melt adhesive, 277
Hot wall CVI, 220–2
HP-PE, *see* Polyethylene fibres
Hybrid composites, fatigue in, 355–7
Hybridisation, 61
Hydroforming, thermoplastic, 90

Impact, aircraft structures, 411–12
Impact testing, 375–9
  post-, and performance, 383–8
Impregnation
  carbon–carbon composites, 217
  firing of precursors, 216–18
  liquid, fibrous pre-forms, 210–18
  reactive, 223
  yarns, of, 213
Impregnation techniques, 57–60
  post-, 59–61
  pre-, 57–8
Induction welding, thermoplastics, 95
Infra-red thermography, 348
Inorganic matrix composites, 197–250
  applications of, 226–44
  limitations of, 202
  processing of, 202–26
  properties of, 199–200
  temperature range for use, 199–200
In situ inorganic composites, 223–5
Interfacial failure criteria, 289
Interlaminar shear strength, 157–9
  test, 323–4
Interlaminar stress, 304–9
Interleafing, 153
Isophthalics, 169–70

JAR guidelines, 413–30
Joints, and anti-corrosion, 192
Joint strength, 287–90

Kevlar, *see* Aramid
Knitted matrix technology, 256–7

Laminate construction, and anti-corrosion, 191–2
Laminate cure, 193
Laminate fracture behaviour, 370–4
Laminators, automated, 265–6
Launch vehicle attachment ring, 292–4
Leaching, and corrosion, 189
Light alloys, 226–44
  for structural applications, 231–6
Liquid impregnation, fibrous pre-forms, 210–18
Low-pressure casting, 214

Machined metal tooling, 259
Matched-mould thermoplastic composite fabrication, 88–90
Materials failure criteria, 287–9
Matrix composites, *see* Continuous fibre reinforced thermoplastic matrix composites
  see also Inorganic matrix composites
Matrix resins, *see* High-temperature thermosetting matrix resins
Mechanical properties, of thermoplastics, 69–74
Melt pre-impregnation, 58
Metal coating, 206–7
Metal fatigue, and organic matrix composites, 360–3
Metal matrix composites, processing, 202–3
Metal pre-preg liquid impregnation, 212
Metal reinforcement, with fibre yarns, 230–1
Metals, powder, 203
Microscopic non-destructive testing, 345–6
Mixed-mode delamination testing, 319–23
  cracked lap shear test, 320–1
  EDT in, 319–20
Mode I delamination testing, 311–15
  DCB, 311–15
Mode II delamination testing, 315–19
  ENF test, 315–17
Molar attraction, 188
Moulding, thermoplastics
  compliance, 90, 92
  matched, 88–90
Multidirectional laminated composite material fatigue, 341–4
Multidirectional laminate fracture behaviour, 372–4
MY720, 11

Napkin ring test, 280–1
Narmco5208, *see* TGDDM
Narmco X5250, *see* Bismaleimide resins
Nicalon (SiC/SiC), 213, 244–5
Nickel electroform tooling, 259
Non-destructive testing, organic matrix composites, 344–8
  infra-red, 348
  microscopy, 345–6
  ultrasonics, 346–7
  X-rays, 347–8
Non-organic matrix composites, *see* Inorganic matrix composites

Organic matrix composite material fatigue, 331–67
  biaxial testing, 357–9
  compressive loading, 352–3
  edge effects, 350–1
  environmental effects, 359–60
  fatigue life prediction, 363–5
  frequency effects, 348–9
  hybrids, 355–7
  metals and, 360–3

Organic matrix composite material fatigue—*contd.*
  multidimensional laminated, 341–4
  NDT, 344–8
  stress concentrators, 351–2
  test methods, 332–8
  unidirectional, 338–41
Orthophthalic resins, 169
Osmotic cracking corrosion, 189

PAN, 128–32
  precursor, 128–9
PBO, *see* Aramid
PBT, 180
PEEK, 45, 48, 180–1, 390–1
  processing, 63–5, 67–8
  see also Polymer chemistry
PEI, 154–5
PEK, *see* Polymer chemistry
PES, 159
PET, 180
Phenol-aralkyl resins, 9–10
Phenolic adhesives, 274
Phenolic laminating resins, and anti-corrosion, 172
  properties of, 173
Phenolic resins, 9–10
  conventional, 9
  phenol-aralkyl, 9–10
Pitch precursor carbon fibre, 129–32
Plastics, reinforced, 163–96
Plastics, thermo-, 43–109
Plastic working, inorganic matrix composites, 207–10
Plastisol adhesives, 277
Ply assembly thermoplastic composite fabrication, 92–4
Ply profiling, automatic, 253–4
PMR resins, 17–18
Poisson's ratio, 305–9
Polyacrylonitrile, *see* PAN
Polyamide fibres, aromic, *see* Aramid
Polyamides, 178, 180
Polybenzimidazole resins, 24, 26
Poly(butylene terephthalate), *see* PBT
Polyesters, saturated, 180
Polyesters, unsaturated, and anti-corrosion, 169–71
Poly(ether ether ketone), *see* PEEK
Polyethersulphones, 33, 35
Polyethylene, 178
Polyethylene fibres, 137–9
  price of, 140–1
Poly(ethylene terephthalate), *see* PET
Polyimide resins, 14–21
  thermal stability of, 15–17
Polymer chemistry, 48–50
Polymer-modified composite toughness, 156–8
Polymer-modified resin toughness, 153–5
Polymers, 48–56
  chemistry of, 48–50
  commercially available, 50–1
  properties, 51–6
Polymer solubility, 187–9
Polymer swelling corrosion, 187
Poly(*p*-phenylenebenzobisoxazole), *see* Aramid
Polyphenylquinoxaline resins, 24–5
Polyphthalocyanine resins, 27–8
Polypropylene, 178
Polystyrylpyridine resins, 21–3, 176
Polyurethane adhesives, 275
Post-impact performance, 383–8
Post-impregnated thermoplastics, 67–9
Powder coating, 60
Powder metallurgy, 203
PPS, processing, 63–5
  see also Polymer chemistry
Pre-impregnated true thermoplastics, 63–7
Pre-impregnation techniques, 57–8
  melt, 58
  solvent-aided, 59
Pre-peg glass fibres, 116–23
  production methods, 122
  unidirectional tape, 118
  weave styles, 123
  woven fabric, 118, 122–3
Propeller blades, bonded structures, 290–1

Pseudothermoplastics, 66–7
PSP, *see* Polystyrylpyridine resins
Pultrusion, 209–10
PVC, 182

Racing cars, bonded structures, 290–1
Rail shear test, 319
Rayon precursor, carbon fibre, 128
Reactive impregnation, 223
Reinforced plastics, and anti-corrosion, 163–96
  design and, 192–4
Resins
  anti-corrosion, 169–82
  fatigue, 340–1
  see also High-temperature thermosetting matrix resins
Resistive bonding thermoplastics, 95
Rheology, and thermoplastic composites, 95–7
Roll-formed thermoplastic composites, 88–90
Rubber-modified thermosetting, 147–50
Rubber toughening, 147–50
Rütapox TGMDA, 11

Saturated polyester, 180
Sealants, thermoplastic composites, 94
S-glass, 183
  see also Fibre glass
Shear testing, 336–7
  thick adherend, 281–2
Short beam test, 323–4
Short fibre composites, 355
Silica-based matrix composites, 242, 244
Silica-based reinforced ceramics, 242–4
Silicon carbide
  fibre, 126
  two-dimensional stacking, 244–5
  whisker reinforced ceramic composites, 228–9
Silicon nitride composites, 225
Solidification, directional, 223–5
Solubility, and corrosion, 187–9
Solvent-aided impregnation, 59
Squeeze casting, 214
Static charge, 193
Static strength, aircraft composite structure, 406–8
Strain rate, 380–1
Stress concentrators, 351–2
Stress cracking corrosion, 187
Stretch-draw thermoplastic composite fabrication, 90
Swelling, 187–9

Tank anti-corrosion, 165
Tape adhesives, 277
Tape laying, 254–5
  thermoplastic composites, 92–3
Tape parts, unidirectional, 252–6
Tape winding, 256
TAST, 281–2
Temperature, thermoplastics and, 78–9
TEN, 12–13
N,N'-Tetraglycidyl-4,4'-diaminodiphenylmethane, *see* TGDDM
TGDDM, 10–11, 172
TGPAP, 12–13
Thermal expansion, controlled, 236–7
Thermal mismatch, 192
Thermid-600, 176
Thermographic (IR) NDT, 348
Thermoplastic matrices, and damage tolerance, 390–1
Thermoplastic modifiers, 32–5
Thermoplastics, and anti-corrosion, 176–82
Thermoplastics, and thermosetting resins, 151–2
  see also Continuous fibre reinforced thermoplastic matrix composites
Thermoset fabrication technology, 252–7
Thermosetting, damage tolerance, 390–1

Thermosetting matrix resins, at high temperature, 7–41
Thermosetting resins, and anti-corrosion, 169–76
Thick adherend shear test, 281–2
Tolerance, to damage, 369–98
Tooling materials, 260–1
  characteristics, 263
Tooling technology, for advanced composite fabrication, 257–62
  cleaning, 262
  future, 262–7
  selection, 258–62
Torayca 3900-2, 33
Toughness enhancement, 145–61
  coated fibres, 152–3
  interleafing, 153
  polymer modified resins, of, 153–5
  rubber, of, 147–51
  thermoplastics and, 151–2
Toughness, of thermoplastics, 74–8
Tow laying, 93
Triglycidyl-*p*-aminophenol, *see* TGPAP
Trisepoxy-novolak, *see* TEN
True thermoplastic, 63
Tube processing, 209
Twaron, 137
Two-dimensional silicon carbide composites, 244–5

Ultem, *see* PEI
Ultrasonic NDT, 346–7
Ultrasonic welding, thermoplastics, 95
Unidirectional composite material fatigue, 339–41
Unidirectional laminate fracture behaviour, 370–2
Unidirectional tape pre-peg, 118
Unitape, 118
Unsaturated polyesters, 169–72

V398, 19, 21
Vapour deposition, 206–7, 218–23
  for boron filaments, 124–5
  gaseous precursors, 220
  light alloy reinforcement, and, 229–44
Vapour infiltration, 218–23
  applications, 222–3
  principles of, 219–22
Vehicle anti-corrosion, 165
Vinyl esters, and anti-corrosion, 172, 174–5
VTBN, 152

Weaving, 256–7
Welding, thermoplastics, 95
Whisker reinforced ceramic composites, 228–9
Wire drawing, 213
Woven fabric pre-peg, 118, 122, 123

X-ray NDT, 347–8

Yarn impregnation, 212
Yarns, 230–1